Food Science
and Food
Biotechnology

FOOD PRESERVATION TECHNOLOGY SERIES

Series Editor
Gustavo V. Barbosa-Cánovas

Innovations in Food Processing
Editors: Gustavo V. Barbosa-Cánovas and Grahame W. Gould

Trends in Food Engineering
Editors: Jorge E. Lozano, Cristina Añón, Efrén Parada-Arias,
and Gustavo V. Barbosa-Cánovas

**Pulsed Electric Fields in Food Processing:
Fundamental Aspects and Applications**
Editors: Gustavo V. Barbosa-Cánovas and Q. Howard Zhang

**Osmotic Dehydration and Vacuum Impregnation:
Applications in Food Industries**
Editors: Pedro Fito, Amparo Chiralt, Jose M. Barat, Walter E. L. Spiess,
and Diana Behsnilian

Engineering and Food for the 21st Century
Editors: Jorge Welti-Chanes, Gustavo V. Barbosa-Cánovas,
and José Miguel Aguilera

Unit Operations in Food Engineering
Albert Ibarz and Gustavo V. Barbosa-Cánovas

Transport Phenomena in Food Processing
Editors: Jorge Welti-Chanes, Jorge F. Vélez-Ruiz,
and Gustavo V. Barbosa-Cánovas

Food Science and Food Biotechnology

Gustavo F. Gutiérrez-López, Ph.D.
Escuela Nacional de Ciencias Biológicas
Instituto Politécnico Nacional
México, DF.

Gustavo V. Barbosa-Cánovas, Ph.D.
Washington State University
Pullman, Washington

CRC PRESS

Boca Raton   London   New York   Washington, D.C.

Dedication

To our families

Contents

Series Preface

I am glad to welcome this book on food science and food biotechnology to the growing CRC Food Preservation Technology Series. This new book significantly expands the scope of the series, opening new directions I would like to pursue. The blend of engineering with food science and food biotechnology is quite unique, as well as complex and intriguing. Wonderful things will come from this association, provided we make the right considerations and identify those products that match expectations.

Biotechnology is vast, without boundaries, and offers great possibilities as well as concerns. Let's use it correctly, let's learn more about it, and let's see how it will contribute to the science of making better foods on a daily basis.

I am very enthusiastic about this book and hope that all readers will feel the same. It has significant new material that will open new horizons; and it offers the opportunity to expand two critical disciplines at the same time: food biotechnology and food science.

Gustavo V. Barbosa-Cánovas

Preface

The relevance of biotechnology in food science has significantly grown in the last few years and this trend will continue in the years to come. For this reason, it is becoming quite relevant to identify in an organized manner efforts toward the use of a very important area of scientific research and application (biotechnology) in a domain of tremendous relevance for the whole population (food science).

This book presents a meaningful and up-to-date review of how food science and biotechnology are interacting and covers important aspects of both subjects as well as their interface. Distinguished scientists from key institutions have contributed chapters that provide deep analysis of their particular subjects and at the same time place each topic within the context of this interface. The premise of this book is that an effective discussion on these subjects (food science, biotechnology) requires an effective coupling to convey in a comprehensive manner the state-of-the-art for the fundamentals and applications of both areas.

The book is mainly directed to academics, undergraduate and postgraduate students (including research students) in food science and technology, biotechnology, and bioengineering, who will find a selection of topics ranging from the molecular basis of food preservation and biotechnology to industrial applications. Professionals working in food and biotechnology research centers also may find this book useful.

The first chapter reviews molecular aspects of food biotechnology, presenting the reader a state-of-the-art introduction to the next 14 chapters. The second chapter covers some of the recent developments in food biotechnology. The reader may note a certain degree of overlap between these two chapters. However, Chapter 2 covers a wide range of food applications. Chapters 3, 4, and 5 describe specific examples of upstream processes and bioseparations.

The following eight chapters cover basic and applied situations of food biotechnology such as the impact of crystallography, stability of enzymes, biotechnology of carotenoids, biogeneration of aroma compounds, potential usage of regional alternative food supplies, and developments in the application of emulsifiers. The last three chapters describe food preservation techniques, strongly stressing alternative food processing technologies as well as drying of biotechnological products.

It is very likely that many of the procedures and techniques described in the book will be used in the food biotechnology industry. It is hoped that this text will constitute a worthy addition to the emerging literature on food biotechnology and that the readers will find in it balanced and organized information, with chapters written with the common goal of stressing the interactions of food science and biotechnology and demonstrating, through specific examples, that the bases for many of the situations discussed in the various sections are very similar.

Gustavo F. Gutiérrez-López
Gustavo V. Barbosa-Cánovas

Editors

Gustavo F. Gutiérrez-López received his bachelor of science in biochemical engineering and master of science in food science and technology from the National Polytechnic Institute of Mexico, and his master of science in food process engineering and Ph.D. in food engineering from the University of Reading, U.K. He is currently professor of food engineering at the National School of Biological Sciences of the National Polytechnic Institute of Mexico and president of the Mexican Society of Biotechnology and Bioengineering.

Gustavo V. Barbosa-Cánovas received his bachelor of science in mechanical engineering from the University of Uruguay and his master of science and Ph.D. in food engineering from the University of Massachusetts at Amherst. He is currently professor of food engineering at Washington State University and director of the Center for Nonthermal Processing of Food.

Acknowledgments

The Editors wish to express their gratitude to the following institutions and individuals who contributed to making this book possible:

National School of Biological Sciences of the National Polytechnic Institute of Mexico (ENCB-IPN), CYTED (Ibero-American Program to promote science and technology) Subprogram XI, Mexican Society of Biotechnology and Bioengineering (SMBB), and Washington State University (WSU) for supporting the preparation of this book.

Our fellow colleagues Lidia Dorantes, María Eugenia Jaramillo, Rosalva Mora, Gloria Dávila, and Humberto Hernández (IPN-ENCB) for their valuable comments and suggestions throughout the preparation of this book.

J. Anderson (WSU) for her professionalism and dedication throughout the entire editorial process.

V. Aguilar-Clark, L. Alamilla, M. Cornejo, A. Ortíz, and J. Chanona, all from ENCB-IPN, and Gipsy Tabilo (WSU) for their decisive participation in helping the editors prepare the final version of this book by revising references, formatting all the manuscripts, and incorporating in the text all the editorial comments.

Contributors

Wendy Dayanara Acosta-Hernández Departamento de Graduados e Investigación en Alimentos, Escuela Nacional de Ciencias Biológicas, Instituto Politécnico Nacional, Carpio y Plan de Ayala, México

Valente B. Alvarez, Department of Food Science and Technology, The Ohio State University, Columbus, Ohio

Ali Asaff-Torres Departamento de Biotecnología y Bioingeniería CINVESTAV–IPN, Instituto Politécnico Nacional, México

Gustavo V. Barbosa-Cánovas Department of Biological Systems Engineering, Washington State University, Pullman, Washington

Gloria Dávila-Ortiz Departamento de Graduados e Investigación en Alimentos, Escuela Nacional de Ciencias Biológicas, Instituto Politécnico Nacional, Carpio y Plan de Ayala, México

Lidia Dorantes-Alvárez Departamento de Graduados e Investigación en Alimentos, Escuela Nacional de Ciencias Biológicas, Instituto Politécnico Nacional, Carpio y Plan de Ayala, México

Miguel A. Duarte-Vázquez Departamento de Investigación y Posgrado en Alimentos, PROPAC, Facultad de Química, Universidad Autónoma de Querétaro, Querétaro, México

Blanca E. García-Almendárez Departamento de Biotecnología, CBS, Universidad Autónoma Metropolitana–Iztapalapa, Delegación Iztapalapa, México

Gustavo Fidel Gutiérrez-López Departamento de Graduados e Investigación en Alimentos, Escuela Nacional de Ciencias Biológicas, Instituto Politécnico Nacional, Carpio y Plan de Ayala, México

Humberto Hernández-Sánchez Departamento de Graduados e Investigación en Alimentos, Escuela Nacional de Ciencias Biológicas, Instituto Politécnico Nacional, Carpio y Plan de Ayala, México

Victor T. Huang General Mills, Minneapolis, Minnesota

Sergio Huerta-Ochoa   Departamento de Biotecnología, CBS, Universidad Autónoma Metropolitana–Iztapalapa, Delegación Iztapalapa, México

María Eugenia Jaramillo-Flores   Departamento de Graduados e Investigación en Alimentos, Escuela Nacional de Ciencias Biológicas, Instituto Politécnico Nacional, Carpio y Plan de Ayala, México

Taehyun Ji   Department of Food Science and Technology, The Ohio State University, Columbus, Ohio

Cristian Jiménez-Martínez   Departamento de Graduados e Investigación en Alimentos, Escuela Nacional de Ciencias Biológicas, Instituto Politécnico Nacional, Carpio y Plan de Ayala, México

Ashok Khare   School of Food Biosciences, The University of Reading, Whiteknights, Reading, United Kingdom

Rosalva Mora-Escobedo   Departamento de Graduados e Investigación en Alimentos, Escuela Nacional de Ciencias Biológicas, Instituto Politécnico Nacional, Carpio y Plan de Ayala, México

Arun S. Mujumdar   Department of Mechanical Engineering, National University of Singapore, Singapore

María Soledad de Nicolás-Santiago   Departamento de Graduados e Investigación en Alimentos, Escuela Nacional de Ciencias Biológicas, Instituto Politécnico Nacional, Carpio y Plan de Ayala, México

Keshavan Niranjan   School of Food Biosciences, The University of Reading, Whiteknights, Reading, United Kingdom

Juan Alberto Osuna-Castro   CINVESTAV, Unidad Irapuato, Libramiento Norte, Carretera México-León, Irapuato, México

Octavio Paredes-López   CINVESTAV, Unidad Irapuato, Libramiento Norte, Carretera México-León, Irapuato, México

Ana María Pilosof   Departamento de Industrias, Facultad de Ciencias Exactas y Naturales, Universidad de Buenos Aires, Buenos Aires, Argentina

Lilia Arely Prado-Barragán   Departamento de Biotecnología, CBS, Universidad Autónoma Metropolitana–Iztapalapa, Delegación Iztapalapa, México

Carlos Regalado-González   Departamento de Investigación y Posgrado en Alimentos, PROPAC, Facultad de Química, Universidad Autónoma de Querétaro, Querétaro, México

José J. Rodríguez   Department of Biological Systems Engineering, Washington State University, Pullman, Washington

Leobardo Serrano-Carreón   Instituto de Biotecnología, Departamento de Bioingeniería, Universidad Nacional Autónoma de México, Morelos, México

Manuel Soriano-García   Departamento de Bioestructura, Instituto de Química, Universidad Nacional Autónoma de México (UNAM), Circuito Exterior, Ciudad Universitaria, Delegación Coyoacán, México

Viviana Taragano   Departamento de Industrias, Facultad de Ciencias Exactas y Naturales, Universidad de Buenos Aires, Buenos Aires, Argentina

Mauricio R. Terebiznik   Departamento de Ingeniería Química, Facultad de Ingeniería, Universidad de Buenos Aires, Buenos Aires, Argentina

Mayra De la Torre-Martínez   Departamento de Biotecnología y Bioingeniería, CINVESTAV–IPN, Instituto Politécnico Nacional, México

Hye Won Yeom   Department of Food Science and Technology, The Ohio State University, Columbus, Ohio

Q. Howard Zhang   Department of Food Science and Technology, The Ohio State University, Columbus, Ohio

Vanessa Zylberman   Departmento de Industrias, Faculta de Ciencias Exactas y Naturales, Universidad de Buenos Aires, Buenos Aires, Argentina

1

Introduction to Molecular Food Biotechnology

Juan Alberto Osuna-Castro and Octavio Paredes-López

CONTENTS

1-56676-892-6/03/$0.00+$1.50
© 2003 by CRC Press LLC

Introduction

Molecular biotechnology is an exciting revolutionary scientific discipline based on the ability of researchers to transfer specific units of genetic information from one organism to another. This conveyance of a gene relies on the techniques of genetic engineering (recombinant DNA technology). The application of science and technology, with molecular biology being one of the more recent developments, has resulted in greatly increased yields per unit of cultivated area, due in great part to production of plants that are resistant to insect predation, fungal and viral diseases, and environmental stresses such as short-term drought, excessive heat, and acidic and alkaline soils. Some biotechnologists now see plants as biofactories or bioreactors that need only water, minerals, sunlight, and the proper combination of genes to produce high-value biomolecules such as enzymes, starches, oils, vitamins, pigments, nutraceuticals, and vaccines for the food and pharmaceutical industries. Biotechnology can also be applied to the production or transformation of food and food ingredients and animal feed by developing microorganisms able to produce chemicals, antibiotics, polymers, amino acids, and various food additives.

Components of Molecular Biotechnology

The first experiments in which DNA fragments were joined *in vitro* and the recombinant molecules reintroduced into living cells were performed nearly 30 years ago (Cohen et al., 1973). The basic information obtained in these early experiments, together with numerous new findings in all fields of bioscience, as well as in chemical, physical, and computer sciences, have led to the development of modern molecular biotechnology. This new field has at least three components: (1) recombinant DNA technology; (2) biomolecular engineering, including metabolic and protein engineering; and (3) molecular bioinformatics, including functional genomics and proteomics (Glick and Pasternak, 1998; Kao, 1999; Ryu and Nam, 2000).

Recombinant DNA Technology

Fundamental steps in recombinant DNA technology, also called gene cloning, molecular cloning, or engineering genetics, are the isolation, enzymatic cleavage, and joining (ligation) of a specific DNA fragment of interest into a cloning vector to make a recombinant DNA molecule (Olmedo-Alvarez, 1999). This construct is then transferred into a host (such as bacteria, yeast, or animal or plant cell), amplified, and maintained within the host. The introduction of DNA into bacterial host cells is termed *transformation* (Glick and Pasternak, 1998; Olmedo-Alvarez, 1999). Those host cells that take up the DNA construct (transformed cells) are identified and selected from those that do not carry the recombinant molecule desired.

Restriction Endonucleases

Recombinant DNA molecules could not be constructed without the use of restriction endonucleases (or restriction enzymes). For molecular cloning, both the source DNA that contains the target sequence and the cloning vector must be consistently cut into discrete and reproducible fragments (Russell, 1998). It was only after restriction endonucleases were discovered that the development of recombinant DNA technology became feasible (Glick and Pasternak, 1998). The important feature of these enzymes is that they are able to cleave double-stranded DNA molecules internally at specific nucleotide pair sequences called *restriction sites*. Restriction enzymes are used to produce a pool of discrete and required DNA fragments to be cloned; they are also utilized to analyze the positioning of restriction sites in a piece of cloned DNA or in a segment of DNA in the genome, in this way allowing us to obtain their restriction maps (Russell, 1998). Over 400 different restriction enzymes have been characterized and purified; thus, endonucleases that cleave DNA molecules at many different DNA sequences are available (Olmedo-Alvarez, 1999).

Plasmid Cloning Vectors

Plasmids are self-replicating, double-stranded circular DNA molecules that are maintained in bacteria as independent extra-chromosomal entities necessary to join or ligate *in vitro* DNA molecules or fragments and perpetuate them in a host cell. The particular plasmids used for cloning experiments are derivatives of naturally occurring bacterial plasmids engineered to have features that facilitate gene cloning. Because they are most commonly used, we focus here on particular features necessary for *Escherichia coli* plasmid vector cloning (Glick and Pasternak, 1998; Russell, 1998; Olmedo-Alvarez, 1999):

1. Small size, which is necessary because the efficiency of transfer of exogenous (foreign) DNA into *E. coli* decreases significantly with plasmids that are more than 15 kb long
2. An origin of replication, which allows the plasmid to replicate in *E. coli* because it provides a sequence recognized by the replication enzymes in the cell
3. Unique restriction endonuclease cleavage sites for several different restriction enzymes (called a *polylinker* or *multiple cloning site*), into which the insert DNA can be cloned
4. One or more dominant selectable markers for identifying recipient cells that carry the construct from cells lacking it

The most commonly used markers are genes mediating resistance to antibiotics, such as ampicillin, chloramphenicol, kanamycin/neomycin, and tetracycline (Glick and Pasternak, 1998; Russell, 1998; Olmedo-Alvarez, 1999).

Polymerase Chain Reaction

The polymerase chain reaction, usually referred to as PCR, is an extremely powerful procedure that allows the amplification of a selected DNA sequence *in vitro*. A three-step cycling process achieves this amplification, which can be more than a million-fold.

A typical PCR process entails a number of cycles for amplifying a specific DNA sequence; each cycle has three successive steps:

1. *Denaturation* — The first step in the PCR amplification system is the thermal denaturation of the DNA sample by raising the temperature within a reaction tube to 95°C.
2. *Annealing* — In the next step, the denatured DNA is annealed to primers by incubating at 35 to 60°C. The ideal annealing temperature depends on the base composition of the primers.

3. *Elongation polymerization or replication* — Taq DNA polymerase is used to replicate the DNA segment between the sites complementary to the oligonucleotide primers. The primers provide the 3′-hydroxyl ends required for covalent extension, and the denatured DNA provides the required template function.

Polymerization reaction is usually carried out at 70 to 72°C (Glick and Pasternak, 1998; Snustad and Simmons, 2000). The products of the first cycle of replication are then denatured, annealed to oligonucleotide primers, and replicated again with Taq DNA polymerase. The procedure is repeated many times until the desired level of amplification is achieved. The amplification occurs in an exponential manner (Glick and Pasternak, 1998; Snustad and Simmons, 2000). All steps and temperature changes required during PCR cycles are usually carried out in an automated programmable block heater known as a *PCR machine* or *thermal cycle* (Snustad and Simmons, 2000).

Polymerase chain reaction technologies provide shortcuts for many cloning and sequencing applications. These procedures permit scientists to obtain definitive structural data on genes and DNA sequences when very small amounts of DNA are available.

Biomolecular Engineering

Metabolic Engineering

The development of recombinant DNA technology has led to the emergence of the field of metabolic engineering, the purposeful and directed modification of intracellular metabolism and cellular properties. Metabolic engineering is generally defined as the redirection of one or more enzymatic reactions to produce new compounds in an organism, to improve the production of existing compounds, or to mediate the degradation of compounds (Jacobsen and Khosla, 1998; Nielsen, 1998; Bailey, 1999; DellaPenna, 2001).

The complete sequencing of genomes from several organisms has resulted in the increased availability of genes for metabolic engineering. The number of databases and computational tools to deal with this information has also increased. This development has stimulated, and will continue to stimulate, advances in metabolic engineering. Specific recent advances include the development of nuclear magnetic resonance (NMR)-based methods for monitoring of intracellular metabolites and metabolic flux and the application of metabolic control analysis and metabolic flux analysis to a variety of systems. It has become possible to perform detailed analyses of cellular functions through both *in vivo* and *in vitro* measurements (Nielsen, 1998; Bailey, 1999; DellaPenna, 2001). NMR has been used for *in vivo* measurement of *in vivo* intracellular metabolites, and, based on experiments with ^{13}C-enriched carbon sources followed by measurements of the fractional enrichment of ^{13}C in cellular amino acids, it is possible to quantify intracellular distribution with high precision (Nielsen, 1998).

Although progress in both pathway gene discovery and the ability to manipulate gene expression in transgenic plants has been most impressive during the past two decades, attempts to use these tools to engineer plant metabolism have met with limited success (DellaPenna, 2001; Lassner and Bedbrook, 2001). Though there are notable exceptions, most attempts at metabolic engineering have focused on modifying (positively or negatively) the expression of single genes affecting routes. In general, the ability to predict experimental outcomes has been much improved when targeting conversion or modification of an existing compound to another, rather than attempting to increase flux through a pathway.

Modification of metabolic storage products or secondary metabolic pathways, which often have relatively flexible roles in plant biology, has also been generally more successful than manipulations of primary and intermediary metabolism (DellaPenna, 2001). Thus, exploiting the full biosynthetic capacity of food crops requires a thorough knowledge of the metabolic routes in plants and the regulatory processes involved in plant biochemistry (Galili et al., 2001). When novel branch-points in plant metabolic pathways are introduced by genetic engineering, the introduced enzyme or enzymes must possess a sufficiently high affinity for their substrate(s) to compete with endogenous enzymes (Jacobsen and Khosla, 1998). In addition, the effects of novel carbohydrates, proteins, or lipids on plant physiology and development may limit the range and quantity of products that can be synthesized. The tissue and/or cellular compartment in which the compound is produced may also limit accumulation of the product; for example, the accumulation of a fructosyltransferase construct that had a vacuolar-targeting signal did not lead to an aberrant phenotype, whereas plants containing a construct with apoplastic-targeting signal sequence exhibited severe necrosis (Goddijn and Pen, 1995; Jacobsen and Khosla, 1998). Therefore, the potentially negative effects of engineered compounds on plant growth or development may be prevented by targeting gene product to appropriate cellular compartments; the same effect could also be obtained by using tissue-specific promoters (Goddijn and Pen, 1995; Lassner and Bedbrook, 2001).

In metabolic engineering, a field in continuous progress, hybrid approaches that involve both directed and evolutionary steps are likely to become increasingly important. For example, genetic engineering can first be used to add a gene or set of genes to an organism; an evolutionary approach, such as selection in a continuous reactor, can then be used to achieve further improvements in factors such as growth rate, regulatory properties, or resistance to toxic metabolites (Jacobsen and Khosla, 1998).

Also, the principle of breeding and *in vitro* evolution can be used to access natural product diversity rapidly and in simple laboratory microorganisms such as *E. coli* and in model plants such tobacco or *Arabidopsis* (Schmidt-

Dannert et al., 2000; Lassner and Bedbrook, 2001). Breeding new biosynthetic pathways may involve mixing and matching genes from different sources, even from unrelated metabolic routes, and at the same time creating new biosynthetic functions by random mutagenesis, recombination, and selection, all in the absence of detailed information on enzyme structure or catalytic mechanism. Because the gene functions introduced into a recombinant organism are not coupled to its survival, this approach can be freely used to explore a wide variety of possible product compounds (Schmidt-Dannert et al., 2000).

Protein Engineering

It is possible with recombinant DNA and PCR technologies to isolate the gene for any protein that exists in nature, to express it in a specific host organism, and to produce a purified product that can be used commercially. However, the physicochemical and functional properties of these naturally occurring proteins are often not well suited for either industrial or food processing applications (Goodenough, 1995; Shewry, 1998). By using a set of techniques that specifically change amino acids encoded by a cloned gene, proteins can be created with properties that are better suited than their natural counterparts for therapeutic and industrial uses (Chen, 2001). Protein engineering provides an excellent opportunity to explore and manipulate the structures, composition, and functional properties of food proteins, with special emphasis on storage proteins from food crops (Goodenough, 1995; Shewry, 1998; Chen, 2001). The final goal is to improve their quality for traditional end uses and to introduce new properties for novel food applications. In the enzyme case, some target properties that we may wish to improve or modify may include thermal tolerance or pH stability, or both, enabling the altered protein version to be used under conditions that would inactivate the native macromolecule (Glick and Pasternak, 1998; Chen, 2001). It may also be desirable to modify the reactivity of an enzyme in nonaqueous solvents so chemical reactions can be catalyzed under nonphysiological conditions. Another desired trait would be to change an enzyme so that a cofactor is no longer required for continuous industrial production processes in which the cofactor must be supplied on a regular basis (Goodenough, 1995; Glick and Pasternak, 1998; Chen, 2001). Finally, one can cite alteration of the allosteric regulation of an enzyme to diminish the impact of metabolite feedback inhibition and to increase product yield (Glick and Pasternak, 1998; Chen, 2001).

Protein engineering is undergoing the most profound and exciting transformation in its history, promising unprecedented expansion in the scope and applications for modified or improved proteins and/or enzymes with desired properties. Two complementary strategies are currently available for achieving these goals: rational design and directed evolution or irrational design (Nixon and Firestine, 2000; Tobin et al., 2000).

Protein Rational Design

It is not a simple matter to produce a new protein with specified predetermined properties; however, it is quite feasible to modify the existing properties of known proteins. Theoretically, these changes can be carried out at either the protein or gene level (Glick and Pasternak, 1998); however, chemical modifications of proteins are generally harsh and nonspecific and, in the case of food proteins, have negative effects on nutritional quality (Glick and Pasternak, 1998; Shewry, 1998). They are also required repeatedly for each batch of protein, so it is preferable to manipulate the DNA sequence of a cloned gene to create an altered protein with novel properties. Unfortunately, it is not always possible to know in advance which individual amino acids or short sequence of those amino acids contribute to a particular characteristic. The process for generating amino acid coding changes at the DNA level is called *directed mutagenesis* (Chen, 2001). Determining which amino acids of a protein should be changed to attain a specific property is easier if the three-dimensional structure of the protein has been well characterized by x-ray crystallographic analysis or other analytical procedures (Goodenough, 1995; Chen, 2001). For most proteins, such detailed information is lacking. In these cases, individual amino acid substitutions or secondary structure engineering have generated enzymes or food proteins with desired properties. Despite these spectacular examples, however, numerous attempts at redesigning proteins have failed (Goodenough, 1995; Shewry et al., 2000; Chen, 2001). These failures might have resulted, to some extent, from an incomplete understanding of the underlying mechanisms or structure–function relationship bases required to enhance the desired enzyme or food functional properties and also because a significant number were based on primary amino acid sequence homologies being the only criterion for amino acid replacements (Goodenough, 1995; Nixon and Firestine, 2000; Chen, 2001).

Protein Irrational Design

Protein irrational design is also called *protein directed evolution* and does not require information about protein structure related to function (Nixon and Firestine, 2000; Tobin et al., 2000; Chen, 2001). This technology accesses an important facet of natural evolution that was lacking in previous formats: the ability to recombine mutations from individual genes similar to natural sexual recombination. This approach employs a random process in which error-prone PCR is used to create a library of mutagenized genes. When coupled to selective pressure or to high-throughput screening, progeny sequences or mutants encoding desirable functions are identified. Desirable clones may be iteratively creating offspring that contain multiple beneficial mutations, until the evolved sequence encoding the desired function is obtained (Nixon and Firestine, 2000; Tobin et al., 2000; Chen, 2001). One limitation of directed evolution is the prerequisite for having a sensitive and efficient method for screening a large number of potential mutants (Chen,

2001). Although either rational design or directed evolution can be very effective, a combination of both strategies will probably represent the most successful route to improve the properties and function of a protein (Nixon and Firestine, 2000).

Molecular Bioinformatics

Functional Genomics and DNA Microarrays

The availability of complete sequence information for many different organisms is driving a revolution in the biological sciences. Unfortunately, the billions of bases of DNA sequence do not tell us what all the genes do, how cells work, how cells form an organism, what goes wrong in disease, or how cells respond to stimuli (Lockhart and Winzeler, 2000). This is where functional genomics comes into play (Kao, 1999; Lockhart and Winzeler, 2000). The goal of functional genomics is not simply to provide a catalog of all the genes and information about their functions, but also to understand how the components work together to comprise functioning cells and organisms (Lockhart and Winzeler, 2000). For the first time, technologies that analyze thousands of genes in parallel are generating comprehensive, high-resolution, and quantitative information on the cellular states of living organisms (Lockhart and Winzeler, 2000; Ryu and Nam, 2000). Thus, the widespread and routine use of functional genomic tools that are based on DNA microarray technology promises to shed light on virtually all scientific arenas, from the fundamental issue of how cells grow to the medical challenge of understanding cancer and other human diseases; breeding plant crops, farm animals, and food-grade microorganisms in order to improve food quality and production; and the industrial goal of developing fermentation processes (Kao, 1999; Ryu and Nam, 2000; Lockhart and Winzeler, 2000). The development of microarray technology is intimately connected with the transition of molecular biology from its classical phase into its postgenomic era. DNA microarray technology promises not only to dramatically speed up the experimental work of molecular biologists but also to make possible an entirely new experimental approach in molecular biology (Kao, 1999; Blohm and Guiseppi-Eli, 2001). Instead of investigating the complexity of biological effects by analyzing single genes of putative importance one after the other, many or even all genes of an organism can now be tested at once: first, to find out which genes are involved in a biological event and, second, to analyze in detail their (inter)actions afterwards. Nucleic acid array works by hybridization of labeled RNA or DNA in solution with DNA molecules attached at specific locations on a surface. The hybridization of a sample to an array is, in effect, a highly parallel search by each molecule for a matching partner on an affinity matrix, with the eventual pairings of molecules on the surface determined by the rules of molecular recognition (Lockhart and Winzeler, 2000).

Traditionally, the arrays have consisted of DNA fragments, often with unknown sequence, spotted on a porous membrane such as nylon material. The arrayed DNA fragments often come from cDNA, genomic DNA, or plasmid libraries, and the hybridized material is often labeled with radioactivity (Lockhart and Winzeler, 2000). Recently, the use of glass as a substrate and fluorescence for detection, together with the development of new technologies for synthesizing or depositing nucleic acids in glass slides at very high densities, have allowed the miniaturization of nucleic acid arrays with concomitant increases in experimental efficiency and information content (Blohm and Guiseppi-Eli, 2001). While making arrays with more than several hundred elements was, until recently, a significant technical achievement, arrays with more than 250,000 different oligonucleotide probes or 10,000 different cDNAs per square centimeter can now be produced in significant numbers (Lockhart and Winzeler, 2000; Blohm and Guiseppi-Eli, 2001). Although it is possible to synthesize or deposit DNA fragments of unknown sequence, the most common implementation is to design arrays based on specific sequence information, a process sometimes referred to as downloading the genome onto a chip (Kao, 1999; Lockhart and Winzeler, 2000; Ryu and Nam, 2000; Blohm and Guiseppi-Eli, 2001). The amounts of RNA or cDNA hybridized to each probe on a microarray can be measured by scanning the blot with an imaging system that measures the amount of radioactivity or fluorescence and analyzing the results with a computer that compares the signals with those produced by known control probes and RNAs or cDNAs (Kao, 1999; Lockhart and Winzeler, 2000; Blohm and Guiseppi-Eli, 2001).

Proteomics

It is becoming increasingly clear that the behavior of gene products is difficult or impossible to predict from gene sequence. Even if a gene is transcribed, its expression may be regulated at the level of translation, and protein products are subject to further control by posttranslational modifications, varying half-lives, and compartmentation in protein complexes (Dove, 1999). Proteomics, more appropriately called *functional proteomics*, is a field that promises to bridge the gap between genome sequence and cellular behavior (Blackstock and Weir, 1999). Proteomics aims to study the dynamic protein products of the genome and their interactions (Blackstock and Weir, 1999). Its rapid emergence in biotechnology is being driven by the development, integration, and automation of large-scale analytical tools, such as two-dimensional gel electrophoresis (2D-PAGE) and tandem mass spectrometry, which have simplified protein analysis and characterization, and the emergence of sophisticated bioinformatics approaches for simplifying complex and interrelated data (Blackstock and Weir, 1999; Patterson, 2000).

Proteomics has its roots in traditional biochemical techniques of protein characterization. Over 25 years ago, high-resolution 2D-PAGE was developed, and its basic principles have remained unchanged (O'Farrell, 1975; Blackstock and Weir, 1999). Essentially, the proteins in a cell or tissue extract are separated, first in one dimension on the basis of charge, and then in a second dimension on the basis of molecular size, resulting in defined spots (O'Farrell, 1975). Early 2D-PAGE experiments focused on comparing the resulting pattern of spots for different tissues, or for bacteria cultured under different growth conditions. Because it was difficult or impossible to determine which proteins the individual spots represented, the usefulness of the system was strictly limited (O'Farrell, 1975). At present, the dominant approach to proteomics combines 2D-PAGE, which separates, maps, and quantifies proteins, with mass spectrometry (MS)-based sequencing techniques, identifying both the amino acid sequences of proteins and their posttranslational appendages (Patterson, 2000). This approach is allied with database search algorithms to sequence and characterize individual proteins. Proteomics relies greatly on genomic databases to facilitate protein identification and in so doing indicates which genes within the database are important in defined circumstances. Consequently, the two fields do not merely focus on complementary levels within the cell but have a synergistic relationship (Dove, 1999; Patterson, 2000).

The use of standardized procedures, robotics, and sophisticated bioinformatics led to systematic evolution of industrialized approaches with an unprecedented level of sensitivity and selectivity (Patterson, 2000). Using the 2D-PAGE/MS approach, the leading proteomics laboratories can separate hundreds of proteins within individual gels at low femtomolar sensitivities and characterize amino acid sequences and posttranslational modifications of proteins at a rate of up to 1000 a week (Patterson, 2000).

This approach to proteomics has limitations. There remain difficulties, for instance, in subjecting hydrophobic and very low molecular weight protein to gel-based analysis. Nevertheless, it represents by far the most powerful and comprehensive means yet devised to screen protein components of biological samples and to compare, for example, healthy and diseased specimens (Blackstock and Weir, 1999; Dove, 1999).

On the other hand, two different approaches are currently being pursued to separate proteins in chips (Dove, 1999; Lueking et al., 1999). In one technique, chips are prepared with different surface chemistries, much like traditional protein chromatography columns, but in a flat array. Proteins can then be separated on the basis of their chemical affinities and identified with MS (Dove, 1999; Patterson, 2000). In the second strategy, which is more similar to the affinity array used in DNA chips, scientists attach antibodies or nucleic acid sequences to the chip surface and then allow antigens or transcription factors to bind to the surface, where they can be detected and identified (Dove, 1999; Lueking et al., 1999).

Applications of Molecular Biotechnology

Plant Biotechnology for Food Production

Plant genetic engineering may be defined as the manipulation of plant development, structure, or composition by insertion of specific DNA sequences (Halford and Shewry, 2000). These sequences may be derived from the same species or even variety of plant. This may be done with the aim of altering the levels or patterns of expression of specific endogenous genes — in other words, to make them more or less active or to alter when and where in the plant they are switched on or off (Halford and Shewry, 2000). Alternatively, the aim may be to change the biological (e.g., regulatory or catalytic) properties of the proteins they encode. However, in many cases, genes are derived from other species, which may be plants, animals, or microbes, and the objective is to introduce novel biological properties or activities (Halford and Shewry, 2000).

To date, numerous transgenic plants have been generated, including many crop and forest species. In the near future, plant biotechnology will have an enormous impact on conventional breeding programs, because it can significantly decrease the 10 to 15 years that it currently takes to develop a new variety using traditional techniques; further, it will also be used to create plants with novel traits (Halford and Shewry, 2000; Lassner and Bedbrook, 2001; Ryals, 2001).

From a biotechnology point of view, there exist two areas in which plant genetic engineering is being applied as a means of enhancing the rational exploitation of plants. In the first, the addition of genes often improves the agronomic performance or quality of traditional crops (Ryals, 2001). Thus, tremendous progress has occurred in the genetic engineering of crop plants for disease-, pest-, stress-, and herbicide-resistance traits, as well as for traits that enhance shelf life and processing characteristics of harvested plant materials (Grierson, 1998; Dunwell, 1999). Some genetically determined agricultural traits, such as recombinant resistance approaches or delay of fruit senescence and ripening, are attractive because they involve only minor changes to the plant (i.e., the introduction of a single heterologous gene). Consequently, the characteristics of commercially successful cultivars are likely to remain unmodified due to genetic improvement (Grierson, 1998; Ryals, 2001).

In the second major area, genetic manipulation is being exploited with the objective of improving the quality of plant products consumed by human beings, and that will affect their nutrition and health. This has already yielded more nutritious grains with modified oil, protein, carbohydrate content, and composition; process-improved flours; designer oilseeds with tailored end-uses; and plants producing high-value biomolecules, such as milk and pharmaceutical proteins, industrial enzymes, vitamins, pigments, nutraceuticals, and edible vaccines for the food and pharmaceutical industries

(DellaPenna, 1999; Dunwell, 1999; Hirschberg, 1999; Kishore and Shewmaker, 1999; Mazur et al., 1999; Fischer and Emans, 2000; Giddings et al., 2000; Ye et al., 2000; Rascón-Cruz et al., 2001). Among the approaches used to produce such transgenic plants with new quality attributes, which represent the second generation of genetically modified plants, are (1) manipulation of plant endogenous metabolic pathways in order to favor the accumulation of important and desired products, and (2) generation of transgenic plants that can act as living bioreactors or biofactories (Goddijn and Pen, 1995; DellaPenna, 1999; Dunwell, 1999; Hirschberg, 1999; Kishore and Shewmaker, 1999; Mazur et al., 1999; Fischer and Emans, 2000; Giddings et al., 2000).

Improvement of Plant Nutritional and Functional Quality

Human beings require a diverse, well-balanced diet containing a complex mixture of both macronutrients and micronutrients in order to maintain optimal health. Macronutrients, carbohydrates, lipids, and proteins make up the bulk of foodstuff and are used primarily as an energy source (Guzmán-Maldonado and Paredes-López, 1999). Modifying the nutritional composition of plant foods is an urgent worldwide health issue, as basic nutritional needs for much of the world population are still unmet (DellaPenna, 1999; Guzmán-Maldonado and Paredes-López, 1999; Kishore and Shewmaker, 1999; Mazur et al., 1999). Large numbers of people in developing countries exist on diets composed mainly of a few staple foods which usually present poor food quality for some macronutrients and many essential micronutrients (DellaPenna, 1999; Guzmán-Maldonado and Paredes-López, 1999; Kishore and Shewmaker, 1999; Mazur et al., 1999).

Seeds and tubers are the most important plant organs harvested by humankind, in terms of their total yield and their use for food, feed, and industrial raw material. This exploitation is possible because they contain rich reserves of storage compounds such as starch, proteins, and lipids. Genetic engineering provides an opportunity to explore and manipulate the structure, nutritional composition, and functional properties of those macromolecules to improve their food quality for traditional end uses and to introduce new properties for novel applications (Goodenough, 1995; Shewry, 1998; Rooke et al., 1999; Chakraborty et al., 2000; Osuna-Castro et al., 2000; Shewry et al., 2000).

Plant Proteins

Because animal proteins are more expensive, people in developing countries virtually depend on seed proteins alone for their entire protein requirement. But, unlike animal proteins such as casein and egg albumin, which are nutritionally more balanced in terms of essential amino acids, plant proteins are generally deficient in some essential amino acids (Guzmán-Maldonado and Paredes-López, 1999). Animals, including humans, are incapable of synthesizing essential amino acids; these must be supplied from outside by food.

Seed storage proteins have been classified into four groups on the basis of their solubility in water (albumins), dilute saline (globulins), alcohol/water mixtures (prolamins), and diluted acid or alkali (glutelins) (Shewry and Casey, 1999). Albumin and globulin storage proteins are also classified according to their sedimentation coefficients into 2S albumin and 7–8S and 11–12S globulin, also known as vicilins and legumins, respectively (Shewry and Casey, 1999). Seeds form the major source of dietary proteins for humans and their livestock. The seed storage globulins of legumes are low in the sulfur-containing amino acids cysteine and methionine, such as in soybeans and common beans, whereas cereal prolamins are normally deficient in lysine and tryptophan, such as in maize and rice (Shewry, 1998; Tabe and Higgins, 1998; Yamauchi and Minamikawa, 1998; Rascón-Cruz et al., 2001). Consequently, diets based on a single cereal or legume species result in amino acid deficiencies (Guzmán-Maldonado and Paredes-López, 1999).

Storage proteins also confer functional properties which allow seeds to be processed in many food systems (Shewry, 1998). These proteins are particularly important in soybeans for texturing and gel (tofu) formation, and in the production of bread, pasta, and other products from wheat (Shewry, 1998). Thus, functional characteristics are largely determined by the amount and properties of the storage proteins.

Processing quality poses some problems, as it may be determined by the biophysical properties of the seed proteins (rather than their composition) and their interactions with other seed components, such as starch, lipids, and non-starch polysaccharides (cell wall components) (Miflin et al., 1999). In this case, it may be necessary to define quality criteria in molecular and structural terms before attempting to make improvements (Shewry, 1998; Shewry et al., 2000). In addition, functional attributes can only be assessed after grain has been broken down into its component parts. Small-scale grain processing facilities suitable for gram quantities of grain are necessary to isolate grain fractions similar to those produced in commercial scale milling and processing plants. These small-scale facilities eliminate the time-consuming seed multiplication step necessary for large-scale assays (Mazur et al., 1999; Rooke et al., 1999). Tiered functional assays, in which high-throughput functional, nutritional, and sensory evaluations are carried out, also increase the efficiency of functionality assessments (Mazur et al., 1999; Rooke et al., 1999).

Due to the great importance of seed storage proteins, they are one of the major targets for creating food crops tailored to provide better nutrition for humans and improved food functional properties (Tabe and Higgins, 1998; Yamauchi and Minamikawa; 1998). Understanding the structural features of seed reserve proteins is important to provide a basis for proposed engineering mutations at suitable sites of these proteins, with the ultimate goal of enhancing nutritional and/or food processing quality, without affecting their folding/assembly, transport, deposition, and packing into protein bodies of plant seeds (Shewry, 1998; Shewry et al., 2000).

Nutritional Quality

The benefits of food crops that produce seed proteins with improved amino acid composition are numerous. One way to reach this goal is to increase the amount of essential amino acids present in seed material used either for human consumption or for livestock feed (Tabe and Higgins, 1998). Currently, when corn is used as animal feed, it must be supplemented with soybean meal or purified lysine, or both (Glick and Pasternak, 1998). The development of high-lysine crops such as corn or soybeans is a good approach to replace the use of expensive lysine supplements.

Lysine is a nutritionally important essential amino acid; its level in plants is largely regulated by the rate of synthesis and catabolism (Galili et al., 2001). Lysine is a member of the aspartate family of amino acids and is produced in bacteria by a branched pathway that also produces threonine, methionine, and isoleucine (Galili et al., 2001). The first step in the conversion of aspartic acid to lysine is phosphorylation of aspartic acid by aspartokinase (AK) to produce β-aspartyl phosphate. The condensation of aspartic β-semialdehyde with pyruvic acid to form 2,3-dihydrodipicolinic acid, which is catalyzed by dihydrodipicolinic acid synthase (DHDPS), is the first reaction in the pathway that is committed to lysine biosynthesis. AK and DHDPS are the two key enzymes in the lysine biosynthesis pathway, which are both feed-inhibited by lysine (Eggeling et al., 1998; Tabe and Higgins, 1998; Yamauchi and Minamikawa; 1998; Mazur et al., 1999; Galili et al., 2001).

Falco et al. (1995) isolated bacterial genes encoding highly insensitive forms of AK and DHSPS from *E. coli* and *Corynebacterium*, respectively. A deregulated form of the plant DHSPS was created by site-specific mutagenesis (Karchi et al., 1993; Mazur et al., 1999). Expression of these genes in tobacco leaves produces high concentrations of free lysine, but no accumulation was observed in tobacco seed with the use of either constitutive or seed-specific promoters (Karchi et al., 1993). It was discovered that the failure to augment lysine concentrations in the seeds was due to the presence of an active catabolic pathway. However, in the seed-specific expression of a feedback-insensitive AK alone resulted in a 17-fold increase in free threonine and a 3-fold increase in free methionine in the seed of transgenic tobacco (Karchi et al., 1993; Tabe and Higgins, 1998).

In soybean and canola seeds, lysine accumulated sufficiently to more than double the total seed lysine content (Table 1.1) (Falco et al., 1995; Mazur et al., 1999). In corn, expression with an endosperm-specific promoter did not lead to lysine accumulation, whereas expression with an embryo-specific promoter gave high levels of lysine sufficient to raise overall lysine concentrations in seed 50 to 100%, with minor accumulation of catabolic products (Table 1.1) (Mazur et al., 1999).

On the other hand, a number of studies have focused on increasing the sulfur (S)-amino acid contents in legume and rapeseed crops and lysine and tryptophan contents in cereal crops (Shewry, 1998). For example, a gene for β-phaseolin, a 7S globulin from *Phaseolus vulgaris*, was modified by addition

TABLE 1.1

Advances in Improvement of Nutritional and Functional Quality of Seed Storage Proteins

Improvement	Protein	Target Seed Crop	Results	Ref.
Nutritional	Engineered high-methionine bean phaseolin	Tobacco	Degraded due to misfolding	Hoffman et al., 1988
Nutritional	Bacterial-insensitive AK and DHDPS	Soybean, canola, and maize	Increase in total seed lysine level	Falco et al., 1995; Mazur et al., 1999
Nutritional	Methionine-rich 2S albumin of Brazil nut	Soybean and canola	Spectacular increase in total methionine content; allergenic in humans	Altenbach et al., 1992; Nordlee et al., 1996
Nutritional	Methionine-rich sunflower 2S albumin	Lupin	Increase of methionine level	Molvig et al., 1997
Nutritional	Essential amino-acid-rich amaranth 2S albumin	Potato	Improved nutritional quality	Chakraborty et al., 2000
Nutritional	Amarantin with elevated levels of essential amino acids	Maize	Improved nutritional quality	Rascón-Cruz et al., 2001
Nutritional	Antisense cruciferin	Canola	Increase in sulfur-containing amino acids and lysine levels	Kohno-Murase et al., 1995
Functional	Modified 7S/11S globulin ratios by cosuppression technology	Soybean	Improved emulsifying, water- and fat-binding, and gel properties	Mazur et al., 1999; Kinney et al., 2001
Functional	HMW subunit of wheat glutenin	Wheat	Massive increase in dough elasticity	Rooke et al., 1999

Abbreviations: AK, aspartokinase; DHDPS, dihydrodipicolinic acid synthase; HMW, high molecular weight.

of a 45-bp nucleotide sequence encoding a methionine-rich region from a maize 15-kDa zein, a seed storage protein (Hoffman et al., 1988). The added peptide was predicted to form an α-helix structure and was inserted into an α-helical region of phaseolin. The modification would increase the number of methionine residues on the protein from three to nine. The modified globulin gene was expressed, under the control of β-phaseolin specific-seed promoter at the same level as the wild-type gene in the seeds of tobacco, as measured by mRNA abundance. However, the engineered, high-methionine phaseolin accumulated to a much lower concentration than the unmodified protein. It was concluded that the high-methionine globulin was unstable in the developing seed (Table 1.1). The three-dimensional structure of phaseolin indicated that the introduction of methionine residues caused misfolding of the modified phaseolin.

Spectacular success was achieved with the methionine-rich 2S storage albumin of Brazil nut, which contains 26% sulfur-containing amino acids. To elevate the methionine content of the rapeseed crop, the cDNA of the 2S albumin from Brazil nut seed was ligated to the promoter region of the phaseolin gene, and this chimeric gene was introduced into canola plants; an increase of up 30% in total seed methionine was obtained (Table 1.1) (Altenbach et al., 1992). When the promoter region of the legumin gene from field bean directed synthesis of the Brazil nut albumin in *Vicia narbonensis* (narbon beans), the content of total protein was raised 6% (Saalbach et al., 1995). This was sufficient to give a three-fold increase in total seed methionine. In addition, it was reported that high expression levels have also been achieved in seeds of transgenic soybean (Nordlee et al., 1996).

Although Brazil nut albumin is unusually rich in cysteine and methionine, it has a major disadvantage for use in food or feed; it is highly allergenic in its purified form and in extracts of transgenic seeds and could therefore result in the development of allergenic reactions in humans or livestock. This has prevented the commercial development of transgenic crops with improved levels of sulfur-containing amino acid (Nordlee et al., 1996; Yamauchi and Minamikawa, 1998). A related 2S albumin from sunflower seeds contains 16% methionine residues and is not an allergen, making it an interesting target for improving nutritional traits in legume crops (Molvig et al., 1997). Its cDNA was ligated to the promoter of the vicilin gene from pea, and then transferred to lupin. Methionine content in seed proteins of transgenic lupin was approximately twofold higher than the non-transgenic lupin (Molvig et al., 1997). Results of a rat feeding trial indicated that the nutritional value of the seeds from the transgenic lupin was significantly higher than that of the untransformed lupin.

A different approach to improving the nutritional quality of rapeseed (*Brassica napus*) has been taken by Kohno-Murase et al. (1995). They transformed plants with an antisense gene for the 11S storage protein cruciferin. This did not significantly affect the total protein content of the seeds, but the nutritionally poor cruciferins were decreased and the 2S albumins increased. The net result was increases of about 32, 10, and 8% in the levels of cysteine, lysine (which is lower in rapeseed [*Brassica napus*] than in soybean), and methionine, respectively (Table 1.1).

Potato is the most important noncereal food crop and ranks fourth in terms of total global food production, besides being used as animal feed and as raw material for manufacture of starch, alcohol, and other food products. The essential amino acids that limit the nutritive value of potato protein are lysine, tyrosine, methionine, and cysteine.

Chakraborty et al. (2000) used an interesting strategy for improving potato nutritional quality. The 2S albumin from amaranth seed (*Amaranthus hypochondriacus*), termed AmA1, is nonallergenic in nature and in its purified form and is rich in all essential amino acids, including those in which potato is deficient; the amino acid composition corresponds well with World Health Organization standards for optimal human nutrition (FAO/WHO, 1991). These authors reported the tuber-specific expression of AmA1 in potato by using granule-bound starch synthase (GBSS). The expression of AmA1 in transgenic tubers resulted in a significant increase in all essential amino acids. Unexpectedly, the transgenic plants also contained more total protein in tubers (35 to 45% increase) compared to control plants (with 1.1 g per 100 g tuber), which was in broad correlation with the increase of most essential amino acids. One highly expressing tuber population, labeled pSB8, showed a significant 2.5- to 4-fold increase in lysine, methionine, cysteine, and tyrosine contents; interestingly, in other highly expressing tubers, their amounts appeared to be 4- to 8-fold (Table 1.1).

These findings are consistent with immunoblot data, wherein the expression of AmA1 was found to be 5- to 10-fold higher in pSB8G tubers than that in pSB8 tubers. A striking increase in the growth and production of tubers was observed in transgenic populations. Hypersensitivity tests in mice did not evoke an IgE response, which negated the possibility that the protein is allergenic (Chakraborty et al., 2000). In addition, the literature has not revealed any allergenicity associated with amaranth grain or amaranth forage. In fact, amaranth grain and its food products, even eaten as fresh vegetables, have been exploited and consumed in Mexico and Latin American countries for many centuries without the development of allergies (Guzmán-Maldonado and Paredes-López, 1999; Segura-Nieto et al., 1999). The pseudocereal amaranth has been identified as a food crop comparable with most potential food and feed resources because of the exceptional nutritional–functional quality of its storage proteins (Guzmán-Maldonado and Paredes-López, 1999; Segura-Nieto et al., 1999). Specifically, its 11S globulin, called amarantin, one of the most important amaranth proteins, contains a good balance of essential amino acids that nearly meets the needs of human protein nutrition in reference to protein requirements established by the World Health Organization (FAO/WHO, 1991; Segura-Nieto et al., 1994, 1999; Barba de la Rosa et al., 1996).

A previous report showed that amarantin cDNA may be synthesized in *E. coli*, exhibiting electrophoretic, immunochemical, and surface hydrophobicity properties similar to those of native amarantin from amaranth seed

(Osuna-Castro et al., 2000). Thus, it would be advantageous to express novel seed proteins such as amarantin in maize grains with the objective of improving amino acid composition.

Very recently, Rascón-Cruz et al. (2001) found that the amarantin cDNA with signal peptide and under control of the tissue-specific promoter from rice glutelin 1 (osGT1) was synthesized, processed correctly, and accumulated specifically in the seed endosperm of tropical maize. In addition, they achieved a remarkable change in respect to total protein content (44 to 75%) in transgenic maize seeds, finding a significant increase in the limiting essential amino acids ranging from 23 to 83%, when compared with untransformed maize.

Thus, the highest expression was achieved in the maize line, which showed a significant 1.3- and 1.6-fold increase in lysine and tryptophan, respectively, which are deficient in ordinary maize, and the third limiting amino acid, isoleucine, also increased around 2-fold (Table 1.1). In addition, the content of other essential amino acids in transgenic maize was improved. Amarantin has an elevated content of essential amino acids and, because maize constitutes an important component of the diet of people in many developing countries and in particular of Mexico, the final target of this research work is to overexpress amarantin in order to more fully exploit its nutritional potential.

Functional Quality

In soybeans, two classes of storage proteins, 7S and 11S globulins, predominate, each with useful functional characteristics. Transgenic lines in which one or the other of these classes was eliminated by cosuppression technology, and lines with altered 7S/11S ratios were then tested at increasing scale for emulsification, water- and fat-binding properties, gel-forming ability, and other parameters relevant to processing, cooking, and food manufacture. From these assessments, specific transgenic soybeans have been selected for testing in milk and meat replacement products (Table 1.1) (Mazur et al., 1999; Kinney and Jung, 2001).

The ability to make leavened bread from wheat flour depends largely on the unusual properties of the gliadin and glutenin proteins, which together constitute gluten; these properties are not shared by the prolamins of barley and wheat, although they are related structurally to the gluten proteins (Shewry, 1995, 1998; Miflin et al., 1999). Gluten storage proteins form a continuous network in dough, conferring the viscoelastic properties necessary to entrap carbon dioxide released during the proofing of leavened bread. The protein network also provides the cohesiveness required for other foods; for example, high elasticity is required for making noodles and pasta, while more extensible doughs are needed for making cakes and biscuits (cookies). The quality of wheat is determined by genetic and environmental factors, with poor quality generally resulting from low gluten elasticity (Shewry, 1998; Shewry et al., 2000).

Wheat gluten is a complex mixture of proteins with over 50 individual components, but one group of proteins, the high-molecular-weight (HMW) subunits of glutenin, which form HMW (above 1×10^6-Da) polymers stabilized by inter-chain disulfide bonds, are particularly important. A range of studies provide strong evidence that the glutenin polymers are responsible for gluten elasticity, and the variation in quality is associated with differences in the number (three, four, or five), amounts (about 6 to 12% of the total protein), and properties of the glutenin polymers and their constituent subunits (Shewry, 1998; Shewry et al., 2000). Shewry's research group has shown that increasing the number of expressed subunits in a wheat line with a poor-quality background from two to three and four results in stepwise increases in dough elasticity, mirroring the effects of manipulating gene dosage by conventional breeding; however, it has been possible to go beyond the gene dosage obtainable by classical breeding. This results in gluten that contains over 20% of HMW subunits and a massive increase in dough elasticity (Table 1.1) (Rooke et al., 1999). The resulting flour is actually too strong to be used for bread making, but may be valuable for blending to fortify poor-quality wheat. This is, however, a coarse approach, and future improvements may require a finer adjustment of gluten structure. This could be achieved by making specific mutations — for example, changing the pattern of disulfide bonds.

Similarly, the amounts and properties of other types of gluten protein could also be manipulated. Protein engineering is a powerful tool for studying the structures and biophysical properties of individual gluten proteins, and it is even possible to determine the functional properties of heterologously expressed proteins by incorporation into dough using a small-scale mixograph (Rooke et al., 1999; Shewry et al., 2000). However, such incorporation experiments are difficult to perform and it is not possible to ensure that incorporated protein will form the same molecular interactions as it would if expressed in the developing wheat grain. The developing of wheat transformation systems, therefore, provides an opportunity to determine the roles of individuals proteins, engineered or not, in wheat gluten structure and functionality. This will eventually allow for the production of a variety of wheats in which the structure and properties of gluten are fine-tuned for various end uses, including bread, pasta and noodles, and other baked products (Shewry, 1998; Shewry et al., 2000).

Lipids

In most commercial seed oils, more than 95% of the mass can be attributed to only a few fatty acids — specifically, lauric, mirystic, palmitic, stearic, oleic, linoleic, and linolenic acids (Murphy, 1996, 1999; MacKenzie, 1999). Plants produce a wide diversity of fatty acids, the majority of which accumulate in the seed as triacylglycerols. Vegetable oils generally are preferred to oils and fats from other sources because of their higher content of mono- and polyunsaturated fatty acids (Figure 1.1). Unsaturated fatty acids are healthier than saturated fatty acids, and the monounsaturated form, oleic acid, is also more stable in frying and cooking applications than are the polyunsaturated forms, linoleic and linolenic (MacKenzie, 1999; Napier et al., 1999).

It is now possible to engineer "designer" oilseeds tailored to specific end uses. Plant lipids have different end uses, from industrial applications such as lubricants and detergents to food and nutrition (Murphy, 1996, 1999). They also have important roles as nutraceuticals and pharmaceuticals. The global market for plant-derived oils is immense; consequently, the major biotechnology companies have driven much of the research.

Saturated Fatty Acids

Laurate, which is used in confectionery, is normally obtained from either coconut or palm oil. Although both plants yield relatively high levels of the fatty acid, they are limited in their agricultural utility. Research work carried out at Calgene Company has demonstrated the feasibility of engineering canola to produce lauric acid by introducing the gene encoding a lauroyl-specific acyl-carrier protein (ACP), thioesterase, from the California bay plant. This enzyme causes premature chain termination of the growing fatty acid, resulting in the accumulation of lauric acid, a 12-carbon saturated fatty acid, rather than normal C18 oils (Figure 1.1). Canola normally contains less than 0.1% lauric acid; however, with the latter approach, Calgene has produced a number of canola lines that accumulate around 40% laurate (Voelker et al., 1992). The novel fatty acids are recovered from transgenic canola by standard processing methods. These oils, trivially named canola laurate, are now marketed as a partially hydrogenated vegetable oil for use in confectionery under the trade name Laurical.

This work has also demonstrated the feasibility of producing large amounts of transgene-modified plant oils to supplement or replace fluctuating natural sources (Murphy, 1996, 1999; MacKenzie, 1999). Saturated fatty acids are relatively uncommon in most plant storage lipids because of the presence of highly active desaturases in developing seeds. The composition of canola oil has been modified by expressing an antisense stearoyl-ACP desaturase cDNA, under the control of a napin or ACP promoter, in seeds of *Brassica rapa* and *B. napus*. The activity and amount of stearoyl-ACP desaturase were greatly reduced, resulting in a marked increase in stearic acid from less than 2% to as much as 40%; oleic acid levels also decreased (Knutzon et al., 1992).

In other experiments, the cloning of the genes encoding each of the soybean fatty acid desaturases enabled the cosuppression of the seed-specific desaturase, causing seed oleic acid levels to rise from 25% in nontransformed lines to 85% in transgenic soybean (Mazur et al., 1999). The cosuppressed, seed-specific, high-oleic-acid, transgenic soybean plants showed excellent agronomic properties.

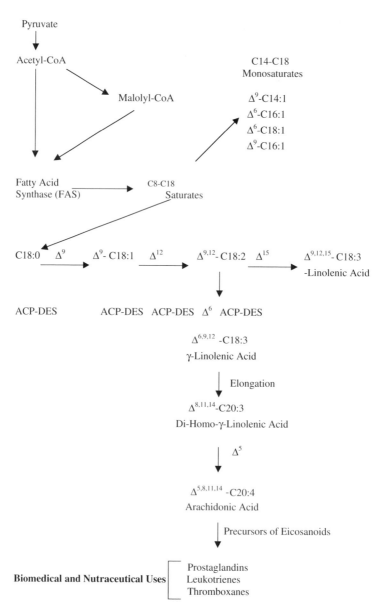

FIGURE 1.1

Generalized scheme for the biosynthesis and metabolism of lipids. Fatty acid precursors, such as pyruvate and malate, are imported into plastid for conversion to acetyl-CoA, then the FAS complex converts Acetyl-CoA and Malolyl-CoA units into C8–C18 saturated acyl-ACPs. Depending on the plant species, C14–C18 saturates may be desaturated by a variety of ACP–DES leading to synthesis of PUFAs. The arachidonic PUFA is subject to further trans-formations to form eicosanoid precursors that present important medical and nutraceutical applications. Abbreviations: ACP, acyl-carrier protein; ACP–DES, acyl-ACP desaturase; FAS, fatty acid synthase; PUFAs, polyunsaturated fatty acids. (Adapted from Murphy et al., 1999; Napier et al., 1999.)

Unsaturated Fatty Acids

It is well documented that the risk of coronary heart disease is positively correlated with elevated levels of high-density lipoprotein (LDL) cholesterol (Kishore and Shewmaker, 1999; MacKenzie, 1999). The C12, C14, and C16 saturated fatty acids are known to induce hypercholesterolemic effects and to increase the proportions of LDLs in the bloodstream, although they differ in their impact (MacKenzie, 1999). The intake of saturated fatty acids has been shown to be closely related to the clotting activity of platelets and their response to thrombin-induced aggregation (MacKenzie, 1999). Consequently, there has been, and continues to be, a parallel research effort to improve the quality of vegetable oils by reducing the total amount of saturated fatty acids (Kishore and Shewmaker, 1999; MacKenzie, 1999).

Unsaturated fatty acids have recently attracted interest as targets for genetic manipulation (Figure 1.1) (Napier et al., 1999). Polyunsaturated fatty acids (PUFAs; long-chain fatty acids containing two or more double bonds introduced by specific desaturase enzymes) have entered the biomedical and nutraceutical areas as a result of the elucidation of their biological role in certain clinical conditions (Napier et al., 1999). Besides pharmaceutical applications, public perceptions of healthy foods and lifestyles have also brought them to the attention of the consumer. Specifically, interest is growing in 18- to 22-carbon PUFAs containing three or more double bonds, as these compounds have emerging therapeutic roles (MacKenzie, 1999; Napier et al., 1999). Current commercial production of PUFAs, largely from seed oils, marine fish, and certain mammals, is considered inadequate for the future PUFA market. However, the seeds of commonly cultivated agricultural plants contain a rather limited range of fatty acids that is truly representative of the diversity of unusual fatty acids accumulated predominantly as storage lipids by plants and fungi (Murphy, 1999). Some of these less common unsaturated fatty acids are highly valuable, although the organisms from which they originate are generally unsuitable for large-scale production and cultivation (Murphy, 1996, 1999). For example, linoleic and linolenic acid PUFAs cannot be synthesized *de novo* by animals; they are therefore classified as essential and must be ingested (MacKenzie, 1999; Napier et al., 1999). Furthermore, for metabolic purposes, the two families cannot be interconverted. Arachidonic acid is synthesized by the elongation of γ-linolenic acid to a longer chain PUFA, a 20-carbon molecule, followed by desaturation at the Δ^5 position (Figure 1.1). Both PUFAs are precursors of a group of short-lived regulatory molecules called the *eicosanoids*. The eicosanoids (prostaglandins, leukotrienes, thromboxanes) have many functions and are particularly important in reproductive function and the regulation of blood pressure (Horrobin, 1990). The ability, therefore, to produce crop plants accumulating fatty acids such as arachidonic acid and other PUFAs is an obvious biotechnological target (Figure 1.1) (Murphy, 1996, 1999; MacKenzie, 1999; Napier et al., 1999).

Transgenic plants expressing genes that encode fatty acid desaturases have been recently generated. Thus, an animal (*Caenorhabditis elegans*) microsomal fatty acid desaturase was expressed in *Arabidopsis*, resulting in increased levels of α-linolenic acid (Spychalla et al., 1997). Moreover, these same transgenic plants desaturated exogenously applied 20-carbon PUFAs, such as arachidonic, indicating that the enzyme was a Δ^{15}/Δ^{17} desaturase (e.g., ω-3) with specificity for both 18- and 20-carbon fatty acids. On the other hand, a *Mortierella alpina* desaturase with an amino-terminal cytochrome b_5-domain was shown to exhibit Δ^5-fatty acid desaturase activity, converting exogenously supplied di-homo-γ-linolenic acid ($\Delta^{8,11,14}$-C20:3); also, transgenic canola was obtained resulting in the expression of a number of unusual, Δ^5-desaturated 18-carbon PUFAs (Michaelson et al., 1998; Napier et al., 1999).

γ-Linolenic is a registered pharmaceutical with wide applications for antiviral and cancer therapies; unfortunately, it only accumulates in a limited number of species, including evening primrose, borage, and amaranth. In 1997, Sayanova et al. were the first to isolate a cDNA from borage plant encoding the Δ^6-fatty acid desaturase responsible for synthesis of this fatty acid and to introduce it into tobacco (used as a model crop). It resulted in the accumulation of γ-linolenic at levels equivalent to or greater than those found in borage (Figure 1.1). The final goal, the production of novel high-value pharmaceutical oils and eicosanoid precursors, may be difficult to achieve but will probably require transgenic oilseed producing both the Δ^5 and Δ^6 desaturases and other as yet undefined enzymes (Napier et al., 1999).

Carbohydrates

Starch is composed of two different glucan chains, amylose and amylopectin. These polymers have the same basic structure but differ in their length and degree of branching, which ultimately affects the physicochemical properties of both polysaccharides (López et al., 1994; Guzmán-Maldonado and Paredes-López, 1999). Amylose is an essentially linear polymer of glucosyl residues joined via α-1,4 glycosidic bonds, whereas amylopectin exists as a branched α-1,4; α-1,6 D-glucan polymer. The physicochemical properties of the α-1,4 glucans are based on the extent of branching and/or polymerization (López et al., 1994; Guzmán-Maldonado and Paredes-López, 1999).

Starch is a very important staple in the diet of the world population and is widely used in the food and beverage industries to produce glucose and fructose syrups as sweetener and to confer functional properties for food processing (López et al., 1994; Schulman, 1999). Starch is primarily used as thickener, but also as binder, adhesive, gelling agent, film former, and texturizer in many snacks.

The relative amounts of amylose and amylopectin are what give polysaccharides their unique physical and chemical properties, which convey specific functionality and are of biotechnological importance (López et al., 1994; Guzmán-Maldonado and Paredes-López, 1999; Slattery et al., 2000).

Therefore, it might be of value to produce polysaccharides with features that are intermediate between amylose and amylopectin, that are more highly branched, or that have a higher molecular weight (Visser and Jacobsen, 1993; Slattery et al., 2000).

The basic pathways of amylose and amylopectin synthesis are well understood and genes readily cloned (Visser and Jacobsen, 1993; Schulman, 1999; Slattery et al., 2000). Thus, the potential exists to produce plant starches with a wider range of structures and properties. The targets are varied but include mutant wheat starches to mimic those from maize, phosphorylated starch (currently obtained from potato), and resistant starches for healthy diets, among others (Guzmán-Maldonado and Paredes-López, 1999; Schulman, 1999).

Most starch is synthesized from sucrose, a route that involves four steps: initiation, elongation, branching, and granule formation. In maize and many other plant species, at least 13 enzymes have been identified in the starch biosynthetic pathway (Visser and Jacobsen, 1993; Schulman, 1999; Slattery et al., 2000). Of these, three enzymes are considered to be key in the synthesis of amylose and amylopectin:

1. ADP-glucose pyrophosphorylase (AGP) is involved in the initiation step and generates the glucosyl precursor ADP-glucose from glucose-1-phosphate.

2. Two distinct classes of starch synthase (involved in elongation and granule formation) are found within the plastids: those bound exclusively to the granule, known as granule-bound starch synthases (GBSSs), and others that can be found in the soluble phase or granule bound, known as starch synthases (SSs).

3. Branching enzyme (involved in branching and granule formation) is a transglycosylase involved in amylopectin synthesis, also known as starch synthase of amylopectin biosynthesis.

Amylose has been shown to be the product of GBSSs. The waxy mutants, which lack GBSS, have only amylopectin but still possess soluble starch synthases, suggesting that different enzymes participate in amylose and amylopectin synthesis (Slattery et al., 2000).

Starch Quality

High-amylose starches have numerous industrial applications including fried snack products to create crisp, evenly browned snacks; as thickeners, as they are strong gelling agents; and, owing to their rapid setting properties, in confectionery (López et al., 1994; Guzmán-Maldonado and Paredes-López, 1999; Schulman, 1999; Slattery et al., 2000). An added bonus of these types of starches is that they hamper the penetration of cooking oils, leading to a decrease in fat intake by the consumer (Guzmán-Maldonado and Paredes-López, 1999; Schulman, 1999; Slattery et al., 2000). Thus, high-amylose starch is in great demand by the starch industry because of its unique functional properties. However, very few high-amylose crop varieties are commercially available. Recently, Schwall et al. (2000) achieved the goal of producing amylose-rich potatoes by simultaneously inhibiting two isoforms of starch branching enzymes (SBE A and SBE B) to below 1% of the wild-type activities through SBE A + B antisense inhibition; that is, to generate potato starch with very high amylose content, antisense SBE B lines were retransformed with an antisense SBE A construct. Starch and granule morphology and composition were noticeably altered. Normal HMW amylopectin was absent, which was also reflected in the strong decrease of short chains compared to wild-type starch, whereas the amylose content was increased to levels comparable to the highest commercially available maize starches, with an apparent amylose level of 90%. In normal potatoes, about 80% of the starch consists of amylopectin, while 20% is amylose.

Because antisense inhibition of AGP in transgenic potatoes abolishes starch formation in tubers, an increase in the quantity of storage starch has been attempted by raising AGP activity. This possibility was investigated by transforming potatoes with a mutant *E. coli* AGP gene (*glgC16*); the bacterial strain accumulated 30% more glycogen than normal (Stark et al., 1992). The results showed a 30% increase in starch content. A different approach to increase starch concentration is the ectopic expression of inorganic pyrophosphatase (PP); thus, *E. coli* PP in the cytosol of potato produced some unexpected results. Both the rates of sucrose degradation and starch synthesis increased, and the transformed tubers accumulated 20 to 30% more starch than wild-type tubers (Geigenberger et al., 1998).

It has previously been cited that in normal potatoes about 80% of the starch consists of amylopectin, while 20% is amylose. Thus, efforts targeted at practical alteration of starch quality, producing amylose-free crops, began with a mutagenesis program in potatoes and the successful creation of a low-amylose line, the result of a single point mutation in the GBSS gene responsible for amylose production. By means of genetic engineering, an amylose-free potato crop has also been produced, expressing an antisense RNA for the GBSS transcript. The pertinent transgenic amylopectin potato is largely free of amylose (Tramper, 2000).

Plants as Bioreactors

As a result of plant genetic engineering, compounds of commercial interest that were previously available only from exotic plant species, from other organisms, or in limited amounts can now be produced in domesticated crops (Goddijn and Pen, 1995). In fact, currently producing biomass by growing crop plants in the field can, in general, compete with any other production system; the process is inexpensive and requires limited facilities to produce bulk quantities (Goddijn and Pen, 1995; Arakawa et al., 1999).

Molecular biotechnology will enable broadening of the range of products and use of transgenic plants as a versatile renewable and low-cost source of novel high-value molecules (Goddijn and Pen, 1995; Arakawa et al., 1999; Dunwell 1999; Fischer et al., 1999; Fischer and Emans, 2000; Giddings et al., 2000). This area of novel commercial exploitation of plants is called *biofarming* or *molecular farming* and involves the crop-plant-based production of industrial or therapeutic biomolecules. In this application, the plant can be considered as a solar-powered bioreactor and an attractive alternative to conventional microbial or animal cell expression systems. Its requirements are simple and inexpensive: sunlight, mineral salts from the soil (or fertilizers), and water (Goddijn and Pen, 1995; Arakawa et al., 1999; Dunwell 1999; Fischer et al., 1999; Fischer and Emans, 2000; Giddings et al., 2000). Similarly, as traditional agriculture takes advantage of these characteristics in the large-scale production of feed and foodstuff items, they can be equally exploited by plant molecular farming in the production of commercially important recombinant proteins and other biomolecules. In addition to economic benefits that plants present when used as biofactories, other very important benefits include reduced risk from human pathogen contamination such as mammalian viruses (human immunodeficiency virus and hepatitis B), blood-borne pathogens, oncogenes, and bacterial toxins (Arakawa et al., 1999; Fischer et al., 1999; Fischer and Emans, 2000; Giddings et al., 2000). The cultivation, harvesting, transport, storage, and processing of transgenic crops would also use an existing infrastructure and require relatively little capital investment, making their commercial production by molecular farming technology an exciting prospect.

Vaccines

The availability of recombinant biopharmaceuticals and the identification of molecules involved in many devastating human diseases and economically important plant diseases have created demand for amounts of safe, inexpensive, recombinant proteins (Fischer et al., 1999). Large-scale production of these molecular medicines, such as recombinant antibodies and vaccines, has the potential to make new therapies and novel diagnostic tests widely available for detecting and combating these diseases (Fischer et al., 1999; Fischer and Emans, 2000).

Vaccination has been one of the greatest advances in medical science, dramatically improving human life expectancy and quality (Fischer and Emans, 2000). It is the most cost-effective form of health care. In 1992, the World Health Organization (Geneva) and a consortium of philanthropic organizations presented a Children's Vaccine Initiative. The focus of this project was to encourage the development of technology that would make vaccines available to developing countries, where they are needed most. Priority areas included lower cost vaccines that could easily be distributed in poor countries lacking refrigeration, healthcare infrastructure, and oral vaccines (Walmsley and Arntzen, 2000).

Subunit vaccines consist of specific macromolecules that induce a protective immune response against a pathogen (Walmsley and Arntzen, 2000). Plants have now been modified on the basis of peptide epitopes of pathogens capable of invoking a protective immune response (vaccinogen) to produce vaccines against a variety of human and animal diseases (Giddings et al., 2000). Traditional vaccines utilize either killed or attenuated whole disease-causing organisms. Plant-based vaccines allow for the use of vaccine-selected specific subunits, avoiding the risk of causing the disease, as is possible with whole organisms.

Plant Vaccines

The first demonstration of expression of a vaccinogen in plants ocurred in 1990 when Curtiss and Cardineau expressed the *Streptococcus mutans* surface protein antigen A (SpaA) in tobacco. After incorporation of the transgenic tobacco tissue into the diet of mice, a mucosal immune response was induced to the SpaA protein; the induced antibodies were demonstrated to be biologically active when they reacted with intact *S. mutans*.

The most widespread of autoimmune diseases is diabetes mellitus, a condition affecting millions worldwide (Dunwell, 1999). Its treatment involves daily injections of insulin, a procedure designed to regulate the levels of glucose in the blood; if glucose levels are too high, then nonspecific protein glycosylation can occur, a process that often leads to nerve damage, blindness, and a range of other life-threatening conditions (Dunwell, 1999). It is known that oral administration of disease-specific autoantigens can prevent or delay onset of autoimmune disease symptoms. In an experiment to test transgenic potatoes that expressed either human insulin (the major autoantigen) alone or a hybrid protein in which insulin was linked to the C-terminus of the cholera toxin B subunit (LT-B), cholera toxin was used as a potent oral antigen and oral adjuvant that induces the production of antitoxin antibodies (Arakawa et al., 1998). Non-obese diabetic mice fed tuber tissue containing microgram amounts of the LT-B–insulin fusion protein showed a significant reduction in pancreatic inflammation and a delay in the onset of diabetic symptoms.

Reports have since followed of expression of a hepatitis antigen in tobacco and lettuce, a rabies antigen in tomato, and a cholera antigen in tobacco and potatoes (Table 1.2). Animal trials demonstrating antigenicity of plant-derived vaccinogens include tobacco- and lettuce-derived hepatitis B surface antigen (HBsAg), a tobacco- and a potato-derived bacterial diarrhea antigen, a potato-derived Norwalk virus antigen, and an *Arabidopsis*-derived foot-and-mouth disease antigen (Table 1.2) (Walmsley and Arntzen, 2000).

Chimeric plant viruses were proven effective as carrier proteins for vaccinogens in 1994, after rabbits raised an immune response against purified recombinant cowpea mosaic virus (CPMV) particles expressing epitopes derived from human rhinovirus 14 and human immunodeficiency virus-1

TABLE 1.2

Transgenic Plant Production of Vaccine- and Antibody-Like Biopharmaceuticals against Some Important Human Diseases

Vaccine/Antibody	Host Plant	Potential Medical Use	Ref.
Human immunodeficiency virus-1 (HIV-1)	Cowpea	AIDS	Porta et al., 1994
Colorectal cancer antigen	Maize, tobacco	Cancer	Verch et al., 1998
E. coli heat-labile toxin B subunit	Potato, tobacco	Cholera	Tacket et al., 1998; Walmsley and Arntzen, 2000
Streptococcus mutans surface protein antigen A (SpaA); *S. mutans* adhesin	Tobacco	Dental caries; protection against oral streptococcal colonization	Curtiss and Cardeneaus, 1990; Ma et al., 1994; 1995; 1998
Heat-labile *E. coli* toxin B subunit	Potato, tobacco	*E. coli* diarrhea	Tacket et al., 1994; 1998; Walmsley and Arntzen, 2000
Norwalk virus capsid protein	Potato, tobacco	Viral diarrhea	Mason et al., 1996
Insulin-cholera toxin B subunit fusion protein	Potato	Diabetes	Arakawa et al., 1998
Hepatitis B surface antigen (HBsAg)	Lettuce, lupin, tobacco	Hepatitis B	Walmsley and Arntzen, 2000; Fischer and Emans, 2000; Giddings et al., 2000
Herpes simplex virus 2	Soybean	Vaginal herpes infection	Zeitlin et al., 1998
Malarial B-cell epitope	Tobacco	Malaria	Turpen et al., 1995
Rabies virus epitopes	Spinach	Rabies	Moldeska et al., 1998

Source: Adapted and modified from Fischer and Emans, 2000; Giddings et al., 2000.

(HIV-1); for both chimerae, virus particle yields were found to be in the range of 1 to 2 g/kg cowpea leaf tissue (Porta et al., 1994). Numerous reports have since verified plant viruses as effective alternative vaccinogen expression vectors (Table 1.2). Antibodies have been stimulated in mice after injection with plant virus-derived HIV-1 epitopes and rabies virus epitopes (Porta et al., 1994).

Modelska et al. (1998) were the first to detect a mucosal immune response after oral induction with a plant virus-derived vaccinogen. A recombinant alfalfa mosaic virus (AMV) was engineered to express two rabies virus epitopes. Mice were immunized intraperitoneally or orally by gastric intubation or by feeding on virus-infected spinach leaves. Interestingly, mice vaccinated through diet produced twice the level of anti-vaccinogen IgA than that detected in intubated animals. It appears that the additional protection against digestion afforded through delivery of the vaccinogen in plant cells (e.g., bioencapsulation within plant cell walls and membrane compartments) allows successful oral immunization. The produced levels of serum IgG and IgA proved capable of improving the clinical symptoms caused by intranasal infection with an attenuated rabies virus strain.

The first human clinical trials for a transgenic plant-derived antigen were approved by the U.S. Food and Drug Administration and performed in 1997. Transgenic potatoes constitutively expressing a synthetic bacterial diarrhea vaccinogen, the B subunit of *E. coli* heat labile toxin (LT-B), were orally delivered to human volunteers (Tacket et al., 1998).

Thus, each participant received potato cubes from a random sample of non-transgenic control or transgenic tubers. Prior to and at multiple time points after ingestion of the potato, serum and fecal samples were taken and analyzed. A significant rise in LT-B antibodies was displayed by 10 of the 11 test participants, whereas no LT-B-specific antibodies were detected in control volunteers. Interestingly, serum antibody levels induced by ingestion of the transgenic potatoes comparable to those measured when participants were challenged with 10^6 virulent enterotoxigenic *E. coli* organisms (Tacket et al., 1998). Also, other human trials are currently in progress using orally-delivered, potato-derived HBsAg as booster for the commercial hepatitis B vaccine and potato-delivered Norwalk virus virus-like particles for a viral diarrhea vaccine (Thanavala et al., 1995; Tacket et al., 1998; Walmsley and Arntzen, 2000).

As well as in direct therapeutic applications, plant-delivered HBsAg is also being tested for use in diagnostic systems. HBsAg has been expressed in tobacco, and the recombinant product was used successfully in the hemagglutination test routinely conducted by the Japanese Blood Center on blood samples for HBsAg-positive donors (Tsuda et al., 1998). The serological results were comparable to those achieved with the standard antigen from *E. coli*.

Antibodies

The alternative to inducing the immune system to produce antibodies is to deliver them directly for use in passive immunization (Giddings et al., 2000). A key breakthrough in making molecular farming in plants a reality was the demonstration of functional antibody expression in tobacco leaves (Hiatt et al., 1989). The importance of this is underscored by the fact that monoclonal antibodies and recombinant antibodies are essential therapeutic and diagnostic tools used in medicine for human and animal health care, as well as the life science and biotechnology fields (Table 1.2) (Fischer et al., 1999). Advances in modern recombinant DNA technology and antibody engineering have made possible the production of novel polypeptides with desirable properties, smaller antibody fragments, or antibody–fusion proteins linked to enzymes, biological response modifiers, or toxins (Table 1.2) (Fischer and Emans, 2000). Antibodies or antibody fragments produced in plants are often referred to as *plantibodies*.

Plantibodies

The production of a recombinant monoclonal antibody in plants was first described by Hiatt et al. (1989). The antibody chosen for study, 6D4, was a mouse IgG_1 antibody that recognizes a synthetic phosphonate ester and can catalyze the hydrolysis of certain carboxylic esters. For the creation of transgenic tobacco plants synthesizing the intact antibody, a multimer of two heavy-chain polypeptides and two light-chain polypeptides covalently linked by disulfide bonds, a two-step strategy was adopted. Thus, genes encoding either the heavy chain or the light chain were expressed in tobacco, followed by sexual crossing of individual plants expressing either a heavy or a light chain.

From the highest overproducing F1 lines, the yield of 6D4 antibody was up to 10 g/kg of the total protein (Hiatt et al., 1989). This represents one of the highest reported levels for a recombinant produced in transgenic plants (Fischer and Emans, 2000; Giddings et al., 2000; Walmsley and Arntzen, 2000). Moreover, in F1 plants expressing the assembled 6D4 antibody, the protein is secreted into the intercellular spaces and accumulated, in cell suspension, at up to 20 mg/L (Hein et al., 1991). This compares favorably with the levels of monoclonal antibody secreted into the medium by cultured hybridoma cells.

In 1994, Ma et al. reported the production of monoclonal antibody Guys 13 in transgenic tobacco; Guys 13 is a mouse IgG_1 immunoglobulin that binds to the 185-kDa cell surface antigen of *Streptococcus mutans*, the main causative agent of dental caries in human beings. The 185-kDa antigen is streptococcal adhesin, which mediates initial attachment of the bacterium to the tooth surface. This monoclonal antibody was first expressed in tobacco by sequentially crossing plants expressing its individual components. This permitted the production of high levels of whole recombinant Guys 13 (500 µg/g of leaf material) (Ma et al., 1995). Three years later, the same group showed that Guys 13 monoclonal plantibody afforded specific protection in human volunteers against oral streptococcal colonization for at least 4 months (Ma et al., 1998).

Expression of an anticancer monoclonal antibody in plants using a tobacco mosaic virus (TMV) gene delivery system was reported by Verch et al. (1998). This well-studied antibody (CO17-1A) against a colorectal cancer antigen is an IgG_1 molecule. Mature IgG_1 antibody was formed; also, the authors report that trials are underway to test the tumor-suppressive activity of this anticancer plantibody (Verch et al., 1998). On the other hand, a pharmaceutical company plans to begin injecting cancer patients with doses of up to 250 mg of antibody-cancer drug purified from maize seeds. It is also cultivating transgenic soybean producing monoclonal antibodies against herpes simplex virus 2 (HSV-2). The plantibodies showed no detectable differences from the mammalian cell-culture-derived version in terms of their molecular weight, stability in human semen and cervical mucus over 24 hours, ability to diffuse in such mucus, and efficacy in preventing vaginal HSV-2 infection in mice models (Zeitlin et al., 1998).

Finally, the banana is an ideal fruit to contain edible vaccines and antibodies. Bananas are grown extensively throughout the developing world, are an inexpensive food widely consumed by infants or children, and, in contrast to potatoes, can be eaten uncooked. In the near future, banana-producing edible biopharmaceuticals such as vaccines and antibodies could be produced against a range of diseases, including polio, diphtheria, yellow fever, HIV, and certain types of viral diarrhea (Giddings et al., 2000).

We predict that plants will be the premier expression system for diagnostic and therapeutic compounds. Plant expression systems have the potential to be as abundant tomorrow as prescription drugs are today. We foresee that molecular farming will provide a basket full of novel medicines for the diseases of the 21st century, just as plants were the source of medicine in Aztec, Mayan, Egyptian, and Greek times.

Milk Proteins

Human milk has long been recognized as the best-balanced diet for infants, supplying high nutritional value in casein and whey proteins as well as providing a source of antimicrobial proteins such as lactoferrin and lysozyme (Arakawa et al., 1999). Casein protein (3 to 3.5 g/L human milk) makes up about 30 to 35% of the total milk protein (Arakawa et al., 1999). There are two subclasses of human casein: (1) β-casein, which is the most abundant form, representing about 70%; and (2) κ-casein, which accounts for around 27% of the total casein in human milk. β-casein is a 25-kDa globular protein that exists in several isoforms depending on its phosphorylation state. Up to five phosphate groups can be attached to serine and threonine residues. The binding of calcium, magnesium, and phosphate ions facilitates formation of large casein aggregates referred to as *micelles*, giving milk its white color (Arakawa et al., 1999). In addition to its emulsification properties, human β-casein has diverse biological effects such as enhancement of absorption of calcium and other divalent cations, an opiate agonist effect (β-casomorphins), immunostimulating and modulating effects, and antibacterial functions (Mitra and Zhang, 1994).

It is generally accepted that human milk is superior to all other milk substitutes for growing infants; however, a large market exists for milk substitutes, as not all mothers are able to nurture their infants by breastfeeding due to adverse physical, health, psychological, or socioeconomic conditions (Arakawa et al., 1999). Infant formulae based on bovine milk or soybeans have been, until recently, traditional substitutes for mother's milk (Chong et al., 1997; Glick and Pasternak, 1998). Although cow's milk has been recognized as an excellent source of proteins, vitamins, and minerals, its consumption is often related to excessive weight gain and cow's milk protein allergy and intolerance (CMPA/CMPI), which may affect the gastrointestinal tract, the respiratory tract, skin, and blood (Chong et al., 1997; Glick and Pasternak, 1998; Arakawa et al., 1999). CMPA is a disease of infancy and usually appears in the first few months of life. Prenatal or early neonatal exposure to cow's milk protein increases the risk, not only of adverse reactions to bovine milk proteins but also of development of allergies to other food, specially soybean and egg (Host et al., 1995). Soy-based infant formulae have been recommended as hypoallergenic alternatives for non-breastfed infants; however, many infants with allergy to bovine milk are also allergic to soy proteins (Arakawa et al., 1999).

Reconstitution of Human Milk Proteins in Food Plants

Production of several human milk proteins in vegetables and fruits could provide a novel source of improved nutrition for children and malnourished human populations in economically emerging countries (Chong et al., 1997). The generation of individual transgenic food plants producing several milk proteins, in addition to vaccine protein antigens, can contribute to a cost-effective and nutritionally superior vegetable-based diet for people in need of more complete nutrition and protection against diseases (Arakawa et al., 1998; Chong and Langridge, 2000; Giddings et al., 2000).

Recently, Chong et al. (1997) succeeded in expressing β-casein, a major component of human milk proteins, in potato plants, becoming the first reported expression of a human milk protein gene in food crops. A β-casein expression level of 0.01% of total soluble protein was detected in leaf and tuber tissue. The plant-synthesized β-casein appears to be a single peptide of approximately 24 kDa, around 1 to 1.5 kDa smaller than the human counterpart isoform, which is phosphorylated at two sites. The reason for the apparent reduction in molecular size is not clearly understood. Recently, human β-casein was introduced into tomato and its presence at the protein level was detected (Arakawa et al., 1999).

Lactoferrin, one of the major whey proteins in human milk, is an iron-binding glycoprotein of around 80 kDa (Anderson et al., 1989; Chong and Langridge, 2000). Although little is known about the function of milk lacto-ferrin *in vivo*, functions ascribed to lactoferritin *in vitro* include promotion of cell growth, antimicrobial activity, and immune-modulating properties (Arakawa et al., 1999; Chong and Langridge, 2000). Antibacterial as well as antiviral activity of lactoferritin has been reported. Human milk lactoferrin contains a specific antimicrobial domain consisting of a loop of 18 amino acid residues. This region significantly inhibits growth of *E. coli*, and, based on its iron-chelating properties, lactoferrin impedes bacterial iron utilization, causing bacteriostasis (Anderson et al., 1989; Chong and Langridge, 2000).

Mitra and Zhang (1994) first reported expression, in a constitutive manner, of human lactoferrin in tobacco cells. The antibacterial properties of transgenic callus extracts were tested against four different phytopathogenic bacterial strains: *Xanthomonas campestris* pv. phaseoli, *Pseudomonas syringae* pv. phaseolicola, *Pseudomonas syringae* pv. syringae, and *Clavibacter flaccumfaciens* pv. flaccumfaciens. In colony-forming unit reduction assays, transgenic calli containing recombinant lactoferrin exhibited substantially higher antibacterial activity than native lactoferrin. In 2000, Chong and Langridge also expressed the human milk lactoferrin gene, but in potato plants and under the control of both the auxin-inducible manopine synthase (*mas*) promoter and the CaMV35S tandem promoter. Auxin activation of the *mas* promoter increased lactoferrin expression levels in transformed tuber and leaf tissues to approximately 0.1% of total soluble plant proteins, which was significantly greater than that driven by the CaMV35S constitutive promoter (around 0.01%). Antimicrobial activity, bacteriostatic and/or bacteriocidal, against different human pathogenic bacterial strains (*E. coli* Migula, *Salmonella paratyphi*, and *Staphylococcus aureus*) were detected in potato tuber tissues. Tuber extract containing more lactoferrin showed consistently stronger antimicrobial effects. This is the first report of synthesis of biologically active human milk lactoferrin in edible crops.

A cDNA encoding human α-lactalbumin has recently been introduced into tobacco plants to produce biologically functional human α-lactalbumin exhibiting an apparent molecular weight identical to human-derived α-lactalbumin (Takase and Hagiwara, 1998). When combined with galactosyltransferase, the tobacco-derived α-lactalbumin was fully active in the synthesis of lactose.

Several recombinant milk proteins have been produced in the milk of different transgenic animal species (Glick and Pasternak, 1998). However, only a few human milk proteins such as human β-casein, lactoferrin, and α-lactalbumin have been synthesized in plants (Mitra and Zhang, 1994; Chong et al., 1997; Takase and Hagiwara, 1998; Arakawa et al., 1999; Chong and Langridge, 2000). Because human milk protein genes are available, reconstitution of human milk proteins, including the caseins (β and κ) and whey proteins (α-lactalbumin, serum albumin, lactoferrin, lysozyme, and immunoglobulins), can now be achieved in transgenic food plants (Mitra and Zhang, 1994; Chong et al., 1997; Takase and Hagiwara, 1998; Arakawa et al., 1999; Chong and Langridge, 2000). Moreover, modification of milk protein properties by protein engineering now provides the opportunity to further improve digestibility; to increase protection against microbial pathogens by production of antimicrobial peptides such as lactoferrin, isracidin, and casecidin; and to enhance the feeling of well-being during periods of stress by producing human α-casein-derived neurotropic peptides (casomorphins) (Arakawa et al., 1999; Chong and Langridge, 2000).

The ultimate objective will be to reconstitute a panel of essential human milk proteins with the goal of enhancing the beneficial properties of milk proteins used for both infant formulae and in dairy food products for simultaneous improvement in human nutrition and the prevention of autoimmune and infectious diseases, so that we can say with certainty that we will never outgrow our need for milk.

Micronutrients

Essential micronutrients in the human diet include 17 minerals and 13 vitamins that are needed at minimum levels to alleviate nutritional disorders (DellaPenna, 1999; Grusak and DellaPenna, 1999; Guzmán-Maldonado and Paredes-López, 1999). Nonessential micronutrients encompass a vast group of unique organic phytochemicals that are not strictly required in the diet but when present at sufficient levels are linked to the promotion of good health (Steinmetz and Potter, 1996; Bliss, 1999).

On the other hand, before attempting to manipulate nutritional components in food crops, careful consideration must be given to the selection of target compounds, their efficacy, and whether excessive dietary intake could have unintended negative health consequences (DellaPenna, 1999). For select mineral targets (iron, calcium, selenium, and iodine) and a limited number of vitamin targets (folate; vitamins E, B_6, and A), the clinical and epidemiological evidence is clear that they play a significant role in maintenance of optimal health and are limited in diets worldwide (DellaPenna, 1999).

Carotenoids

Carotenoids comprise a group of natural pigments that are ubiquitous throughout nature. Over 600 different carotenoids with diverse chemical structures have been identified in bacteria, fungi, algae, and plants (Shewmaker et al., 1999; Mann et al., 2000). Their colors range from yellow to red, with variations of brown and purple; in addition, carotenoids as colorants take advantage of their good pH stability and their insensitivity to reducing agents such as ascorbic acid (Mann et al., 2000). Not to be outdone, humanity through the ages has learned to exploit the pleasing visual properties of carotenoid pigments by supplementing feedstocks and incorporating carotenoid pigments into cosmetics and foods (Delgado-Vargas et al., 2000; Jez and Noel, 2000). As precursors of vitamin A, they are fundamental components in our diet and play additional important roles in human health (Delgado-Vargas et al., 2000; Van den Berg et al., 2000; Ye et al., 2000). Because animals are unable to synthesize them *de novo*, they must obtain them by dietary means.

Other outstanding industrial uses of carotenoids include pharmaceuticals and nutraceuticals. In fact, they are important nutraceutical compounds and natural lipophilic antioxidants, whose sale as food and feed supplements is estimated to be approximately U.S.$500 million, and the market is expanding (Albrecht et al., 2000). All this is due in part to the discovery that these natural products can play a role in the prevention of cancer and chronic disease (mainly because of their antioxidant properties) and, more recently, that they exhibit significant tumor suppression activity as a result of specific interactions with cancer cells (Sandmann et al., 1999; Albrecht et al., 2000; Van den Berg et al., 2000).

The commercial demand for carotenoids is mainly met by chemical synthesis and, to a minor extent, by extraction from natural sources or microbial fermentation (Sandmann et al., 1999; Shewmaker et al., 1999; Sandmann, 2001). Moreover, although a wide range of natural carotenoid derivatives is known to date, most of these are biosynthetic intermediates that accumulate only in trace amounts, making it very difficult to extract sufficient material for purification (Albrecht et al., 2000). Some important dietary carotenoids are not abundant in the human diet. Zeaxantin, for instance, is a rare carotenoid, which together with lutein is the essential component of the macular pigment in the eye (Delgado-Vargas et al., 2000; Van den Berg et al., 2000). Low levels of intake increase the risk of age-related macular degeneration. Marigold extracts from *Tagetes erecta* or the dried flowers themselves are well known as supplements for chicken feed to color the eggs and the chicken skin (Delgado-Vargas et al., 2000). Interestingly, marigold flowers contain high concentrations of lutein as the major pigment (Delgado-Vargas and Paredes-López, 1997).

During ingestion of carotenoids, the efficiency of their absorption depends largely on the type of food, its processing, and the amount of dietary fat or oil. Whether the presence of a carotenoid in the food matrix might facilitate its bioavailability is still not known (Van den Berg et al., 2000; Sandmann, 2001). As a result of this, the number of carotenoids available for assessing their biological function and pharmaceutical and nutraceutical potential by *in vivo* and *in vitro* assay systems is very limited.

Carotenoids are a large family of C_{40} isoprenoid pigments. Their colorant and biological action, such as antioxidant activity, are related to the number and location of conjugated double bonds within their structure, cyclization of the ends of the molecules, and their modification by oxygen-containing R groups such as hydroxyl, keto, and epoxi groups (Albrecht et al., 2000; Mann et al., 2000; Schmidt-Dannert et al., 2000).

The first committed step in carotenoid biosynthesis is the condensation of two geranyl-geranyl diphosphate (GGDP) molecules to form the C_{40} backbone, the colorless phytoene (Figure 2.2). Phytoene desaturases from bacteria can introduce four double bonds, yielding red carotenoid lycopene, whereas plants utilize two desaturase enzymes to complete this conversion. Phytoene desaturase (PDS) catalyzes the first two desaturations (phytoene to phytofluene to ζ-carotene), whereas the conversion of ζ-carotene to lycopene via neurosporene is performed by ζ-carotene desaturase (ZDS). The cyclization of lycopene, by lycopene cyclase, forms either α- or β-carotene, and subsequent hydroxylation reactions produce the xanthophylls, lutein, and zeaxanthin (Sandmann, 2001).

Different approaches have been followed to modify the carotenoid content in plants to enhance their nutritional value: (1) modification of carotenoid products in tomato, (2) increasing the amounts of preexisting carotenoids in rapeseed (*Brassica napus*), and (3) engineering a carotenogenic pathway in tissue that is completely devoid of carotenoids, such as rice endosperm (Table 1.3) (Shewmaker et al., 1999; Romer et al., 2000; Schmidt-Dannert et al., 2000; Ye et al., 2000).

During lycopene deposition in tomato fruit ripening, the activity of phytoene synthase is the major controlling factor of the route; therefore, this enzyme should be an ideal target for the genetic manipulation of the carotenoid composition of tomato fruit (Fray et al., 1995). The constitutive high-level expression of tomato phytoene synthase-1 in transgenic tomato has resulted in carotenoid-rich seed coats, cotyledons, and hypocotyls, but also in reduced levels of carotenoids in ripe tomato fruit due to gene silencing with the endogenous gene and dwarfism due to redirection of GGDP into

TABLE 1.3

Selected Essential Micronutrients for Human Diet, Their Daily Allowances, Manipulation by Plant Biotechnology, and Potential Applications

Nutrient	Maximum Adult RDA[a]	Engineered Plant	Result	Potential Application	Ref.
		Vitamin			
Vitamin A	1 mg RE[b]	Tomato, rapeseed, rice	Increased levels of β-carotene	Provitamin A deficiency nutraceutical	Shewmaker et al., 1999; Romer et al., 2000; Ye et al., 2000
Vitamin E	10 mg TE[c]	*Arabidopsis*	Elevated content of α-tocopherol and reduced γ-tocopherol content	Vitamin E deficiency nutraceutical	Shintani and DellaPenna, 1998
		Minerals			
Iron	15 mg	Tobacco, rice	Improved Fe content	Anemia nutraceutical	Goto et al., 1998, 1999
Zinc	15 mg	—	—	—	—
Calcium	1200 mg	—	—	—	—
Phosphorus	1000 mg	Tobacco, rapeseed	Reduced phytic acid levels	Improvement in mineral bioavailability	Pen et al., 1993; Tramper, 2000

[a] Recommended dietary allowances per day; values represent the highest RDA either for male or female adults, except for pregnant or lactating women.
[b] Vitamin A activity is expressed in retinol equivalent (RE). One RE is equal to 1 mg of all-*trans*-retinol, 6 mg of all-*trans*-β-carotene, or 12 mg of other provitamin A carotenoids.
[c] One TE (α-tocopherol equivalent) is equal to 1 mg (R,R,R)-α-tocopherol.
Source: Adapted and modified from Shintani and DellaPenna, 1998; DellaPenna, 1999; Guzmán-Maldonado and Paredes-López, 1999.

the gibberellin pathway (Fray et al., 1995). The resultant plants were reduced in size. This work illustrates how problems arise when a balanced metabolism is disturbed (Sandmann, 2001). On the other hand, elevation of the provitamin A (β-carotene) content in transgenic tomato plants was achieved by manipulation of the desaturation activity. Romer et al. (2000) overexpressed a single carotenoid gene encoding PDS, which converts phytoene into lycopene, from *Erwinia uredovora*, under the control of a constitutive promoter and with the protein being targeted to the plastid by pea ribulose biphosphate carboxylase small subunit transit sequence. These researchers found that the expression of that gene in transformed tomatoes did not elevate total carotenoid levels. However, the β-carotene content increased about threefold, representing up to 45% of the total carotenoid level. The transgenic tomato fruit contained approximately 5 mg all-*trans*-β-carotene or 800 retinal equivalents (Table 1.3). Thus, 42% of the RDA is contained in a single provitamin A tomato fruit, as compared to 23% of the control fruit. The advantage of β-carotene instead of retinol (vitamin A) in the diet is that it is nontoxic and can be stored by the body. The alteration in carotenoid content of these transgenic plants did not affect growth and development and their phenotype was stable and reproducible over at least four generations.

Also, the genetic manipulation of canola seeds to increase the carotenoid content to high levels was a tremendous success (Shewmaker et al., 1999). Overexpression of a bacterial phytoene synthase gene extended with a plastid-targeting sequence under a seed-specific promoter increased the carotenoid content of mature canola seed by up to 50-fold. In the transformant, the embryos were bright orange, as compared to the green embryos in control canola. In the transgenic seeds, concentrations of carotenoids (mainly α- and β-carotene) of more than 1 mg/g fresh weight accumulated, yielding oil with 2 mg carotenoids per g oil (Table 1.3).

Other unexpected results were obtained upon transformation of tobacco with an algal β-carotene ketolase gene. Mann et al. (2000) expanded upon the metabolic framework to redirect metabolic flux of the tobacco carotenoid biosynthetic pathway to produce the marine compound astaxanthin (Figure 1.2). Introducing β-carotene ketolase from unicellular algae into tobacco, astaxanthin could be synthesized using the endogenous pool of β-carotene in tobacco flowers. Tissue-specific synthesis was accomplished by linking a gene promoter for flower petal expression to a fused gene encompassing a transit peptide sequence for plastid localization and the algal ketolase coding sequence. Expression was high in flowers as visualized by the red nectar pigmentation caused by astaxanthin and other ketocarotenoids (Mann et al., 2000). Total carotenoid levels were increased to 140% compared with the wild type. It is important to note that the astaxanthin produced in the transgenic plants had the same chirality as the natural astaxanthin found in marine organisms. In contrast, the synthetic astaxanthin that is currently used as fish feed is a mixture of stereoisomers, of which 75% have an unnatural chiral structure (Hirschberg, 1999). These results demonstrate the

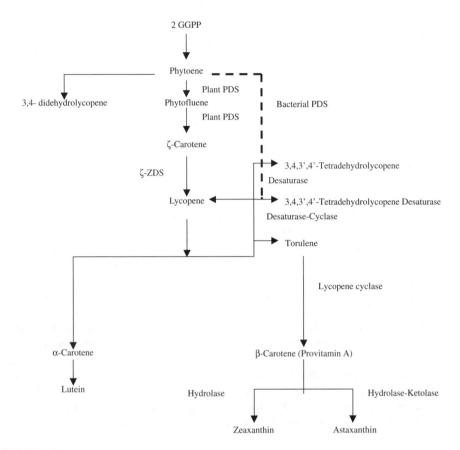

FIGURE 1.2
Biosynthetic pathway of α- and β-carotene, their oxo derivaties, and schematic view of some
the transformations of novel carotenoids obtained by using molecular breeding and an *in vitro*
evolution approach. Production of β- carotene is universal in plants, fungi, and bacteria. Other
carotenoids of biotechnological, nutraceutical, and pharmaceutical interest are lutein, zeaxanthin, and astaxanthin, including the novel and engineered carotenoids (3,4-didehydrolycopene,
3,4,3′,4′-tetradehydrolycopene, and torulene), which present improved antioxidant properties.
Abbreviations: GGDP, geranyl-geranyl diphosphate; PDS, phytoene desaturase; ζ-ZDS, ζ-carotene desaturase. (Adapted and modified from Jez and Noel, 2000; Sandmann, 2001.)

prospect of genetically engineering carotenoid biosynthesis toward the production of naturally and commercially valuable compounds in plants.

Rice, a major staple food, is usually milled to remove the oil-rich aleurone
layer that turns rancid upon storage. The endosperm, the remaining edible
part of rice grains, lacks several essential nutrients, such as provitamin A.
In fact, rice in its milled form contains neither β-carotene nor any of its
immediate precursors. Thus, predominant rice consumption promotes vitamin A deficiency, a serious public health problem in at least 26 countries,
including highly populated areas of Asia, Africa, and Latin America (Kishore
and Shewmaker, 1999; Ye et al., 2000; Potrykus, 2001).

Immature rice endosperm synthesizes the carotenoid precusor GGDP. To convert GGDP to β-carotene, Ye et al. (2000) programmed the endosperm to carry out the necessary additional enzymatic reactions leading to the formation of cyclic carotenoids (β-carotene) in the rice endosperm. Transformation was carried out with a plasmid containing a plant phytoene synthase gene and a bacterial phytoene desaturase gene, which together should mediate the synthesis of lycopene from GGDP (Figure 1.2). Both reading frames were extended with transit sequences for targeting the endosperm plastids. One was under control of the endosperm-specific glutelin and the other under one constitutive promoter.

Surprisingly, such transgenic plants did not accumulate lycopene as predicted. Instead, these plants produced essentially the same end products (β-carotene, lutein, and zeaxanthin). The authors speculate that the enzymes necessary to convert lycopene into β-carotene, lutein, and zeaxanthin are constitutively expressed in normal rice endosperm or are induced when lycopene is produced. Co-transformation with another construct that carried the third gene of interest, lycopene β-cyclase, increased the β-carotene content of the rice endosperm to a maximum level of 1.6 μg/g dry weight (Ye et al., 2000). The resulting yellow-colored endosperm, containing provitamin A (β-carotene) and other carotenoids of nutritional importance, could provide additional health benefits. This type of transformed rice accumulating large levels of β-carotene is known as *golden rice*, and as little as 300 g of the cooked golden rice, a typical Asian diet, should provide almost the entire daily vitamin A requirement (Table 1.3). Golden rice exemplifies the best that agricultural biotechnology has to offer a world whose population is predicted to reach 7 billion by 2013 (Potrykus, 2001).

Vitamins

Tocopherols, the lipid-soluble antioxidants collectively known as vitamin E, are essential ingredients in human nutrition (Traber and Sies, 1996). Several epidemiological studies have indicated that vitamin E supplementation (100 to 400 international units [IU], or approximately 250 mg of α-tocopherol daily) results in decreased risk for cardiovascular disease and cancer, aids in immune function, and prevents or slows down a number of degenerative diseases associated with aging, such as cataracts, arthritis, and disorders of the nervous system caused by cumulative damage to tissues mediated by reactive oxygen species (Grusak and DellaPenna, 1999; Hirschberg, 1999).

The four naturally occurring tocopherols (α-, β-, γ-, and δ-tocopherol) differ only in the number and position of methyl substituents on the aromatic ring (Shintani and DellaPenna, 1998). Of tocopherol species present in foods, natural single (R,R,R)-α-tocopherol is the most important to human health, has the highest vitamin E activity, and occurs as a single isomer (Grusak and DellaPenna, 1999). Although all tocopherols are absorbed equally during digestion, only single (R,R,R)-α-tocopherol is preferentially retained and distributed throughout the body; even though synthetic α-tocopherol is employed as a vitamin E supplement, it is a racemic mixture of eight different stereoisomers. Most of these isomers are less efficacious than the (R,R,R) isomer (Traber and Sies, 1996; Grusak and DellaPenna, 1999). Tocopherols exist at different concentrations in various plant species and also vary among tissues. Whereas leaves of most common plants contain low levels of tocopherols (10 to 50 μg/g fresh weight), they can accumulate to high concentrations (500 to 2000 μg/g) in seeds (Hirschberg, 1999). However, in most seed crops, including those from which the major edible oils are derived (soybean, corn, canola, cottonseed, and oil palm), α-tocopherol is present only as a minor component because its immediate biosynthetic precursor γ-tocopherol predominates; for example, in soybean oil, α- and γ-tocopherol account for 7 and 70%, respectively, of the total tocopherol pool. Although other major oilseeds have similar patterns, seed oils still represent the major source of naturally derived dietary α-tocopherol (Grusak and DellaPenna, 1999).

The most recent U.S. recommended daily allowance (RDA) suggests that up to 10 mg TE (α-tocopherol equivalent) be consumed every day (Table 1.3) and, because of the abundance of plant-derived components in most diets, this RDA is often met in the average human diet (Shintani and DellaPenna, 1998; DellaPenna, 1999).

Substantial increases in the α-tocopherol content of the major food crops are necessary to supply the public with dietary sources of vitamin E that can approach the desired therapeutic levels and benefit, because doing so is nearly impossible from the average diet, unless a concerted effort is made to ingest large quantities of specific food enriched in that vitamin (Shintani and DellaPenna, 1998; DellaPenna, 1999; Grusak and DellaPenna, 1999). γ-Tocopherol is methylated to α-tocopherol in a reaction catalyzed by γ-tocopherol methyltransferase (γ-TMT). These observations suggest that γ-TMT activity is likely limiting in the seeds of most agriculturally important oilseed crops and may be responsible for the low proportion of α-tocopherol synthesized and accumulated. As such, γ-TMT is a prime molecular target for manipulation of α-tocopherol content in crops (Shintani and DellaPenna, 1998; Hirschberg, 1999). An exquisite and very significant example of metabolic engineering in this direction has been reported by Shintani and DellaPenna (1998), who overexpressed a γ-TMT cDNA in *Arabidopsis* seeds under control of a seed-specific carrot promoter.

Untransformed *Arabidopsis* seeds contain around 370 ng of total tocopherols per milligram, mostly composed of γ-tocopherol. In the transgenic seeds, 85 to 95% of the tocopherol pool was α-tocopherol, representing an 80-fold increase in α-tocopherol levels compared with the wild-type control (Table 1.3). Interestingly, the total seed tocopherol amount was not altered in these plants, indicating either a lack of stringent feedback regulation (by γ-tocopherol) of the route in seeds or that α-tocopherol functions by the same mechanism as γ-tocopherol. This qualitative change in tocopherol composition reflects a 9-fold rise in total vitamin E activity of the seed oil due to the different vitamin E potency of α- vs. γ-tocopherol. Similar increases in vitamin E activity in commercially important oilseed crops could be achieved (DellaPenna, 1999; Grusak and DellaPenna, 1999).

Minerals

In developing countries, cereal grains and some legumes are the primary and least expensive sources of calcium, iron, and zinc; however, their intake does not satisfy the mineral requirements of the populations of these countries (Table 1.3) (Guzmán-Maldonado et al., 2000).

Guzmán-Maldonado et al. (2000) cited that the percentage of anemic subjects in developing countries (26%) was higher than that observed in Europe (10.9%) and the U.S. (8%); data revealed that anemia was predominantly caused by iron deficiency. They also cited that 40% of iron intake derived from legumes and cereals. Recent reports indicate that iron deficiency is the most prevalent micronutrient problem in the world, affecting over 2 billion people globally, many of whom depend on beans as their staple food (Goto et al., 1999; Guzmán-Maldonado et al., 2000).

Some crops, such as spinach and legumes, are known for their iron content; however, these plants usually contain oxalic acid and phytate-like substances that decrease its bioavailability (Bliss, 1999). Iron in crops has been improved by increasing the iron concentration of the hydroponic culture media or soil (Bliss, 1999; Goto et al., 1999); however, this method is costly and cannot be used to target iron accumulation to a desirable part of the plant.

Higher plants utilize one of two strategies for iron acquisition (Grusak and DellaPenna, 1999; Curie et al., 2001). Strategy 1 involves an obligatory reduction of ferric iron, usually as a Fe(III) compound, prior to membrane influx of Fe^{2+}; all dicotyledonous plants and non-grass monocots use this approach. Strategy 2 (used by grasses) employs ferric chelators called *phytosiderophores* that are released by roots and chelate ferric iron in the rhizosphere. The Fe(III)–phytosiderophore is absorbed intact via a plasmalemma transport protein. When plants using either approach are challenged with Fe-deficiency stress, the processes associated with one or the other strategy are upregulated in the plant root system (Grusak and DellaPenna, 1999; Curie et al., 2001).

Modifying seeds to store the excess Fe chelated to peptides or in heme-containing enzymes might further enhance the seed Fe nutritional quality by improving bioavailabity (Grusak and DellaPenna, 1999). In the first case, one target might be the *maize yellow stripe 1* (YS1) mutant, which is deficient in Fe(III)–phytosiderophore uptake, suggesting that YS1 is an Fe(III)–phytosiderophore transporter. In 2001, Curie et al. showed that YS1 is a membrane protein that mediates iron uptake and whose expression in a yeast iron uptake mutant restores growth specifically on Fe(III)–phytosiderophore media.

In the second approach, we have the ferritin, which is an iron-storage protein found in animals, plants, and bacteria (Goto et al., 1999). Recent studies show that both plants and animals use ferritin as the storage form of iron and that, when orally administered, it can provide a source of iron for treatment of rat anemia (Beard et al., 1996). These findings suggest that increasing the ferritin content of cereals by genetic engineering may help to solve the problem of dietary iron deficiency. To achieve this goal, Goto et al. (1999) introduced a soybean ferritin gene in rice seed, under the control of a rice seed-storage glutelin promoter, to mediate the accumulation of iron specifically in the grain. Their results indicated that soybean ferritin was overexpressed and accumulated in the rice endosperm tissue. Interestingly, transgenic rice stored up to three times (31.8 μg/g dry weight) more iron than untransformed seeds (11.2 μg/g dry weight).

This achievement suggests that it may be feasible to produce ferritin rice as an iron supplement in the human diet. The iron content in a meal-size portion of ferritin rice (5.7 mg Fe per 150 g dry weight) would be sufficient to supply 30 to 50% of the daily adult iron requirement (around 13 to 15 mg Fe) (Table 1.3). Also, this group constitutively expressed soybean ferritin in tobacco which resulted in a maximum iron content in transgenic leaves that was about 1.3 times that of control leaves (Goto et al., 1998). This increase seems to be low compared with the rice system; however, it is interesting to note that increments in absolute amounts of iron in these two systems are very similar (24.3 μg/g dry weight in tobacco vs. 27 μg/g dry weight in rice) (Goto et al., 1998, 1999). Although a positive correlation can be observed between exogenous ferritin and iron content, it is possible that the amount of iron accumulation is restricted by transport of iron to the ferritin molecule, rather than simply by the level of ferritin protein. Thus, it may be possible to store larger amounts of iron in the exogenous ferritin molecule by cointegrating into the target plant genome, at the same time, the ferritin gene and the gene Fe(III)–phytosiderophore transporter, such as the membrane protein of the maize yellow stripe 1 mutant (Goto et al., 1998, 1999; Curie et al., 2001).

Numerous studies have led to the conclusion that phytic acid and tannins may bind proteins and some essential dietary minerals (calcium, iron, and zinc), thus making them unavailable or only partially available for absorption (Bliss, 1999; Guzmán-Maldonado et al., 2000). Zinc is essential for normal growth, appetite, and the immune function, being an essential component of more than 100 enzymes involving digestion, metabolism, and wound healing (Bliss, 1999; Guzmán-Maldonado et al., 2000). While iron deficiency has long been considered a major nutritional problem, zinc deficiency has only recently been recognized as a public health problem (Guzmán-Maldonado et al., 2000).

On the other hand, approximately 60 to 65% of the phosphorus present in cereal, legume, and oilseed crops exists as phytic acid (*myo*-inositol-hexaphosphate) which, accordingly, represents the major storage form of phosphate in plants (Tramper, 2000; Ward, 2001). However, in this form, the phosphate remains largely unavailable to monogastrics as these species are devoid of sufficient, suitable, endogenous phosphatase activity that is capable of liberating the phosphate groups from the phytate core structure.

The animal inability to degrade phytic acid has a number of nutritional consequences; phytic acid chemically complexes to zinc, iron, and calcium, preventing their assimilation by the animal (Bliss, 1999; Guzmán-Maldonado et al., 2000; Tramper, 2000; Ward, 2001). Methods to reduce the levels of these phytochemicals, such as phytic acid and tannins, should enhance the bioavailability of the micronutrients they affect (Table 1.3) (Pen et al., 1993; Bliss, 1999; DellaPenna, 1999; Grusak and DellaPenna, 1999; Tramper, 2000; Ward, 2001). One strategy to improve mineral bioavailability is to reduce phytic acid by adding phytase to the diet or to increase the level of endogenous phytase in the dietary components. Thus, an *Aspergillus niger* phytase gene was transferred to tobacco plants, and transgenic seeds accumulated phytase protein up to 10 g/kg total soluble protein. When samples of milled transgenic tobacco seed were mixed with standard poultry feed under conditions that stimulated the crop and stomach of the chicken, inorganic phosphate was released from the fodder (Pen et al., 1993). Also, the Dutch company Plantzyme succeeded in producing phytase in rapeseed and accumulating it in the seeds. The advantage is that the seeds can be directly added to the diet, without having to isolate the enzyme first (Tramper, 2000).

Manipulation of Fruit Ripening

A major problem in fruit marketing is premature ripening and softening during transport. These changes are part of the natural aging (senescence) process of the fruit (Gray et al., 1992). When compared to other plant organs, fruits exhibit a high metabolic activity even in their postharvest life (Giovannoni et al., 1998). It is precisely this biochemical activity that is one of the main causes of the high perishability of these commodities, resulting in short shelf life. Some postharvest problems have been solved for many commercially important plants, especially those grown in temperate climates, by harvesting before they ripen on the plant and/or storing at low temperatures or in modified or controlled atmospheres (Gomez-Lim, 1999); however, these approaches have had limited success when applied to fruits of tropical origin. For instance, some of these fruits fail to ripen properly, developing an unpleasant taste, if they are harvested at a green or immature stage (Gomez-Lim, 1999).

Fruit ripening represents a biological process unique to plant species in which developmental and hormonal signaling systems orchestrate a variety of biochemical and physiological changes which, in summation, result in the ripe stage of fruit maturation (Giovannoni et al., 1998). These changes lead to a soft, edible fruit. Some of these changes include synthesis of metabolites related to the development of flavor and aroma, synthesis of pigments, degradation of chlorophyll, alterations in organic acids and cell wall metabolism, and softening of the fruit tissue. In so-called climacteric fruits such as tomato, cucurbit, banana, apple, mango, and many others (Giovannoni et al., 1998; Gomez-Lim, 1999), the initiation of ripening is characterized by a dramatic increase in respiration and biosynthesis of the gaseous hormone ethylene. Inhibition of ethylene biosynthesis or ethylene perception via exogenous application of inhibitors, or endogenous expression of transgenes, has been shown to have profound inhibitory effects on ethylene-mediated plant processes, including climacteric fruit ripening (Gray et al., 1992; Giovannoni et al., 1998; Grierson, 1998).

Cell-wall-metabolizing enzyme polygalacturonase (PG) catalyzes the hydrolysis of polygalacturonic acid chains in unmethylated regions of pectin (Grierson, 1998). One gene appears to be responsible for the endopolygalacturonase that is synthesized *de novo* during tomato ripening, although three different polypeptide forms of PG are found in tomato; the difference is possibly due to carbohydrate content (Bird et al., 1988; Grierson, 1998).

The inhibition of PG was first achieved in transgenic tomato using antisense genes driven by a constitutive promoter; both PG mRNA and enzyme activity were reduced by 90% (Sheehy et al., 1988; Smith et al., 1990). Interestingly, this gene silencing was stably inherited and the PG activity had no effect on other ripening attributes such as color change and ethylene synthesis (Sheehy et al., 1988; Smith et al., 1990). Cell wall pectin, which normally decreases in molecular weight during ripening, was shown to retain a high molecular weight during ripening of fruits with a silenced PG gene (Smith et al., 1990). No significant change in firmness could be detected, at least in some varieties, and this and other observations led various researchers to question whether PG was in fact involved in softening (Smith et al., 1990). It is now clear from biochemical studies and experiments with transgenic plants, however, that PG has a distinct effect on the textural quality of tomato (Grierson, 1998). Inhibiting its expression by antisense genes is the basis for the Flavr Savr™ tomato and, in 1994, the U.S. Food and Drug Administration ruled that the Flavr Savr tomato is as safe as tomatoes that are bred by conventional approaches. In 1996, the genetically modified puree of that transgenic tomato was marketed in the United Kingdom. Thus, we have two different opportunities to exploit low PG tomatoes. For the fresh market, low PG makes the fruit less susceptible to cracking, splitting, and mechanical damage; therefore, fruit can be left to ripen longer on the vine before harvesting. While these same advantages may also be significant for processing varieties, the main benefit is in the longer chain pectin and perhaps the larger size of cell clumps in puree (Grierson, 1998).

The plant growth hormone ethylene induces the expression of a number of genes involved in fruit ripening and senescence (Theologis, 1992). This gas is largely responsible for fruit and vegetable spoilage, thus possibilities have been explored aimed at modifying the ethylene formation or content in plants and fruits (Theologis, 1992). Ethylene is synthesized from S-adenosylmethionine via the intermediate ACC (aminocyclopropane-1-carboxylic acid). The enzyme ACC synthase catalyzes the formation of ACC. The second step leading to ethylene production is catalyzed by ACC oxidase, also known as the ethylene-forming enzyme (EFE) (Figure 1.3) (Theologis, 1992). Thus, two small multigene families encoding ACC synthase and (aminocyclopropane-1-carbolic acid) oxidase control the biosynthesis of ethylene (Theologis, 1992; Grierson, 1998).

Treatment of plants with chemical compounds that block ethylene formation with sequestrants of it delay both fruit ripening and senescence (Gomez-Lim, 1999). Thus, premature fruit ripening might be prevented by inhibiting the ability of the plant to synthesize ethylene. Almost complete inhibition of ACC synthase, and therefore of ethylene generation, using antisense genes inhibited the change in color and texture of tomato fruits (Oeller et al., 1991). Ripening changes could be restored by adding ethylene or ethylene-related compounds. A recent reexamination of the properties of these transgenic tomatoes has revealed, however, that they still generate sufficient ethylene to induce PG gene expression (Sitrit and Bennet, 1997).

Inhibiting the expression of a specific cDNA by antisense strategy in transgenic tomatoes first identified the genes for ACC oxidase. Inhibiting ACC oxidase by 95% permitted normal development of ripening attributes of fruit attached to the plant but prevented the extreme softening, cracking, and spoilage normally associated with over-ripening so that the fruit lasted for several weeks (Picton et al., 1993). The transgenic plants synthesized a lower level of ethylene than did normal plants, and again the fruit of transgenic tomatoes had a significantly longer shelf life. Interestingly, if the fruits were picked at the mature-green stage, they never ripened fully; adding ethylene externally did stimulate color development, but over-ripening and deterioration did not occur (Picton et al., 1993). Another approach for reducing

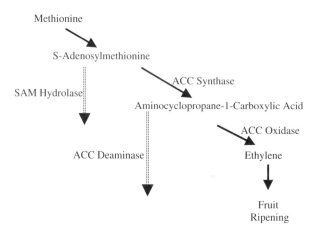

FIGURE 1.3
Plant ethylene biosynthesis and its control by genetic engineering affects fruit ripening process. Abbreviations: ACC, aminocyclopropane-1-carboxylic acid; SAM, S-adenosylmethionine. A bold and full arrow indicates normal ethylene biosynthesis pathway; a broken arrow indicates catabolism of ethylene intermediates, which have also been used to manipulate fruit ripening. (Adapted and modified from Theologis, 1992.)

ethylene formation is to overexpress bacterial genes for S-adenosylmethion-ine hydrolase or ACC deaminase, thus depleting plant cells of the required precursor (Figure 1.3) (Grierson, 1998).

While most reports of transgenic plants in which ethylene amounts have been lowered in an effort to delay fruit ripening have focused on tomato plants (Theologis, 1992; Picton et al., 1993; Giovannoni et al., 1998; Grierson, 1998), one research group published the generation of low-ethylene trans-genic cantaloupe melon with reduced ACC oxidase gene expression by anti-sense technology (Ayub et al., 1996). Melons of the cantaloupe Charenties type were chosen because of their good eating quality but poor storage capability. ACC oxidase activity in fruit was virtually undetectable and eth-ylene production was diminished below 1% of the control level. Transgenic melons developed a functional abscission layer but remained on the plant for longer. Pigment production in the flesh followed the normal ripening pattern, but the peel remained green while controls turned yellow. After 10 days of storage at 25°C, the transgenic melons were still green and retained their shape, whereas the wild-type fruit had a shriveled yellow peel, showed signs of fungal infections, and had soft flesh and a squashed shape. Applying ethylene to the transgenic fruits restored the yellowing phenotype to the peel (Ayub et al., 1996). The latter results suggest that those molecular approaches could be effective in a range of different fruits, offering new opportunities for enhancing their quality and nutritional value.

Microbial Biotechnology in Industry

Microorganisms are important for many reasons, particularly because they produce things that are of value to us (Demian, 2000a,b). These can be very large materials such as proteins, nucleic acids, carbohydrates, food polymers, or even cells, or they can be smaller molecules that are usually separated into metabolites that are essential for vegetative growth (primary) and those that are inessential (secondary) (Demian, 2000a,b).

Although microbes are extremely good at producing an amazing array of valuable products, they usually produce them only in the amounts they need for their own benefit; thus, they tend not to overproduce their metabolites (Demian, 2000b). By contrast, the industrial microbiologist screens for waste-ful strains that overproduce and excrete a particular compound that can be isolated and marketed (Chotani et al., 2000; Demian, 2000a,b). After a desired strain has been found, a development program is initiated to improve titers by modification of culture conditions using mutation and recombinant DNA techniques (Chotani et al., 2000; Demian, 2000b). The microbiologist is actu-ally modifying the regulatory controls remaining in the original culture so that its inefficiency can be further increased and the microorganism will excrete tremendous amounts of these high-value products into the medium.

The main reason for the use of microorganisms to produce compounds that can otherwise be isolated from plants and animals or synthesized by chemists is the ease of increasing production by environmental and genetic manipulation. Thousand-fold increases have been reported for small metabolites. Of course, the higher the specific level of production, the simpler the job of product isolation (Chotani et al., 2000; Demian, 2000b).

Microbial Metabolites

Primary metabolites are the small molecules of all living cells that are intermediates or end products of the pathways of intermediary metabolism, or are building blocks for essential macromolecules, or are converted into coenzymes (Glick and Pasternak, 1998; Chotani et al., 2000; Demian, 2000a,b). Primary metabolites used in the food and feed industries include alcohols, amino acids, flavor nucleotides, organic acids, polyols, polysaccharides, sugars, and vitamins (Glick and Pasternak, 1998; Chotani et al., 2000; Demian, 2000a,b).

Microbially produced secondary metabolites are extremely important for health and nutrition. As a group that includes antibiotics, other medicinals, toxins, biopesticides, and animal and plant growth factors, they have tremendous economic importance (Glick and Pasternak, 1998; Chotani et al., 2000; Demian, 2000a,b). Secondary metabolites have no function in the growth of the producing cultures (although, in nature, they are essential for the survival of the producing organism); they are produced by certain restricted taxonomic groups of organisms and are usually formed as mixtures of closely related members of a chemical family.

Microbial Production of Small High-Value Molecules

To date, molecular biotechnology research has focused largely on the production of a range of different proteins; however, recombinant DNA technology can also be used to enhance the production of a range of low-molecular-weight compounds (Glick and Pasternak, 1998). With efficient expression systems, it is relatively straightforward to clone and express a particular target protein. The expressed protein is either the final product (e.g., a restriction enzyme) or a catalyst for a specific chemical reaction (Glick and Pasternak, 1998). Sometimes, a novel catalytic activity is introduced into a microorganism by genetic manipulation and used to produce *in vivo* low-molecular-weight metabolites. In these cases, the host microorganism is engineered to become a factory for the production of useful metabolites (Glick and Pasternak, 1998; Chotani et al., 2000; Demian, 2000a,b).

Vitamins

Vitamin C (L-ascorbic acid) is used on a large scale as an antioxidant in food, animal feed, beverages, pharmaceutical formulations, and cosmetic applications (Chotani et al., 2000). The current world market of ascorbic acid is 60,000 to 70,000 metric tons per year and generates annual revenues in excess of U.S.$500 million (Glick and Pasternak, 1998; Chotani et al., 2000). An expensive process that includes one microbial fermentation step and a number of chemical steps starting with D-glucose currently synthesizes ascorbic acid commercially. The last step in this process is the acid-catalyzed conversion of 2-keto-L-gulonic acid (2-KLG) (Chotani et al., 2000). Biochemical studies for the metabolic routes of a number of different microorganisms have shown that it may be possible to synthesize 2-KLG by an alternative pathway (Glick and Pasternak, 1998; Chotani et al., 2000). For example, some bacteria (*Acetobacter, Gluconobacter,* and *Erwinia*) can convert glucose to 2,5-diketo-D-gluconic acid (2,5-DKG) and others such as *Corynebacterium, Brevibacterium,* and *Arthrobacter* have the enzyme 2,5-DKG reductase, which converts 2,5-DKG to 2-KLG.

Thus, a novel process involves the use of a genetically engineered *Erwinia* sp. strain containing a gene encoding 2,5-DKG reductase from *Corynebacterium* sp. (Anderson et al., 1985) The transformed *Erwinia* cells were able to convert D-glucose directly to 2-KLG. The endogenous *Erwinia* enzymes, localized in the inner membrane of the bacterium, converted glucose to 2,5-DKG, and the cloned 2,5-DKG reductase, localized in the cytoplasm, catalyzed 2,5-DKG to 2-KLG. The engineered organism transforms glucose to 2-KLG in a single-step fermentation, which can be easily converted by acid or base to L-ascorbic acid (Pramik, 1986). Using this elegant approach, Genencor International Company has produced up to 120 g of 2-KLG per liter in less than 120 hours of fermentation time in 14-L fermenters (Chotani et al., 2000; Demian, 2000b).

The goal of manufacturing vitamin C directly by fermentation has remained elusive. By employing a metabolic selection strategy, Genecor International Company has now identified a 2-KLG to ascorbic acid activity in two yeast species, namely *Candida blankii* and *Cryptococcus dimmnae*. Another direct route from D-glucose to L-ascorbic acid in microalgae has been developed (Chotani et al., 2000). Additional bioengineering is required to advance toward the direct fermentation of glucose to vitamin C without the need to isolate 2-KLG.

Vitamin B_2 (riboflavin) was produced commercially for many years by both fermentation and chemical synthesis, but today fermentation is the main route. Overproducers include two yeast-like molds, *Eremothecium ashbyii* and *Ashbya gossypii*, which synthesize riboflavin in concentrations up to 20 g/L (Demian, 2000a). New processes using *Candida* sp. or recombinant *Bacillus subtilis* strains that produce greater than 30 g riboflavin per liter have been recently developed (Perkins et al., 1999; Demian, 2000a).

Vitamin B_{12} (cyanocobalamin) is produced industrially with *Propionibacterium shermanii* or *Pseudomonas denitrificans* (Demian, 2000a). Such strains make about 100,000 times more vitamin B_{12} than they need for their own growth. The early stage of *P. shermanii* fermentation is conducted under anaerobic conditions in the absence of the precursor 5,6-dimethylbenzimidazole. These conditions prevent vitamin B_{12} synthesis and allow for the accumulation of the intermediate, cobinamide (Demian, 2000a,b). The culture is then aerated and dimethylbenzimidazole is added, converting cobinamide to the vitamin. In the *P. denitrificans* fermentation, the entire process is carried out under low oxygen. A high level of oxygen results in an oxidizing intracellular environment that represses the formation of the early enzymes in the pathway. Of major importance for the *P. denitri-ficans* fermentation is the addition of betaine (Kusel et al., 1984). Production of vitamin B_{12} has reached levels of 150 mg/L, with a world market value of U.S.$71 million. However, while vitamin B_{12} overproduction is totally dependent upon betaine, the mechanism of control is unknown (Kusel et al., 1984; Demian, 2000b).

D-biotin is a vitamin required for human health care and use as an animal feed additive (Masuda et al., 1995). It is industrially produced by chemical synthesis, but more economical manufacturing methods have been developed by recombinant DNA technology (Demian, 2000a). During production of biotin, feedback repression is caused by the acetyl-coenzyme A carboxylase biotin holoenzyme synthase, with biotin 5-adenylate acting as corepressor. Recently, *Serratia marcescens* has been engineered as a biotin-hyperproducing strain. A recombinant plasmid carrying the mutated biotin operon has been constructed and is introduced into a D-biotin-producing *Serratia* strain that also contains an additional copy of the mutated biotin operon (Masuda et al., 1995). The strain of *S. marcescens* obtained and selected for resistance to biotin antimetabolites is able to produce 600 mg of D-biotin/L in the presence of high levels of sulfur and ferrous iron, in contrast to the 80 mg of D-biotin/L obtained without an additional source of these minerals.

Lactic acid bacteria have very limited biosynthetic capability for the production of vitamins; however, certain exceptions have been noted. The yogurt bacterium *Streptococcus thermophilus* has been observed to produce folic acid, which, in fact, stimulates the growth of the other yogurt bacterium, *Lactobacillus bulgaricus*. *L. lactis* also produces substantial amounts of folic acid during fermentation (Hugenholtz and Kleerebezem, 1999). Many of the gene codings for the route of folic acid biosynthesis have been identified in the genome of this microorganism. Also, genes for riboflavin and biotin biosynthesis have been reported in *L. lactis*. This would make it possible to engineer the production of these vitamins in these food-grade bacteria, as recently reported for *B. subtilis*. Vitamin production processes by lactic acid bacteria would have huge advantages over the currently used processes, as they could also be implemented for *in situ* production, such as food fermentations (Hugenholtz and Kleerebezem, 1999). This work has just started.

Lactic Acid

L-lactic acid has an ancient history of use as a food preservative and food flavoring compound (Hugenholtz and Kleerebezem, 1999). Recently, lactic acid has received attention because it can be condensed into a biodegradable polymer that could become a major bioplastic in the future (Hugenholtz and Kleerebezem, 1999; Chotani et al., 2000; Demian, 2000b). The market for lactic acid is growing rapidly, exceeding several hundred million dollars annually (Hugenholtz and Kleerebezem, 1999; Chotani et al., 2000; Demian, 2000b).

Under non-energy-limiting batch fermentation conditions, homofermentative bacteria predominantly produce lactic acid as their end product (Hugenholtz and Kleerebezem, 1999). Lactic acid yields are highest during glycolysis via the homolactic acid fermentative pathway. Although free lactic acid is preferred for most of industrial processes, anaerobic fermentation for the production of the organic acid operates optimally at pH values where the salt of the organic acid rather than free acid is formed (Hugenholtz and Kleerebezem, 1999; Chotani et al., 2000). To obtain lactic acid in its free form, the fermentation process must be carried out at or below its pKa of 3.87. An intelligent approach was recently published. By insertion of the bovine lactate dehydrogenase A (LDH-A) gene into a *Kluyveromyces lactis*, pyruvate flux toward ethanol production was fully replaced by lactic acid production (1.19 mol lactate per mole of glucose) (Porro et al., 1999). Transferring the process to a 14-L fermenter gave a titer of 109 g/L with productivity of 0.8 g/L/h at pH 4.5. A doubling of yield as well as titer was achieved in a fermentation carried out with a strain of *S. cerevisiae* overexpressing the lactate-proton symporter (Chotani et al., 2000).

Amino Acids

Amino acids are used extensively in the food industry as flavor enhancers, antioxidants, and nutritional supplements; in agriculture, as feed additives; in medicine, in infusion solutions for postoperative treatment; and in the chemical industry, as starting materials for the manufacture of polymers and cosmetics (Glick and Pasternak, 1998). For the most part, amino acids are commercially produced either by extraction from protein hydrolysates or as fermentation products of either *Corynebacterium* or *Brevibacterium* spp., which are both nonsporulating Gram-positive soil bacteria (Glick and Pasternak, 1998). Traditionally, the productivity of these organisms has been improved by mutagenesis and subsequent screening for strains that overproduce certain amino acids (Glick and Pasternak, 1998; Demian, 2000b). However, this way of developing new strains is slow and sometimes inefficient.

Some preliminary progress has been made in increasing the amino acids in *C. glutamicum*. Thus, the synthesis of tryptophan by *C. glutamicum* was enhanced by introducing into wild-type *C. glutamicum* cells a second copy of the gene encoding anthranilate synthetase, which is the rate-limiting enzyme in the normal tryptophan biosynthetic pathway (Glick and Pasternak, 1998). On the other hand, one mutant unable to grow on a minimal medium unless anthranilic acid was added, and which did not produce tryptophan, was transformed with a copy of the anthranilate synthetase construct; the cloned gene did indeed restore most of the capacity of the mutant to synthesize tryptophan. However, the effect of adding this gene to wild-type *C. glutamicum* was much more dramatic, with the synthesis of this amino acid being increased by approximately 130%. This level of overproduction reflects more efficient utilization of available precursor material (Glick and Pasternak, 1998).

Most cereals are deficient in the essential amino acid L-lysine (Shewry, 1998; Tabe and Higgins, 1998). In *E. coli,* the lysine route is controlled very tightly and includes three AKs that are each regulated by a different end product. In addition, after each branch point, their respective final products inhibit the initial enzymes, and no overproduction occurs. However, in lysine fermentation in organisms such as mutants of *C. glutamicum* and its relatives, only a single AK exists, and it is regulated via concerted feedback inhibition by threonine and lysine (Eggeling et al., 1998). By the genetic removal of homoserine dehydrogenase, a glutamate-producing wild-type *Corynebacterium* was converted into a lysine-overproducing mutant that cannot grow unless methionine and threonine are added to the medium (Eggeling et al., 1998). As long as the threonine supplement is kept low, the intracellular concentration of this amino acid is limiting and feedback inhibition of AK is bypassed. Thus, *E. coli* and *Serratia marcescens* have been engineered with plasmid-bearing amino acid biosynthetic operons (Demian, 2000b). Plasmid transformation has been accomplished in *C. glutamicum,* so that recombinant DNA is now used to improve these commercial amino-acid-producing strains (Eggeling et al., 1998). The major manipulations have involved gene cloning to increase the levels of feedback-resistant AK and DHDPS. As a result, lysine industrial production yields 170 g/L and 0.54 moles of L-lysine per mole of glucose used. L-lysine is produced at an annual rate of 300,000 tons with a market of U.S.$600 million (Demian, 2000b).

Carotenoids

The enormous progress in the cloning of carotenogenic genes offers the opportunity of modifiying and engineering the carotenoid pathways in either noncarotenogenic or carotenogenic microorganisms (Sandmann et al., 1999; Sandmann, 2001). Carotenoids have been successfully synthesized in noncarotenogenic yeast *Candida utilis,* which has systematically been genetically modified as a producer host for lycopene, β-carotene, and astaxanthin (Misawa and Shimada, 1998). The foreign bacterial carotenoid biosynthesis gene was altered according to the codon usage for *C. utilis* and expressed under the control of a constitutive promoter derived from the host. This engineered strain yielded around 8 mg of lycopene per gram of dry weight (Misawa and Shimada, 1998).

On the other hand, Schmidt-Dannert et al. (2000) used DNA shuffling (also called sexual PCR) to generate diversity in both phytoene desaturases and lycopene cyclases by recombination of homologous DNA sequences.

Phytoene desaturase gene from *Erwinia herbicola* and *E. uredovora* were subjected to one round of DNA shuffling. *Escherichia coli* engineered to produce high levels of GGDP was transformed with a library of phytoene desaturase variants and screened for their ability to synthesize carotenoids of varying colors. In contrast to the four double bonds introduced by the parental enzymes, variants synthesizing between two and six double bonds were found in this library. One gene variant that synthesizes the fully conjugated compound with six double bonds is 3,4,3',4'-tetradehydrolycopene, a novel compound not known to be synthesized in nature (Figure 1.2) and yielding pink colonies. It is interesting to note that this carotenoid has greater antioxidant activity than lycopene.

Lycopene cyclase does not utilize 3,4,3',4'-tetradehydrolycopene as a substrate. DNA shuffling of the lycopene cyclase genes from *E. herbicola* and *E. uredovora*, however, resulted in the evolution of a cyclase capable of forming rings at both ends of 3,4,3',4'-tetradehydrolycopene to form torulene (Figure 1.2). A related strategy employed by Albrecht et al. (2000) combined carotenogenic genes from different bacteria employing unique pathways that maintain altered product specificities in *E. coli* expressing the biosynthetic machinery for phytoene production. In conjunction with the four-step phytoene desaturase that yields lycopene, a five-step desaturase was used to produce 3,4-didehydrolycopene (Figure 1.2). Further, diversification of the C40 skeleton resulted in the formation of numerous novel carotenoids, including 1-hydroxylated acyclic structures with up to 13 conjugated double bonds. Their antioxidative potential is superior to other related carotenoids. These experiments show that it is possible to use the molecular breeding approach to create novel carotenoids that have greater antioxidant activity than the carotenoids found in nature and hold the promise of creating improved nutraceuticals.

Acknowledgment

We thank Fidel Guevara-Lara from CINVESTAV-IPN for critical review of the manuscript.

References

Albrecht, M., S. Takaichi, S. Steiger, Z. Wang, and G. Sandmann (2000) Novel hydroxycarotenoids with improved antioxidative properties produced by gene combination in *Escherichia coli*, *Nat. Biotechnol.*, 18, 843–846.

Altenbach, S.B., C.C. Kuo, L.C. Staraci, K.W. Pearson, C. Wainwright, and A. Georges-cu (1992) Accumulation of a Brazil nut albumin in seeds of transgenic canola results in enhanced levels of seed protein methionine, *Plant Mol. Biol.*, 18, 235–245.

Anderson, B.F., H.M. Baker, G.E. Norris, D.W. Rice, and E.N. Baker (1989) Structure of human lactoferrin: crystallographic structure analysis and refinement at 2.8 A resolution, *J. Mol. Biol.*, 209, 711–734.

Anderson, S., C.B. Marks, R. Lazarus, J. Miller, K. Stafford, J. Seymour, D. Light, W. Rastetter, and D. Estell (1985) Production of 2-keto-L-gulonato, an intermediate in L-ascorbic synthesis by a genetically modified *Erwinia herbicola*, *Science*, 230, 144–149.

Arakawa, T., D.K.X. Chong, J. Hough, P.C. Engel, and W.H.R. Langridge (1998) A plant-based cholera toxin B subunit–insulin fusion protein protects against the development of autoimmune diabetes, *Nat. Biotechnol.*, 16, 934–938.

Arakawa, T., D.K.X. Chong, C.W. Slattery, and W.H.R. Langridge (1999) Improvements in human health through production of human milk proteins in transgenic food plants, in *Chemicals via Higher Plant Bioengineering*, F. Shahidi, P. Kolodziejcyk, J. Whitaker, A. López Munguia, and G. Fuller, Eds., Kluwer Academic, New York, pp. 149–159.

Ayub, R., M. Guis, M. Ben Amor, L. Gillot, J.P. Roustan, A. Latché, M. Bouzayen, and J.C. Pech (1996) Expression of ACC oxidase antisense gene inhibits ripening of cantaloupe melon fruits, *Nat. Biotechnol.*, 14, 862–866.

Bailey, J.E. (1999) Lessons from metabolic engineering for functional genomics and drug discovery, *Nat. Biotechnol.*, 17, 616–618.

Barba de la Rosa, A.P., A. Herrera-Estrella, S. Utsumi, and O. Paredes-López (1996) Molecular characterization, cloning and structural analysis of a cDNA encoding an amaranth globulin, *J. Plant Physiol.*, 149, 527–532.

Beard, J.L., J.W. Burton, and E.C. Theil (1996) Purified ferritin and soybean meal can be sources of iron for treating iron deficiency in rats, *J. Nutr.*, 126, 154–160.

Bird, C.R., C.J.S. Smith, J.A. Ray, P. Moureau, M.W. Bevan, A.S. Bird, S. Hughes, P.C. Morris, D. Grierson, and W. Schuch (1988) The tomato polygalacturonase gene and ripening specific expression in transgenic plants, *Plant Mol. Biol.*, 11, 651–662.

Blackstock, W.P. and M.P. Weir (1999) Proteomics: quantitative and physical mapping of cellular proteins, *Trends Biotechnol.*, 17, 121–127.

Bliss, F.A. (1999) Nutritional improvement of horticultural crops through plant breeding, *HortScience*, 34, 1163–1167.

Blohm, D.H. and A. Guiseppi-Eli (2001) New developments in microarray technology, *Curr. Opin. Biotechnol.*, 12, 41–47.

Chakraborty, S., N. Chakraborty, and A. Datta (2000) Increased nutritive value of transgenic potato by expressing a nonallergenic seed albumin gene from *Amaranthus hypochondriacus*, *Proc. Natl. Acad. Sci. USA*, 97, 3724–3729.

Chen, R. (2001) Enzyme engineering: rational redesign versus directed evolution, *Trends Biotechnol.*, 19, 13–14.

Chong, D.K.X. and W.H. Langridge (2000) Expression of full-length bioactive antimicrobial human lactoferrin in potato plants, *Transgenic Res.*, 9, 71–78.

Chong, D.K.X., W. Roberts, T. Arakawa, K. Illes, G. Bagi, C.W. Slattery, and W.H.R. Langridge (1997) Expression of a human milk protein, β-casein in transgenic potato plants, *Transgenic Res.*, 6, 289–296.

Chotani, G., T. Dodge, A. Hsu, M. Kumar, R. LaDuca, D. Trimbur, W. Weyler, and K. Sanford (2000) The commercial production of chemicals using pathway engineering, *Biochim. Biophys. Acta*, 1543, 434–455.

Cohen, S., A.C. Chang, H.W. Boyer, and R.B. Helling (1973) Construction of biologically functional bacterial plasmids *in vitro*, *Proc. Natl. Acad. Sci. USA*, 70, 3240–3244.

Curie, C., Z. Penaviene, C. Loulergue, S.L. Dellaporta, J.F. Briat, and E.L. Walker (2001) Maize *yellow stripe* 1 encodes a membrane protein directly involved in Fe(III) uptake, *Nature*, 409, 346–349.

Curtiss, R.I. and C.A. Cardineau (1990) Oral immunisation by transgenic plants, world patent application, WO 90/02484.

Delgado-Vargas, F. and O. Paredes-López (1997) Effects of enzymatic treatments of marigold flowers on lutein isomeric profiles, *J. Agric. Food Chem.*, 45, 1097–1102.

Delgado-Vargas, F., A.R. Jiménez, and O. Paredes-López (2000) Natural pigments: carotenoids, anthocyanins, and betalains — characteristics, biosynthesis, processing, and stability, *Crit. Rev. Food Sci. Nutr.*, 40, 173–289.

DellaPenna, D. (1999) Nutritional genomics: manipulating plant micronutrients to improve human health, *Science*, 285, 375–379.

DellaPenna, D. (2001) Plant metabolic engineering, *Plant Physiol.*, 125, 160–163.

Demian, A.L. (2000a) Microbial biotechnology, *Trends Biotechnol.*, 18, 26–31.

Demian, A.L. (2000b) Small bugs, big business: the economic power of the microbe, *Biotechnol. Adv.*, 18, 499–514.

Dove, A (1999) Proteomics: translating genomics into products?, *Nat. Biotechnol.*, 17, 233–236.

Dunwell, J.M (1999) Transgenic crops: the next generation, or an example of 2020 vision, *Ann. Bot.*, 84, 269–277.

Eggeling, L., S. Oberle, and H. Sahm (1998) Improved L-lysine yield with *Corynebacterium glutamicum*: use of dapA resulting in increased flux combined with growth limitation, *Appl. Microbiol. Biotechnol.*, 1, 24–30.

Falco, S.C., T. Guida, M. Locke, J. Mauvais, C. Sanders, R.T. Ward, and P. Weber (1995) Transgenic canola and soybean seeds with increased lysine, *Biotechnology*, 13, 577–582.

FAO/WHO (1991) Protein quality evaluation: report of a joint FAO/WHO expert consultation, in *FAO Food and Nutrition Paper*, Food and Agriculture Organization of the United Nations, Rome, Italy.

Fischer, R. and N. Emans (2000) Molecular farming of pharmaceutical proteins, *Transgenic Res.*, 9, 279–299.

Fischer, R., J. Drossard, U. Commandeur, S. Schillberg, and N. Emans (1999) Towards molecular farming in the future: moving from diagnostic protein and antibody production in microbes to plants, *Biotechnol. Appl. Biochem.*, 30, 101–108.

Fray, R.G., A. Wallace, P.D. Fraser, D. Valero, P. Heddon, P. Bramley, and D. Grierson (1995) Constitutive expression of a fruit phytoene synthase gene in transgenic tomatoes causes dwarfism by redirecting metabolites from the Gibberellin pathway, *Plant J.*, 8, 693–701.

Galili, G., G. Tang, X. Zhu, and B. Gakiere (2001) Lysine catabolism: a stress and development super-regulated metabolic pathway, *Curr. Opin. Plant Biol.*, 4, 261–266.

Geigenberger, P., M. Hajirezaei, M. Geiger, U. Deiting, U. Sonnewald, and M. Sttit (1998) Overexpression of pyrophosphatase leads to increased sucrose degradation and starch synthesis, increased activities of enzymes for sucrose–starch interconversions, and increased levels of nucleotides in growing potato tubers, *Planta*, 205, 428–437.

Giddings, G., G. Allison, D. Brooks, and A. Carter (2000) Transgenic plants as factories for biopharmaceuticals, *Nat. Biotechnol.*, 18, 1151–1155.

Giovannoni, J.J., P. Kannan, S. Lee, and H.C. Yen (1998) Genetic approaches to manipulation of fruit development and quality in tomato, in *Genetic and Environmental Manipulation of Horticultural Crops*, K.E. Cockshull, D. Gray, G.B. Seymour, and A. Thomas, Eds, CAB International, Oxford, pp. 1–15.

Glick, B.R. and J.J. Pasternak (1998) *Molecular Biotechnology: Principles and Applications of Recombinant DNA*, American Society for Microbiology Press, Washington, D.C.

Goddijn, O.J.M. and J. Pen (1995) Plants as bioreactors, *Trends Biotechnol.*, 13, 379–387.

Gomez-Lim, M.A. (1999) Physiology and molecular biology of fruit ripening, in *Molecular Biotechnology for Plant Food Production*, O. Paredes-López, Ed., Technomic, Lancaster, PA, pp. 303–342.

Goodenough, P. (1995) A review of protein engineering for the food industry, *Int. J. Food Technol.*, 30, 119–139.

Goto, F., T. Yoshihara, and H. Saiki (1998) Iron accumulation in tobacco plants expressing soyabean in the seeds, *Transgenic Res.*, 7, 173–180.

Goto, F., T. Yoshihara, N. Shigemoto, S. Toki, and F. Takaiwa (1999) Iron fortification of rice seed by the soybean ferritin gene, *Nat. Biotechnol.*, 17, 282–286.

Gray, J., S. Picton, J. Shabbeer, W. Schuch, and D. Grierson (1992) Molecular biology of fruit ripening and its manipulation by antisense genes, *Plant Mol. Biol.*, 19, 69–87.

Grierson, D. (1998) GCRI/Bewley lecture: applications of molecular biology and genetic manipulation to understand and improve quality of fruits and vegetables, in *Genetic and Environmental Manipulation of Horticultural Crops*, K.E. Cockshull, D. Gray, G.B. Seymour, and A. Thomas, Eds., CAB International, Oxford, pp. 31–39.

Grusak, M.A. and D. DellaPenna (1999) Improving the nutrient composition of plants to enhance human nutrition and health, *Annu. Rev. Plant Physiol. Plant Mol. Biol.*, 50, 133–161.

Grusak, M.A., N. Pearson, and E. Marentes (1999) The physiology of micronutrient homeostasis in field crops, *Field Crops Res.*, 60, 41–56.

Guzmán-Maldonado, S.H. and O. Paredes-López (1999) Biotechnology for the improvement of nutritional quality of food crop plants, in *Molecular Biotechnology for Plant Food Production*, O. Paredes-López, Ed., Technomic, Lancaster, PA, pp. 553–620.

Guzmán-Maldonado, S.H., J. Acosta-Gallegos, and O. Paredes-López (2000) Protein and mineral content of a novel collection of wild and weedy common bean (*Phaseolus vulgaris* L.), *J. Sci. Food Agric.*, 80, 1874–1881.

Halford, N.G. and P.R. Shewry (2000) Genetically modified crops: methodology, benefits, regulation and public concerns, *Br. Med. Bull.*, 56, 62–73.

Hein, M., Y. Tang, D.A. Mcleod, K.D. Janda, and A. Hiatt (1991) Evaluation of immunoglobulins from plant cells, *Biotechnol. Prog.*, 7, 455–461.

Hiatt, A., R. Cafferkey, and K. Bowdish (1989) Production of antibodies in transgenic plants, *Nature*, 342, 76–78.

Hirschberg, J. (1999) Production of high-value compounds: carotenoids and vitamin E, *Curr. Opin. Biotechnol.*, 10, 186–191.

Hoffman, L.M., D.D. Donaldson, and E.M. Herman (1988) A modified storage protein is synthesized, processed, and degraded in the seeds of transgenic plants, *Plant Mol. Biol.*, 11, 717–729.

Horrobin, D.F. (1990) Gamma linolenic acid: an intermediate in essential fatty acid metabolism with potential as an ethical pharmaceutical and as a food, *Rev. Contemp. Pharmacotherapy*, 1, 1–45.

Host, A., H.P. Jacobsen, S. Halken, and D. Holmenlund (1995) The natural history of cow's milk protein allergy/intolerance, *Eur. J. Clin. Nutr.*, 49(suppl.), S13–S18.

Hugenholtz, J. and M. Kleerebezem (1999) Metabolic engineering of lactic acid bacteria: overview of the approaches and results of pathway rerouting involved in food fermentations, *Curr. Opin. Biotechnol.*, 10, 492–497.

Jacobsen, J.R. and C. Khosla (1998) New directions in metabolic engineering, *Curr. Opin. Chem Biol.*, 2, 133–137.

Jez, J.M. and J.P. Noel (2000) A kaleidoscope of carotenoids, *Nat. Biotechnol.*, 18, 825–826.

Kao, C.M (1999) Functional genomic technologies: creating new paradigms for fundamental and applied biology, *Biotechnol. Prog.*, 15, 304–311.

Karchi, H., O. Shaul, and G. Galili (1993) Seed-specific expression of a bacterial desensitized aspartate kinase increases the production of seed threonine and methionine in transgenic tobacco, *Plant J.*, 3, 721–727.

Kinney, A.J. and R. Jung (2001) Cosuppression of the α-subunits of β-conglycinin in transgenic soybean seeds induces the formation of endoplasmic reticulum-derived protein bodies, *Plant Cell*, 13, 1165–1178.

Kishore, G.M. and C. Shewmaker (1999) Biotechnology: enhancing human nutrition in developing and developed worlds, *Proc. Natl. Acad. Sci. USA*, 96, 5968–5972.

Knutzon, D.S., G.A. Thompson, S.E. Radke, W.B. Johnson, V.C. Knauf, and J.C. Kridl (1992) Modification of *Brassica* seed oil by antisense expression of a stearoyl-acyl carrier protein desaturase gene, *Proc. Natl. Acad. Sci. USA*, 89, 2624–2628.

Kohno-Murase, J., M. Murase, H. Ichikawa, and J. Imamura (1995) Improvement in the quality of seed storage protein by transformation of *Brassica napus* with an antisense gene for cruciferin, *Theor. Appl. Genet.*, 91, 627–631.

Kusel, J.P., Y.H. Fa, and A.L Demian (1984) Betaine Stimulation of vitamin B_{12} biosynthesis in *Pseudomonas denitrificans* may be mediated by an increase in activity of δ-aminolaevulinic acid synthase, *J. Gen. Microbiol.*, 130, 835–841.

Lassner, M. and J. Bedbrook (2001) Directed molecular evolution in plant improvement, *Curr. Opin. Plant Biol.*, 4, 152–156.

Lockhart, D.J. and E.A. Winzeler (2000) Genomics, gene expression and DNA arrays, *Nature*, 405, 827–836.

López, M.G., L.A. Bello-Pérez, and O. Paredes-López (1994) Amaranth carbohydrates, in *Amaranth: Biology, Chemistry and Technology*, O. Paredes-López, Ed., CRC Press, Boca Raton, FL, pp. 103–131.

Lueking, A., M. Horn, H. Eickhoff, K. Bussow, H. Lehrach, and G. Walter (1999) Protein microarrays for gene expression and antibody screening, *Anal. Biochem.*, 270, 103–111.

Ma, J.K., A. Hiatt, N.D. Vine, P. Wand, P. Stabila, C. Van Dollerweerd, K. Mostov, and T. Lehner (1995) Generation and assembly of secretory antibodies in plants, *Science*, 268, 716–719.

Ma, J.K., B.Y. Hikmat, K. Wycoff, N.D. Vine, D. Chargelegue, L. Yu, M.B. Hein, and T. Lehner (1998) Characterization of a recombinant plant monoclonal secretory antibody and preventive immunotherapy in humans, *Nat. Med.*, 4, 601–606.

Ma, J.K., T. Lehner, P. Stabila, C.I. Fux, and A. Hiatt (1994) Assembly of monoclonal antibodies with IgG1 and IgA heavy chain domains in transgenic tobacco plants, *Eur. J. Immunol.*, 24, 131–138.

MacKenzie, S.L. (1999) Chemistry and engineering of edible oils and fats, in *Molecular Biotechnology for Plant Food Production*, O. Paredes-López, Ed., Technomic, Lancaster, PA, pp. 525–551.

Mann, V., M. Harker, I. Pecker, and J. Hirschberg (2000) Metabolic engineering of astaxanthin production in tobacco flowers, *Nat. Biotechnol.*, 18, 888–892.

Mason, H.S., J.M. Ball, J.J. Shi, X. Jiang, M.K. Estes, and C.J. Arntzen (1996) Expression of Norwalk virus capsid protein in transgenic tobacco and potato and its oral immunogenicity in mice, *Proc. Natl. Acad. Sci. USA*, 93, 5335–5340.

Masuda, M., K. Takahashi, N. Sakurai, K. Yanagiya, S. Komatsubara, and T. Tosa (1995) Further improvements of D-biotin production by a recombinant strain of *Serratia marcescens*, *Proc. Biochem.*, 30, 553–562.

Mazur, B., E. Krebbers, and S. Tingey (1999) Gene discovery and product development for grain quality traits, *Science*, 285, 372–375.

Michaelson, L.V., C.M. Lazarus, G. Griffiths, J.A. Napier, and A.K. Stobart (1998) Isolation of a Δ^5-fatty acid desaturase gene from *Mortierella alpina*, *J. Biol. Chem.*, 273, 19055–19059.

Miflin, B., J. Napier, and P.R. Shewry (1999) Improving plant product quality, *Nat. Biotechnol.*, 17(suppl.), BV13–BV14.

Misawa, N. and H. Shimada (1998) Metabolic engineering for production of carotenoids in non-carotenogenic bacteria and yeasts, *J. Biotechnol.*, 59, 169–181.

Mitra, A. and Z. Zhang (1994) Expression of a human lactoferrin cDNA in tobacco cells produces antibacterial protein(s), *Plant Physiol.*, 106, 977–981.

Modelska, A., B. Dietzschold, N. Sleysh, Z.F. Fu, K. Steplewski, D.C. Hooper, H. Kiprowski, and V. Yusibov (1998) Immunization against rabies with a plant-derived antigen, *Proc. Natl. Acad. Sci. USA*, 95, 2481–2485.

Molvig, L., L.M. Tabe, B.O. Eggum, A.E. Moore, S. Craig, D. Spencer, and T.J.V. Higgins (1997) Enhanced methionine levels and increased nutritive value of seeds of transgenic lupins (*Lupinus angustifolius*) expressing a sunflower seed albumin gene, *Proc. Natl. Acad. Sci. USA*, 94, 8393–8398.

Murphy, D.J. (1996) Engineering oil production in rapeseed and other oil crops, *Trends Biotechnol.*, 14, 206–213.

Murphy, D.J. (1999) Production of novel oils in plants, *Curr. Opin. Biotechnol.*, 10, 175–180.

Napier, J., L.V. Michaelson, and A.K. Stobart (1999) Plant desaturases: harvesting the fat of the land, *Curr. Opin. Plant Biol.*, 2, 123–127.

Nielsen, J. (1998) Metabolic Enginering: Techniques for Analysis of Targets for Genetic Manipulations, *Biotechnol. Bioeng.*, 58, 125–132.

Nixon, A.E. and S.M. Firestine (2000) Rational and irrational design of proteins and their use in biotechnology, *IUBMB Life*, 49, 181–187.

Nordlee, J.A., S.L. Taylor, J.A. Townsend, L.A. Thomas, and R.K. Bush (1996) Identification of a Brazil-nut allergen in transgenic soybeans, *N. Eng. J. Med.*, 334, 688–692.

Oeller, P.W., L.M. Wong, L.P. Taylor, D.A. Pike, and A. Theologis (1991) Reversible inhibition of tomato fruit senescence by antisense 1-aminocyclopropane-1-carboxylate synthase, *Science*, 254, 427–439.

O'Farrell, P.H. (1975) High-resolution two-dimensional gel electrophoresis of proteins, *J. Biol. Chem.*, 250, 4007–4021.

Olmedo-Alvarez, G. (1999) Molecular biology research procedures, in *Molecular Biotechnology for Plant Food Production*, O. Paredes-López, Ed., Technomic, Lancaster, PA, pp. 55–87.

Osuna-Castro, J.A., Q. Rascón-Cruz, J. Napier, R. Fido, P.R. Shewry, and O. Paredes-López (2000) Overexpression, purification, and *in vitro* refolding of the 11S globulin from amaranth seed in *Escherichia coli*, *J. Agric. Food Chem.*, 48, 5249–5255.

Patterson, S.D. (2000) Proteomics: The industrialization of protein chemistry, *Curr. Opin. Biotechnol.*, 11, 413–418.

Pen, J., T.C. Verwoerd, P.A. Van Paridin, R.F. Beudeker, P.J.M. Van den Elzen, K. Greerse, J.D. Van der Klis, H.A. Versteegh, A.J. Van Ooyen, and A. Hoekema (1993) Phytase-containing transgenic seeds as a novel feed additive for improved phosphorus utilization, *Biotechnology*, 11, 811–814.

Perkins, J., A. Sloma, T. Hermanm, K. Thieriault, E. Zachgo, T. Erdenberger, N. Hannet, N. Chatterjee, V. Williams, G. Rufo, R. Hatch, and J. Pero (1999) Genetic engineering of *Bacillus subtilis* for the commercial production of riboflavin, *J. Indus. Microbiol. Biotechnol.*, 22, 8–12.

Picton, S., S.L. Barton, M. Bouzayen, A.J. Hamilton, and D. Grierson (1993) Altered fruit ripening and leaf senescence in tomatoes expressing an antisense ethylene forming enzyme transgenic, *Plant J.*, 3, 469–481.

Porro, D., M.M. Bianchi, L. Brambilla, R. Menghini, D. Bolzani, V. Carrera, J. Lievense, C.L. Liu, B.M. Ranzi, L. Frontali, and L. Alberghina (1999) Replacement of a metabolic pathway for large-scale production of lactic acid from engineered yeasts, *Appl. Environ. Microbiol.*, 65, 4211–4215.

Porta, C., V.E. Spall, J. Loveland, J.E. Johnson, P.J. Barker, and G.P. Lomonossoff (1994) Development of cowpea mosaic virus as a high-yielding system for the presentation of foreign peptides, *Virology*, 202, 949–952.

Potrykus, I. (2001) Golden rice and beyond, *Plant Physiol.*, 125, 1157–1161.

Pramik, M.J. (1986) Genentech develops recombinant technique for producing vitamin C, *Gen. Eng. News*, 2, 9–12.

Rascón-Cruz, Q., J.A. Osuna-Castro, N. Bohorova, and O. Paredes-López (2001) Transformation of tropical maize with an amaranth globulin gene (submitted to *Nat. Biotechnol.*).

Romer, S., P.D. Fraser, J.W. Kiano, C.A. Shipton, N. Misawa, W. Schuch, and P.M. Bramley (2000) Elevation of the provitamin A content of transgenic tomato plants, *Nat. Biotechnol.*, 18, 666–669.

Rooke, L., F. Békés, R. Fido, F. Barro, P. Gras, A.S. Tatham, P. Barcelo, P. Lazzeri, and P.R. Shewry (1999) Overexpression of a gluten protein in transgenic wheat results in greatly increased dough strength, *J. Cereal Sci.*, 30, 115–120.

Russell, P.J. (1998) *Genetics*, The Benjamin/Cummings Publishing Company, Menlo Park, CA.

Ryals, J. (2001) Plant biotechnology: is there light at the end of the tunnel?, *Curr. Opin. Plant Biol.*, 4, 151.

Ryu, D.D. and D.H. Nam (2000) Recent progress in biomolecular biotechnology, *Biotechnol. Prog.*, 16, 2–16.

Saalbach, I., T. Pickardt, D.R. Waddell, S. Hillmer, O. Schieder, and K. Muntz (1995) The sulphur-rich Brazil nut 2S albumin is specifically formed in transgenic seeds of the grain legume *Vicia narbonensis*, *Euphytica*, 85, 181–192.

Sandmann, G. (2001) Genetic manipulation of carotenoid biosynthesis: strategies, problems and achievements, *Trends Plant Sci.*, 6, 14–17.

Sandmann, G., M. Albrecht, G. Schnurr, O. Knorzer, and P. Boger (1999) The biotechnological potential and design of novel carotenoids by gene combination in *Escherichia coli*, *Trends Biotechnol.*, 17, 233–237.

Sayanova, O., M.A. Smith, P. Lapinskas, A.K. Stobart, G. Dobson, W.W. Christie, P.R. Shewry, and J. Napier (1997) Expression of a borage desaturase cDNA containing an N-terminal cytochrome b_5 domain results in the accumulation of high levels of Δ^6-desaturated fatty acids in transgenic tobacco, *Proc. Natl. Acad. Sci. USA*, 94, 4211–4216.

Schmidt-Dannert, C., D. Umeno, and F.H. Arnold (2000) Molecular breeding of carotenoid biosynthetic pathway, *Nat. Biotechnol.*, 18, 750–753.

Schulman, A.H. (1999) Chemistry, biosynthesis, and engineering of starches and other carbohydrates, in *Molecular Biotechnology for Plant Food Production*, O. Paredes-López, Ed., Technomic, Lancaster, PA, pp. 493–523.

Schwall, G.P., R. Safford, R.J. Westcott, R. Jeffcoat, A. Tayal, Y.G. Shi, M.J. Gidley, and S.A. Jobling (2000) Production of very-high-amylose potato starch by inhibition of SBE A and B, *Nat. Biotechnol.*, 18, 551–554.

Segura-Nieto, M., P.R. Shewry, and O. Paredes-López (1994) Biochemistry of amaranth proteins, in *Amaranth: Biology, Chemistry and Technology*, O. Paredes-López, Ed., CRC Press, Boca Raton, FL, pp. 76–95.

Segura-Nieto, M., P.R. Shewry, and O. Paredes-López (1999) Globulins of the pseudocereals: amaranth, quinoa and buckwheat, in *Seed Proteins*, P.R. Shewry and R. Casey, Eds., Kluwer Academic Publishers, Dordrecht, pp. 453–475.

Sheehy, R.E., M. Kramer, and W.R. Hiatt (1988) Reduction of tomato fruit polygalacturonase activity in tomato fruit by antisense RNA, *Proc. Natl. Acad. Sci. USA*, 85, 8805–8809.

Shewmaker, C.K., J.A. Sheehy, M. Daley, S. Colburn, and D.Y. Ke (1999) Seed-specific overexpression of phytoene synthase: increase in carotenoids and other metabolic effects, *Plant J.*, 20, 401–412.

Shewry, P.R. (1995) Plant storage proteins, *Biol. Rev.*, 70, 375–426.

Shewry, P.R. (1998) Manipulation of seed storage proteins, in *Transgenic Plant Research*, K. Lindsey, Ed., Harwood Academic Publishers, Reading, U.K., pp.135–149.

Shewry, P.R. and R. Casey (1999) Seed proteins, In: *Seed Proteins*, P.R. Shewry and R. Casey, Eds., Kluwer Academic Publishers, Dordrecht, pp. 1–10.

Shewry, P.R., A.S. Tatham, J. Greenfield, N. Halford, S. Thompson, D.H.L. Bishop, F. Barro, P. Barcelo, and P. Lazzeri (2000) Wheat protein molecular biology and genetic engineering: implications for quality improvement, *Special Publ. Royal Soc. Chem.*, 212, 199–205.

Shintani, D. and D. DellaPenna (1998) Elevating the vitamin E content of plants through metabolic engineering, *Science*, 282, 2098–2100.

Sitrit, Y. and A.B. Bennet (1997) Regulation of tomato fruit polygalacturonase mRNA accumulation by ethylene: a re-examination, *Plant Physiol.*, 116, 1145–1150.

Slattery, C.J., I.H. Kavakli, and T.W. Okita (2000) Engineering starch for increased quantity and quality, *Trends Plant Sci.*, 5, 291–298.

Smith, C.J.S., C. Watson, P.C. Morris, C.R. Bird, G.B. Seymour, J.E. Gray, C. Arnold, G.A. Tucker, W. Schuch, S. Harding, and D. Grierson (1990) Inheritance and effect on ripening antisense polygalacturonase genes in transgenic tomatoes, *Plant Mol. Biol.*, 14, 369–379.

Snustad, D.P. and M.J. Simmons (2000) *Principles of Genetics*, John Wiley & Sons, New York.

Spychalla, J.P., A.J. Kinney, and J. Browse (1997) Identification of an animal omega-3 fatty acid desaturase by heterologous expression in *Arabidopsis*, *Proc. Natl. Acad. Sci. USA*, 94, 1142–1147.

Stark, D.M., K.P. Timmerman, G.F. Barry, J. Preiss, and G.M. Kishore (1992) Regulation of the amount of starch in plant tissues by ADP glucose pyrophosphorylase, *Science*, 258, 287–292.

Steinmetz, K.A. and J.D. Potter (1996) Vegetables, fruits and cancer prevention: a review, *J. Am. Diet. Assoc.*, 96, 1027–1039.

Tabe, L. and T.J.V. Higgins (1998) Engineering plant protein composition for improved nutrition, *Trends Plant Sci.*, 3, 282–286.

Tacket, C.O., R.H. Reid, E.C. Boedeker, G. Losonsky, J.P. Nataro, H. Bhagat, and R. Edelman (1994) Enteral immunization and challenge of volunteers given enterotoxigenic *E. coli* CFA II encapsulated in biodegradable microspheres, *Vaccine*, 12, 1270–1274.

Tacket, C.O., H.S. Mason, G. Losonsky, J.D. Clements, M.M Levine, and C.J. Arntzen (1998) Immunogenicity in humans of a recombinant bacterial antigen delivered in a transgenic potato, *Nat. Med.*, 4, 607–609.

Takase, K. and K. Hagiwara (1998) Expression of human alpha-lactalbumin in transgenic tobacco, *J. Biochem.*, 123, 440–444.

Thanavala, Y., Y.F. Yang, P. Lyons, H.S. Mason, and C. Arntzen (1995) Immunogenicity of transgenic plant-derived hepatitis B surface antigen, *Proc. Natl. Acad. Sci. USA*, 92, 3358–3361.

Theologis, A. (1992) One rotten apple spoils the whole bushel: the role of ethylene in fruit ripening, *Cell*, 70, 181–184.

Tobin, M.B., C. Gustafsson, and G.W. Huisman (2000) Directed evolution: the rational basis for irrational design, *Curr. Opin. Struct. Biol.*, 10, 421–427.

Traber, M.G. and H. Sies (1996) Vitamin E in humans: demands and delivery, *Annu. Rev. Nutr.*, 16, 321–347.

Tramper, J. (2000) Modern biotechnology: food for thought, in *Food Biotechnology*, S. Bielecki, J. Tramper, and J. Polak, Eds., Elsevier Science, Amsterdam, pp. 3–12.

Tsuda, S., K. Yoshioka, T. Tanaka, A. Iwata, A. Yoshikawa, Y. Watanabe, and Y. Okada (1998) Application of human hepatitis B virus core antigen from transgenic tobacco plants for serological diagnosis, *Vox Sanguinis*, 74, 148–155.

Turpen, T.H., S.J. Rein, Y. Charoenvit, S.L. Hoffman, V. Fallarme, and L.K. Grill (1995) Malarial epitopes expressed on the surface of recombinant tobacco mosaic virus, *Biotechnology*, 13, 53–57.

Van den Berg, H., R. Faulks, H.F. Granado, J. Hirschberg, B. Olmedilla, G. Sandmann, S. Southon, and W. Stahl (2000) The potential for the improvement of carotenoid levels in foods and the likely systemic effects, *J. Sci. Food Agric.*, 80, 880–912.

Verch, T., V. Yusibov, and H. Koprowski (1998) Expression and assembly of a full length monoclonal antibody in plants using a plant virus vector, *J. Immunol. Methods*, 220, 69–75.

Visser, R.G.F. and E. Jacobsen (1993) Towards modifying plants for altered starch content and composition, *Trends Biotechnol.*, 11, 63–68.

Voelker, T.A., A.C. Worrell, L. Anderson, J. Bleibaum, C. Fan, D.J. Hawkins, S.E. Radke, and H.M. Davies (1992) Fatty acid biosynthesis redirected to medium chain in transgenic oilseed plants, *Science*, 257, 72–74.

Walmsley, A.M. and C.J. Arntzen (2000) Plants for delivery of edible vaccines, *Curr. Opin. Biotechnol.*, 11, 126–129.

Ward, K.A. (2001) Phosphorus-friendly transgenics, *Nat. Biotechnol.*, 19, 415–416.

Yamauchi, D. and T. Minamikawa (1998) Improvement of the nutritional quality of legume seed storage proteins by molecular breeding, *J. Plant Res.*, 111, 1–6.

Ye, X., S. Al-Babili, A. Kloti, J. Zhang, P. Lucca, P. Beyer, and I. Potrykus (2000) Engineering provitamin A (β-carotene) biosynthetic pathway into (carotenoid-free) rice endosperm, *Science*, 287, 303–305.

Zeitlin, L., S.S. Olmsted, T.R. Moench, M.S. Co, B.J. Martinell, V.M. Paradkar, D.R. Russell, C. Queen, R.A. Cone, and K.J. Whaley (1998) A humanized monoclonal antibody produced in transgenic plants for immunoprotection of the vagina against genital herpes, *Nat. Biotechnol.*, 16, 1361–1364.

2

Recent Developments in Food Biotechnology

Humberto Hernández-Sánchez

CONTENTS

Introduction

Defining Biotechnology

Out of the several definitions of biotechnology, perhaps one of the broadest is the use of living cells, microorganisms, or enzymes for the manufacture of chemicals, drugs, or foods or for the treatment of wastes (Jenson, 1993). A simpler approach defines biotechnology as the use of biological organisms or processes in any technological application (Riley and Hoffman, 1999). Biotechnology has had a tremendous impact on the food industry. It has provided high-quality foods that are tasty, nutritious, convenient, and safe, and it has the potential for the production of even more nutritious, palatable, and stable food (John Innes Centre, 1998).

Traditional Biotechnology

The roots of biotechnology can be found in the ancient processes of food and beverage fermentation. These traditional technologies are present in almost every culture in the world and have evolved over many years without losing their traditional essence. Examples of these processes include the production of some well-known foods, such as bread, wine, yogurt, and cheese. These products, like many others (such as ripened sausages [salami], pickles, sauerkraut, soy sauce, vinegar, beer, and cider), are produced using the natural processes of living organisms (e.g., fermentation) — in other words, by using biotechnology. Some relatively new developments in these traditional products include bio-yogurts, or biogurts, which contain extra bacteria (usually probiotic organisms) that are not found naturally in the original food. These probiotic bacteria most often include *Lactobacillus acidophilus* and *Bifidobacterium bifidum*. Another traditional biotechnology technique is the production of mycoprotein Quorn™ as an alternative to meat, which was developed much more recently. Because these techniques are considered conventional, they have not caused public concern (John Innes Centre, 1998).

Enzymes are also widely used in the food and beverage industries. For economic reasons they are used in a relatively crude form or in a reusable form, usually achieved by immobilization. The dairy industry uses primarily rennins and lactases; the brewing industry, proteases and amylases. High-fructose corn syrup is produced from starch by using α-amylase, amyloglucosidase, and glucose isomerase, and the beverage industry consumes a great amount of pectinases (Smith, 1988).

Molecular (Modern) Biotechnology

"Modern" or molecular biotechnology, in contrast with "traditional" biotechnology, also includes the use of techniques of genetic engineering — that is, techniques for altering the properties of biological organisms. This allows characteristics to be transferred between organisms to give new combinations of genes and improved varieties of plants or microorganisms for use in agriculture and industry (John Innes Centre, 1998). Some consumers are concerned about the safety of using the techniques of modern biology, although in countries such as Japan and the U.S. consumers remain optimistic about biotechnology. They are generally willing to purchase foods developed through these techniques, and the food and agricultural applications are as acceptable as are new medicines (Hoban, 1999).

As stated above, recombinant DNA techniques have revolutionized the fields of biology, biochemistry, and biotechnology, as they have made research of genomes more possible than ever. For example, molecular cloning and the polymerase chain reaction (PCR) have been used in order to obtain the large number of DNA copies required for DNA sequencing methods (McKee and McKee, 1999).

Molecular Cloning

In this technique, a fragment of DNA isolated from a donor cell (e.g., bacteria, yeast, or any animal or plant cell) is incorporated into a vector (e.g., plasmids or phages), by which the gene of interest can be introduced into a host cell. The formation of the recombinant DNA molecule requires a restriction endonuclease to cut and open the vector DNA. After the sticky ends of the vector have been annealed with those of the donor DNA, a DNA ligase joins the two molecules covalently. The recombinant molecule is then inserted into bacterial cells by any suitable method such as electroporation. It is necessary that the recombinant vectors contain regulatory regions recognized by the bacterial enzymes.

Polymerase Chain Reaction

Polymerase chain reaction, considered one of the most significant DNA technologies developed, is a method for amplifying very small amounts of DNA that copies part of a genome for subsequent sequencing or other analyses (Garrison and dePamphilis, 1994). It is an *in vitro* laboratory method to amplify a specific segment of a genome DNA by using a pair of specific primers (oligonucleotides used to start the DNA replication) to allow a section of the DNA to be repeatedly copied. Template strands are separated by heating and then cooled to allow primers to anneal to the template. The temperature is raised again to allow primer extension (template strand copying) by means of a thermostable DNA polymerase. The procedure is repeated 30 to 40 times with an exponential amplification of the DNA concentration (Hill, 1996). With the availability of this technique, enormous quantities of genes have now been sequenced for a wide range of organisms. The genomes of several bacteria and small organisms have already been fully sequenced, and the genomic sequences of many higher organisms such as plants, animals, and humans (around 90% complete) have been published (Institute of Food Technologists, 2000).

Modern Biotechnological Approaches to Traditional Processes

Phage Resistance

Bacteriophages are viruses that infect and kill bacteria, including cheese and fermented milk starter cultures, which leads to low production, low-quality products, and economic losses. The discovery that some starter strains have phage resistance mechanisms encoded on plasmids has resulted in the development of molecular cloning techniques to develop new resistant strains, especially in *Lactococcus lactis*. Currently, research in this subject is divided among three main areas: (1) the study of the mechanisms of infection (Valyasevi et al., 1991; Monteville et al., 1994) and resistance (McKay and Baldwin, 1984; Sing and Klaenhammer, 1990; Garvey et al., 1995, 1996); (2) the study of restriction and modification systems in lactic acid bacteria (Sing and Klaenhammer, 1991; Su et al., 1999); and (3) the study of bacteriophage genomes (Brown et al., 1994; Djordjevic and Klaenhammer, 1997). Something similar is also being developed in the case of yogurt for *Streptococcus thermophilus* (Brussow et al., 1994).

Nisin Resistance

Potential food-grade selectable markers from the lactococci include genes associated with carbohydrate metabolism, bacteriophage resistance, and nisin production or resistance (Froseth and McKay, 1991). Nisin is an important antimicrobial agent produced by some *Lactococcus lactis* subsp. *lactis* strains and is very effective against a variety of Gram-positive bacteria (Froseth et al., 1988). In one of the studies related to the cloning of nisin resistance, the objective was to obtain a genetically modified strain of *L. lactis* subsp. *lactis* lac(–) for potential use in accelerating the ripening rate of cheese. The lactococcal strain DRC3 was chosen as a DNA donor. This strain is a diacetyl producer and shows resistance to nisin and bacteriophage c2 aside from having a known plasmid pattern. An electroporation procedure for the plasmid-mediated genetic transformation of intact cells of *L. lactis* subsp. *lactis* LM0230 was developed. The DNA from DRC3 was isolated by alkaline lysis and purified in a CsCl–ethydium bromide gradient. Competent cells were obtained by culturing the lac(–) strain in glucose-M17 medium with 1.5% glycine and washing the harvested cells with 20% glycerol. Electroporation was performed by a single pulse at 12.5 kV/cm after mixing the cells with the purified plasmid DNA. These conditions produced a transformant that showed limited resistance to nisin. A plasmid with a molecular weight between 1 and 1.8 MDa and absent in the donor strain was observed by electrophoresis. The plasmid encoding resistance to nisin in DRC3 is very large (40 MDa) and is known as pNP40, so it was logical to assume that fragmentation occurred during the electroporation and that only a part of it could penetrate. The transformant strain proved to be suitable for use in starter cultures containing nisin producers and lac(+) lactococci to accelerate cheese ripening (Hernández-Sánchez, 1994).

In a second paper (Froseth et al., 1988), the nisin resistance determinant and an origin of replication on pNP40 was cloned on a 7.6-kb *Eco*RI fragment. When self-ligated, this fragment existed as an independent replicon (pFM011) and contained a 2.6-kb *Eco*RI fragment encoding nisin resistance. This vector was also used to clone a 6.3-kb *Eco*RI fragment coding for bacteriophage insensitivity. This new plasmid, designated pFK012, conferred nisin resistance and an abortive type of phage insensitivity when introduced into *L. lactis* subsp. *lactis* LM0230. The plasmid pFK012 is then a potential food-grade vector formed only by lactococcal DNA (Hughes and McKay, 1992).

A very efficient new food-grade cloning system for industrial strains of *Lactococcus lactis* has just been developed that allows the overexpression of many technologically important cloned genes in industrial strains. This new vector (pFG200) has many advantages, including (1) easy cloning of genes into a versatile polylinker region, (2) a small and stable multicopy vector that is introduced by electroporation into *Lactococcus* with high efficiency, and (3) a selection system allowing selection and maintenance in milk, which allows genetic modification, cloning, and overexpression of *Lactococcus* DNA in a food-grade manner. The plasmids generated with this vector are stably maintained in the host cells for more than 35 generations in media, including milk (Sørensen et al., 2000).

Recombinant Chymosin

It is a fact that good-quality cheese can only be made with good-quality rennet, derived from the stomach of unweaned calves. Chymosin is the proteolytic enzyme in rennet that catalyzes the milk clotting activity, but it is often contaminated with other enzyme activities and microorganisms that may cause quality problems. The availability of this enzyme is limited and can be costly. The calf chymosin gene has been isolated by several research groups and cloned in microorganisms such as *Escherichia coli* (Beppu, 1983), *Kluyveromyces marxianus*, and *Aspergillus niger*. The *K. marxianus* chymosin is excreted into the fermentation broth and is easily purified. The final product is free of other activities or microorganisms and has been on the market now for some time with excellent results (Jenson, 1993).

Other Applications

Biotechnology is the method of choice, in the case of dairy industry starter cultures, to increase the efficiency of substrate conversion, regulate the production of flavor-enhancing metabolites, and increase the ability to produce natural inhibitory substances (bacteriocins) and proteolytic enzymes (Coffey et al., 1994).

Agricultural Biotechnology

Using traditional techniques such as selective breeding, significant crop and animal improvements have been produced for hundreds of years. The typical crop improvement cycle takes 10 to 15 years to be completed and includes germplasm manipulations, genotype selection and stabilization, variety testing, variety increase, proprietary protection, and crop production stages. However, in recent years, a new approach, agricultural biotechnology, has emerged and is a discipline that can contribute to most of these crop improvement stages (Pauls, 1995). This powerful tool has resulted in further improvement in both crop production and crop quality. Originally, the main task of crop breeding was to achieve a high and stable yield potential. Of course, this potential is linked to other characteristics such as resistance to diseases, plagues, drought, etc. (Knorr, 1987). In a similar way, the first projects in agricultural biotechnology were aimed at obtaining varieties with improved agronomic characteristics such as herbicide-tolerant crops and insect-protected crops (Liu, 1999). Two general strategies have been used:

1. Gene addition, in which cloning is used to alter the characteristics of a plant by providing it with one or more new genes
2. Gene subtraction, in which genetic engineering procedures are used to inactivate one or more of the genes already present in the plant (Brown, 1995)

Following are some examples of the application of the gene addition approach:

- *Crops with herbicide tolerance.* The main research is aimed at developing crops with tolerance to broad-spectrum herbicides such as RoundUp™, Liberty™, and imidazolinone. RoundUp is a rapidly degradable herbicide with low toxicity to animals and humans. The active ingredient in this broad-spectrum herbicide is glyphosate, which binds specifically to the enzyme 5-enolpyruvyl-shikimate-3-phosphate synthase (EPSPS), which is responsible for the synthesis of aromatic amino acids in plants. The Monsanto company research team has successfully expressed in different plants a gene encoding a glyphosate-insensitive EPSPS, conferring tolerance to RoundUp and providing in this way superior weed control with less total herbicide use. The bioengineered plants are known as RoundUp-ready crops, and the first plants to be introduced on the market were soybeans, corn, cotton, and canola (Liu, 1999).
- *Insect-protected crops.* The soil bacterium *Bacillus thuringiensis* produces an insecticidal protein generically known as Bt protein. The protein binds to specific receptors located on the gut membrane of

certain insects, mainly caterpillars, interfering with the ion transport system so the insect cannot feed and eventually dies. The main advantage of this Bt protein is that it is innocuous to humans, animals, birds, and beneficial insects such as honeybees. The insertion of the Bt gene in plants allows them to produce this protein and to have permanent protection against sensitive insects. This technique has been so successful that varieties of insect-protected cotton, potato, corn, and sweet corn are already present in the market.

- *Virus-resistant crops.* Viral diseases are one of the leading causes of crop losses. China and the U.S. are the pioneers in the area of preventing such losses, having commercialized virus-resistant tobacco and tomato and later mosaic-virus-resistant squash and watermelon.

- *Other improved crops.* A variety of corn tolerant to severely alkaline soils has been developed by Garst Seed Company, and different types of resistance to microbial pathogens have also been reported (Pauls, 1995).

In addition to crops with improved agronomic traits, a limited number of genetically modified varieties have improved quality traits. The modifications are directed at the main components (oil, proteins, and carbohydrates), although modification in vitamin content, texture, and color has also been done.

Modification of Lipid Metabolism

Vegetable oils are among the world's most important plant products. An oil has three main quality attributes: nutritive value, oxidative stability, and functionality. No oil is considered to be optimal in all three attributes, so modification of an oil's composition is one of the targets of biotechnology. The two main strategies are alteration of a major fatty acid level and creation of an unusual fatty acid (Liu, 1999).

Alteration of a Major Fatty Acid Level

The predominant fatty acids of vegetable oils found in nature consist of just six or seven structures that have chain lengths of 16 or 18 carbons and one to three double bonds. These fatty acids are synthesized from acetyl-CoA by a series of reactions. The assembly of fatty acids and the introduction of the first double bond occurs while these structures are attached to an acyl-carrier protein (ACP). The fatty acids are released from the ACP by the action of a specific thioesterase and can cross the envelope membrane of the plastid in which the reactions take place. After this, the fatty acids are reesterified to CoA. Further reactions such as desaturation to introduce additional double bonds and triacylglycerol formation are believed to occur by means of membrane-bound enzymes in the endoplasmic reticulum (Ohlrogge, 1994).

An example of oil modification by this technique is the development of high-oleic-acid soybean oil by DuPont through antisense suppressing and/or cosuppression of oleate desaturase. The new oil has an oleic acid content of 80% or higher, compared with 24% in normal soybean oil (Liu, 1999). The crop looks so promising that about 50,000 acres were planted in 1998. Because it is more stable, this oil does not require hydrogenation for use in frying or spraying, which reduces processing costs and also avoids the formation of *trans* fatty acids, which are associated with high cholesterol levels. In addition, this new oil has a longer useful life, which is desirable in the fastfood industry (Riley and Hoffman, 1999). In the case of sunflower, the modified crop is known as mid-oleic sunflower, which has a modified fatty acid profile. It was grown on 100,000 acres in the U.S. in 1998. The seed produces low saturated fat oils with 60 to 75% oleic acid, compared with 16 to 20% in normal oil. This product has the potential to replace cottonseed and partially hydrogenated soybean oils in frying and salad oils. Because the mid-oleic sunflower has higher yields than the standard or high-oleic varieties (77 to 89% oleic acid), this type is expected to be preferred in the future (Riley and Hoffman, 1999).

The above examples also address the dietary goal of reducing saturated fatty acid intake. Vegetable oils generally contain far less saturated fatty acids than the 40 to 50% found in animal fats and are considered adequate for reducing cholesterol levels. However, most vegetable oils still contain 10 to 20% saturated fatty acids. One strategy to reduce this level of fat is based on increasing the levels of the enzyme that catalyzes the elongation of palmitoyl-ACP to stearoyl-ACP. The overexpression of 3-ketoacyl-ACP-synthase II in transgenic *Brassica napus* resulted in reduced levels of palmitic acid. A second strategy of reducing the activity of the acyl-ACP thioesterase has been done in soybeans by means of cosuppression. Transformation of soybeans with an additional acyl-ACP thioesterase gene resulted in a reduction of this enzyme activity and a reduction of 50% in the levels of saturated fatty acids in somatic embryos. A third approach is to transform plants with additional membrane-bound desaturases that can convert saturated fatty acids to unsaturated. This approach has been successful in the case of tobacco by introducing genes from rat or yeast in the cells (Ohlrogge, 1994).

The opposite case is that of trying to increase the level of saturated fatty acids. In many cases, the purpose of this transformation is the creation of an alternative to vegetable oil hydrogenation in the manufacture of margarines and shortenings. An increased stearic acid content has been achieved in canola and soybean oils by using a strategy similar to that used in the case of high-oleic soybean (antisense expression of the stearoyl-ACP-desaturase gene). The new oil would contain more than 30% stearate, compared with about 2% in regular canola oil and around 4% in normal soybean oil. It is targeted to replace hydrogenated oils in margarine and liquid shortenings because it contains no *trans* derivatives (Liu, 1999).

Production of Unusual Fatty Acids

Unusual fatty acids include short-, medium-, and very-long-chain fatty acids and those having double bonds at an unusual position or carrying a hydroxy or epoxy group. The introduction of these fatty acids in conventional vegetable oils would turn them into valuable products with high prices. The strategy in this case is to transfer the genes encoding the key biosynthetic enzymes into oilseeds. A classic example of this approach is Laurical™, a high-lauric canola developed by Calgene. Normal canola oil contains no lauric acid. By introducing the acyl-ACP thioesterase gene from the California bay laurel tree into the canola cells, a new modified oil with 38% or more laurate can be obtained. This novel oil could replace the more expensive coconut or palm kernel oils as sources of this fatty acid and as an alternative to cocoa butter (Liu, 1999; Riley and Hoffman, 1999).

An interesting fatty acid modification that reduces the need for hydrogenation and at the same time increases unsaturation in diets is the production of petroselinic acid-rich vegetable oils. Petroselinic acid is an isomer of oleic acid in which the position of the double bond has changed, resulting in a shift in the melting point from 12 to 33°C. This property means that the oils rich in this isomer are unsaturated oils that are solid at room temperature and therefore are ideal substitutes for margarines and shortenings. The strategy in this case is more complex, as apparently three genes are involved in the production of high levels of this fatty acid in the case of coriander and other Umbelliferae; however, studies are in progress (Ohlrogge, 1994).

Protein Modification

The main modifications in this case are directed to increasing the content of an essential amino acid and improving the functionality of a protein in some crops, although soybeans with improved animal nutrition that bolster the protein and amino acid content of soybean meal are near commercial introduction. The increase in the content of essential amino acids is not easy because the information for seed storage protein genes is encoded in several genes in a complex way. A possible strategy involves the transfer of a gene encoding a protein rich in methionine (in the case of legumes) or lysine (in the case of cereals) from other species. In this way, DuPont is developing transgenic soybeans by expressing a methionine-rich zein protein from corn with an 80 to 100% increase in methionine.

Carbohydrate Modification

This type of modification can also produce interesting varieties. For example, high-sucrose beans that have a better taste (less "beany") and greater digestibility were introduced recently, and about 25,000 acres were planted in the U.S. in 1998 (Riley and Hoffman, 1999). Starch modification in potatoes is also an interesting application of plant biotechnology. One of the key enzymes in starch synthesis is ADP glucose pyrophosphorylase. This enzyme is subjected to feedback regulation but, if a bacterial gene that encodes for a feedback-insensitive enzyme is cloned in the potato, a transgenic plant with an average increase of 20 to 30% in starch content is obtained. This new potato provides french fries with more potato flavor, improved texture, reduced energy content, and a less greasy taste (Stark et al., 1996).

The Gene Subtraction Approach

This modification actually does not involve the removal of a gene, but merely its activation. The main procedure to achieve this goal is antisense technology. This technique is a powerful tool that can limit or eradicate the expression of specific genes by sequence-directed targeting of messenger RNA. The two main strategies for this are antisense RNA and peptide nucleic acid (PNA) procedures. In the first case, the gene to be cloned is ligated into the vector in reverse orientation, so when this cloned "gene" is transcribed the RNA that is synthesized is the reverse complement of the mRNA produced from the normal version of the gene. This reverse complement is known as antisense RNA, or asRNA, and is able to prevent synthesis of the product of the gene it is directed against (Brown, 1995). In the second case, PNA is a DNA mimic in which the nucleotides are attached to a pseudopeptide backbone. PNA is able to hybridize with complementary DNA or RNA and the complex is very stable, thus PNA is a good option for use in antisense technology (Good and Nielsen, 1998). Some examples of the application of this technology follow.

Improving Food Quality

The first whole-food product of modern technology to go on the market (in the U.S. in 1994) was a genetically modified tomato, the FlavrSavr™ tomato produced by Calgene, Inc. The first product to arrive in U.K. supermarkets (in 1996) was a tomato puree manufactured from genetically modified tomatoes (increased-pectin tomatoes) produced by Zeneca Plant Science. Tomatoes are usually harvested while they are still green and later treated with ethylene for ripening to occur. This approach makes sure that the fruits remain firm during transportation and storage, thus avoiding softening and loss of product. The tomatoes turn soft, and finally rot when the enzyme polygalacturonase breaks down the pectin that holds the cell walls together. Both of these types of genetically modified tomatoes have been produced by switching off most of the polygalacturonase production, and the tomatoes soften more slowly. Antisense technology is the strategy used in this case. Because the rate of softening has been slowed, the modified tomatoes can remain longer on the plant to develop their full flavor and color but stay firm enough to be transported to the market. In the case of the tomato puree, the need for the addition of thickeners is minimized during its manufacture (John Innes Centre, 1998).

Another example of the application of antisense technology is the prevention of enzymatic browning in potatoes. The development of brown discoloration in a wide range of fruit and vegetables reduces consumer acceptability and is thus of significant economic importance to the farmer and to the food processor. In one study (Bachem et al., 1994), potato internode explants were transformed with *Agrobacterium tumefaciens* containing antisense-polyphenol oxidase (PPO) Ti plasmid constructs. It was shown that antisense inhibition of PPO gene expression abolishes discoloration after bruising of potato tubers in individual transgenic lines grown under field conditions. Using appropriate promoters to express antisense PPO RNA, melanin formation can be specifically inhibited in potatoes. This lack of bruising sensitivity in transgenic potatoes and the absence of any apparent detrimental side effects open up the possibility of preventing enzymatic browning in a wide variety of food crops without having to use treatments such as heating or the use of antioxidants.

Other Improvements

A new crop known as low-phytate or low-phytic-acid corn, providing increased availability of phosphorus, will be marketed this year. Researchers claim nutraceuticals (also called *functional foods*) could conceivably provide immunity to disease or improve the health characteristics of traditional food, and they can be produced by plant biotechnology. An example would be canola oil with a high β-carotene content (Riley and Hoffman, 1999).

The Benefits of Biotechnology

In summary, we can say that the many benefits offered by biotechnology include:

- *Agricultural benefits.* It is possible to identify specific genetic characteristics, isolate them, and transfer them to valuable crop plants.
- *Environmental benefits.* The need for pesticide application is reduced because the plants have the ability to protect themselves from certain pests and diseases. Water usage, soil erosion, and greenhouse gas emissions are reduced through more sustainable farming practices. The productivity of marginal cropland is improved.
- *Food quality benefits.* Biotechnology allows for enhancements in the quality of foods, from increasing crop yields to delaying ripening for better transportability. In the future, consumers may enjoy better taste and nutrition through reduction of undesirable

characteristics such as saturated fats in cooking oils, elimination of allergens and toxicants, and increases in vitamins and other nutrients; nutraceuticals could even help reduce the risk of chronic diseases (Monsanto, 2000).

Despite the benefits, the subject of biotech foods is still a very controversial one and will have to be addressed soon, as it involves a variety of complex concepts and legal principles (Korwek, 2000).

Transformation Techniques in Plant Biotechnology

Recent advances in transformation technology have resulted in the routine production of transgenic plants for an increasing number of crop species. The three main cloning vector systems for higher plants are: (1) the Ti plasmid of *Agrobacterium tumefaciens*, (2) plant viruses such as the caulimoviruses and gemniviruses, and (3) direct gene transfer using DNA fragments not attached to a plant cloning vector (Brown, 1995).

Cereals comprise a commercially valuable group of plants that could benefit from the introduction and expression of foreign genes, so research has been done directed toward the development of regulatory systems for use in cereal transformation. Among these, the reporter genes used in cereal transformation to analyze gene expression are very important. Some of the currently used reporter genes are β-glucuronidase from *Escherichia coli* (one of the most popular reporter genes), luciferase from the firefly, and anthocyanin regulators from corn. The choice depends on the particular experiment, as all three genes have advantages and disadvantages (McElroy and Brettell, 1994).

Rice was the first major monocotyledonous crop species to be transformed and regenerated. Initially, rice transformation was limited to the *Oryza sativa* subsp. *japonica* cultivars. Subsequently, a number of *indica* and *javanica* cultivars have also been transformed and regenerated into fertile transgenic plants. Most transformation studies in rice have used direct DNA uptake into protoplasts, induced by polyethylene glycol (PEG) treatment or electroporation. Also, other transformation methods have been developed that are less genotype dependent, such as microprojectile bombardment of cell suspensions and immature embryos. *Agrobacterium*-mediated gene delivery has also been successful in producing stable transformants. An advantage of working with rice is its relatively small genome, 4.2×10^8 bp per haploid genome. This is in marked contrast to other cereals such as corn and wheat, which have 4×10^9 and 1.7×10^{10} bp per haploid genome, respectively (Ayres and Park, 1994).

Barley has been one of the most recalcitrant crop species in regard to this technology. A prerequisite for successful transformation of barley leading to improved quality, tolerance, or resistance is that the desired characteristic, as stated before, must be coded by a single gene. Examples of genes for quality traits are the gene *egl*1 of *Trichoderma reesei*, coding for a thermostable β-glucanase, or a hybrid gene of *Bacillus amyloliquefaciens* and *B. macerans*, also coding for β-glucanase functional at high temperatures. This transformation has been successfully done so the gene was under germination-specific control. The most common reporter genes used for barley are the ones for β-glucuronidase, kanamycin resistance, chloramphenicol resistance, phosphinotricin (herbicide) resistance, and luciferase.

The gene transfer methods that have been successful for barley transformation are *Agrobacterium* infection, particle bombardment, PEG treatment of protoplasts, electroporation of protoplasts, and laser perforation. All these methods have produced stable gene expression, though particle bombardment has proven to be the best method in cereal transformation, including barley. Using these methodologies, it is possible to deliver foreign genes to barley and to regenerate fertile transgenic barley plants. The foreign genes are stably integrated into the barley genome and are inherited following the laws of Mendel (Mannonen et al., 1994).

In the case of legumes, *Agrobacterium* infection and particle bombardment have been the most successful transformation methods. The first method has been used in chickpeas, lentils, alfalfa, peas, and cowpeas; the second method has been preferred for peanuts, soybeans, and common beans. In most cases, a transgenic plant has been obtained (Christou, 1994).

Transgenic Animals

Although many researchers had previously studied the behavior of animal cells *in vitro*, the first application of such cells which led to a useful product was the production in 1949 of the polio virus. Today, many products are generated from cultured animal cells, including virus vaccines, cellular chemicals (interferons, interleukin-2, thymosin, urokinase), immunobiologicals (monoclonal antibodies), hormones (e.g., growth hormone, prolactin, ACTH), and virus predators (insecticides), among others. While it is unlikely that animal-cell-based processes for the generation of live virus vaccines will be superseded by processes based on genetically engineered bacteria, it is clear that some animal cell products such as insulin and interferons can be produced from modified bacteria (Spier, 1987). However, transgenic animals are becoming more common and useful tools because they provide an *in vivo* look at the capabilities and impact of foreign gene expression in a biological system. Expression of the gene of interest is controlled by DNA promoter elements that direct where and when the gene product will be expressed in the animal. For example, the expression of a transgene in the mammary gland of an animal requires the use of a promoter and regulatory regions of a milk protein gene (i.e., sequences that direct gene expression only in the mammary gland and only during lactation). In this particular case, most of the work has been concerned with the biological production of important and active proteins (such as pharmaceuticals) in the milk of a transgenic animal with the intent of recovering the protein of interest from the milk.

Transgenic technology can also be used to alter the functional and physical properties of milk, resulting in novel properties useful from a nutritional or technological point of view. Of the several candidates for altering the properties of milk, one of them is human lysozyme. If this protein were present in bovine milk at a significant level, it could help to reduce the overall level of bacteria in this product and, because of its positive charge, it could interact with the negatively charged caseins to produce a reduction of the rennet clotting time. Another candidate could be κ-casein. The addition of more bovine κ-casein to the milk could also affect the physical properties of this food. It could increase the thermal stability of casein aggregates and act to decrease the size of the casein micelles. A smaller diameter would lead to a larger available surface area, which would result in a more consistent and firmer curd as well as an increase in cheese yield. Human lactoferrin could be also a good candidate for improving the properties of milk. Lactoferrin is the major iron-binding protein in milk and is responsible for the high bioavailability of milk iron. It also inhibits the growth of bacteria in the mammary gland and the intestines of infants. Lactoferrin can also act as an antioxidant, inhibiting the formation of toxic oxygen radicals. The presence of human lactoferrin could also enhance the association with caseins, increasing rennet gel strength and possibly cheese yield. It can be seen that transgenic livestock has a promising future (Maga and Murray, 1995).

Conclusion

From all of the above, it becomes clear that food biotechnology is an old tradition and a significant challenge for the future. Many genetically modified foods are under development and soon will be on the market. It is the role of the researchers to ensure that all these new food biotechnology products are safe for the environment and to animal and human health. Finally, and according to the reports found in scientific journals and meetings proceedings, for example, the trends in research in the field of food biotechnology include: DNA sequencing of the genomes of several organisms, developing of food-grade vectors to be used in microorganisms of food interest (lactic acid bacteria, bifidobacteria, yeasts, etc.), development of new biotech crops, and, of course, recombinant DNA biotechnology-derived foods as part of the continuing efforts to improve the food supply.

References

Ayres, N.M. and Park, W.D. (1994) Genetic transformation of rice, *Crit. Rev. Plant Sci.*, 13(3), 219–239.

Bachem, C.W.B., Speckmann, G.J., van der Linde, P.C.G., Verheggen, F.T.M., Hunt, M.D., Steffens, J.C., and Zabeau, M. (1994) Antisense expression of polyphenol oxidase genes inhibits enzymatic browning in potato tubers, *Bio/Technology*, 12, 1101–1105.

Beppu, T. (1983) The cloning and expression of chymosin (rennin) genes in microorganisms, *Trends Biotechnol.*, 1(3), 85–89.

Brown, J.C.S., Ward, L.J.H., and Davey, G.P. (1994) Rapid isolation and purification of lactococcal bacteriophage DNA without the use of caesium chloride gradients, *Lett. Appl.. Microbiol.*, 18, 292–293.

Brown, T.A. (1995) *Gene Cloning: An Introduction*, 3rd ed., Stanley Thornes Publishers, Cheltenham, U.K., pp. 295–311.

Brussow, H., Fremont, M., Bruttin, A., Sidoti, J., Constable, A., and Fryder, V. (1994) Detection and classification of *Streptococcus thermophilus* bacteriophages isolated from industrial milk fermentation, *Appl. Environm. Microbiol.*, 60, 4537–4543.

Christou, P. (1994) The biotechnology of crop legumes, *Euphytica*, 74, 165–185.

Coffey, A.G., Daly, C., and Fitzgerald, G. (1994) The impact of biotechnology on the dairy industry, *Biotechnol. Adv.*, 12, 625–633.

Djordjevic, G.M. and Klaenhammer, T.R. (1997) Genes and gene expression in *Lactococcus* bacteriophages, *Int. Dairy J.*, 7, 489–508.

Froseth, B.R. and McKay, L.L. (1991) Molecular characterization of the nisin resistance region of *Lactococcus lactis* subsp. *lactis* biovar *diacetylactis* DRC3, *Appl. Environm. Microbiol.*, 57(3), 804–811.

Froseth, B.R., Herman, R.E., and McKay, L.L. (1988) Cloning of nisin resistance determinant and replication origin on 7.6-kilobase *Eco*RI fragment of pNP40 from *Streptococcus lactis* subsp. *diacetylactis* DRC3, *Appl. Environm. Microbiol.*, 54(8), 2136–2139.

Garrison, S.J. and dePamphilis, C. (1994) Polymerase chain reaction for educational settings, *Am. Biol. Teacher*, 56(8), 476–481.

Garvey, P., Fitzgerald, G.F., and Hill, C. (1995) Cloning and DNA sequence analysis of two abortive infection phage resistance determinants from the lactococcal plasmid pNP40, *Appl. Environm. Microbiol.*, 61, 4321–4328.

Garvey, P., Hill, C., and Fitzgerald, G.F. (1996) The lactococcal plasmid pNP40 encodes a third bacteriophage resistance mechanism, one which affects phage DNA penetration, *Appl. Environ. Microbiol.*, 62, 676–679.

Good, L. and Nielsen, P.E. (1998) Antisense inhibition of gene expression in bacteria by PNA targeted to mRNA, *Nat. Biotechnol.*, 16, 355–358.

Hernández-Sánchez, H. (1994) Obtainment by Electroporation of Nisin-Resistant Mutants of *Lactococcus lactis* subsp. *lactis* LM0230, Ph.D. thesis, Escuela Nacional de Ciencias Biológicas, IPN, México.

Hill, W.E. (1996) The polymerase chain reaction: applications for the detection of foodborne pathogens, *Crit. Rev. Food Sci. Nutr.*, 36(1/2), 123–173.

Hoban, T.J. (1999) Consumer acceptance of biotechnology in the U.S. and Japan, *Food Technol.*, 53(5), 50–53.

Hughes, B.F. and McKay, L.L. (1992) Deriving phage-insensitive lactococci using a food-grade vector encoding phage and nisin resistance, *J. Dairy Sci.*, 75, 914–923.

Institute of Food Technologists (2000) IFT expert report on biotechnology and foods, *Food Technol.*, 54(8), 124–136.

Jenson, I. (1993) Biotechnology and the food supply: food ingredient products of biotechnology, *Food Australia*, 45(12), 568–571.

John Innes Centre (1998) *Biotech Bytes: Food Biotechnology*, http://www.jic.bb-src.ac.uk/exhibitions/bio-future/.

Knorr, D. (1987) Food biotechnology: its organization and potential, *Food Technol.*, 41(4), 95–100.

Korwek, E.L. (2000) Labeling biotech foods: opening Pandora's box?, *Food Technol.*, 54(3), 38–42.

Liu, K. (1999) Biotech crops: products, properties and prospects, *Food Technol.*, 53(5), 42–49.

Maga, E.A. and Murray, J.D. (1995) Mammary gland expression of transgenes and the potential for altering the properties of milk, *Bio/Technology*, 13, 1452–1457.

Mannonen, L., Kauppinen, V., and Eneri, T.M. (1994) Recent developments in the genetic engineering of barley, *Crit. Rev. Biotechnol.*, 14(4), 287–310.

McElroy, D. and Brettell, R.I.S. (1994) Foreign gene expression in transgenic cereals, *Trends Biotechnol.*, 12, 62–68.

McKay, L.L. and Baldwin, K.A. (1984) Conjugative 40-megadalton plasmid in *Streptococcus lactis* subsp. *diacetylactis* DRC3 is associated with resistance to nisin and bacteriophage, *Appl. Environ. Microbiol.*, 47, 68–74.

McKee, T. and McKee, J.R. (1999) *Biochemistry: An Introduction*, 2nd ed., McGraw-Hill, Boston, MA, pp. 608–610.

Monsanto (2000) *An invitation for Dialogue: Biotechnology and the Food Industry*, http://www.fooddialogue.com.

Monteville, M.R., Ardestani, B., and Geller, B.R. (1994) Lactococcal phages require a host cell wall carbohydrate and a plasma membrane protein for adsorption and ejection of DNA, *Appl. Environm. Microbiol.*, 60, 3204–3211.

Ohlrogge, J.B. (1994) Design of new plant products: engineering of fatty acid metabolism, *Plant Physiol.*, 104, 821–826.

Pauls, K.P. (1995) Plant biotechnology for crop improvement, *Biotechnol. Adv.*, 13(4), 673–693.

Riley, P.A. and Hoffman, L. (1999) Value-enhanced crops: biotechnology's next stage, *Agricultural Outlook*, March, 18–25.

Sing, W.D. and Klaenhammer, T.R. (1990) Characteristics of phage abortion conferred in lactococci by the conjugal plasmid pTR2030, *J. Gen. Microbiol.*, 136, 1807–1815.

Sing, W.D. and Klaenhammer, T.R. (1991) Characterisation of restriction-modification plasmids from *Lactococcus lactis* ssp. *cremoris* and their effects when combined with pTR2030, *J. Dairy Sci.*, 74, 1133–1144.

Smith, J.E. (1988) Biotechnology, 2nd ed., *New Studies in Biology,*: Edward Arnold, London, pp. 44–59.

Sørensen, K.I., Larsen, R., Kibenich, A., Junge, M.P., and Johansen, E. (2000) A food-grade cloning system for industrial strains of *Lactococcal lactis*, *Appl. Environm. Microbiol.*, 66, 1253–1258.

Spier, R.E. (1987) Processes and products dependent on cultured animal cells, in *Basic Biotechnology*, J. Bu'lock and B. Kristiansen, Eds., Academic Press, London, pp. 509–524.

Stark, D.M., Barry, G.F., and Kishore, G.M. (1996) Improvement of food quality traits through enhancement of starch biosynthesis, in *Engineering Plants for Commercial Products and Applications*, G.B. Collins and R.J. Shepherd, Eds., *Annals of the New York Academy of Sciences*, 792, 26–36.

Su, P., Im, H., Hsieh, H., Kang'a, S., and Dunn, N.W. (1999) *Lla*FI, a type III restriction and modification system in *Lactococcus lactis*, *Appl. Environm. Microbiol.*, 65, 686–693.

Valyasevi, R., Sandine, W.E., and Geller, B.L. (1991) A membrane protein is required for bacteriophage c2 infection of *Lactococcus lactis* subsp. *lactis* C2, *J. Bacteriol.*, 173, 6095–6100.

3

Bioprocess Design

Ali Asaff-Torres and Mayra De la Torre-Martínez

CONTENTS

1-56676-892-6/03/$0.00+$1.50
© 2003 by CRC Press LLC

Introduction

Most bioprocess design problems are complex, as they usually involve batch, semi-batch, continuous, and cyclic operations. Also, bioprocess performance is affected by biological variability, highly interactive unit operations, multiple processing steps and options, and complex feedstream physical properties. In some instances, design decisions are made on the basis of rules of thumb or heuristics, while in others design information exists so that a unique design solution may be obtained. Often, the design solution will represent a compromise based, for example, on a target selling price where a trade-off between yield and purity must be sought. Identifying the best design and also determining the relative sensitivity of the solution to the variables governing a system are not trivial tasks (Woodley et al., 1996). The variable, or the most important aspect to consider in synthesis and bioprocess design, is always the primary design issue; however, a unique solution does not exist, because possible solutions depend upon the intrinsic characteristics of the biological system, the selling price, consumer demand, or product specifications.

This chapter discusses the steps followed during bioprocess development and synthesis, together with the most important considerations regarding technical and economic constraints and possibilities. Because literature about equipment design and cost estimation is widely available, these aspects will not be analyzed in detail in this chapter. At the end of the chapter, a case study illustrates bioprocess synthesis.

Food Biotechnology Products

Nowadays, traditional and modern biotechnology processes are used to produce a wide variety of foods and food additives. Ethanol, fermented milk products, amino acids, vitamins, enzymes, polysaccharides such as xanthan, single-cell protein, colorants, flavors and fragrances, and various acids such as citric, lactic, acetic, butyric, and glutamic are some of the products obtained by fermentation processes. The scale of manufacture ranges from billions of gallons of ethanol produced at a cost of one to two dollars per gallon to high-value products such as flavors, fragrances, and vitamins with manufacturing costs of hundreds to thousands of dollars per kilogram.

The actual process employed commercially depends on a variety of factors, including economics, process and product patents, raw material availability, and manufacturing equipment. In some cases, traditional processes are employed along with technological advances — for example, in the manufacture of cheese and fermented milk products, beer, and wine. Obtaining products regulated by food and drug agencies may change the manufacturing processes proposed and may involve costly clinical trials. These processes are often not optimized from a production standpoint (Blanch and Clark, 1997).

Design Generalities

Currently, a large amount of literature exists on systematic process synthesis and design methods for chemical processing. The same abundance of literature does not exist for bioprocess synthesis techniques (Zhou and Titchener-Hooker, 1999; Steffens et al., 2000). Generally, process synthesis and design procedures, can be classified as:

- Heuristic methods
- Algorithmic techniques
- Expert systems

Heuristic methods use the rules acquired through the knowledge and experience gained from similar processes or previous designs and are most often employed for bioprocess development; such methods are not always the best for optimal design. Algorithmic techniques and expert systems are not often used for synthesis and bioprocess design, because biological systems are highly complex. Typical flowsheeting packages are often inappropriate for bioprocess operations, which are complicated by the mixture of batch and semi-continuous modes of processing and by strong interactions among unit operations (Gristis and Titchener-Hooker, 1989; Bulmer et al., 1996), in addition to other obstacles such as:

- Unit operations commonly used for downstream purification generally separate components *non-sharply*. A non-sharp separation is one in which all of the feed components are distributed into the effluent streams.

- A relatively large range of unit operations to be used for a particular separation is available, thus making decisions difficult.
- Biological streams generally contain a large number of compounds.

The above-mentioned situations lead to a relatively large number of alternative, feasible flowsheets and, therefore, a correspondingly large search space for the synthesis algorithm (Fraga, 1996; Lienqueo et al., 1996). Synthesis problems of this nature are difficult to solve using numerical optimization techniques; however, important advances in this field have occurred in the past few years. Petrides (1994), Petrides et al. (1995), Lienqueo et al. (1996), Gregory et al. (1996), Zhou et al. (1997), and Steffens et al. (2000) have carried out interesting work using expert knowledge and exploiting differences in physical properties to select unit operations and synthesize economically favorable processes and flow sheets.

The success of biochemical compound manufacturing companies depends on the ability to design an efficient economical optimal manufacturing process (Leser and Asenjo, 1992). New product development requires an interdisciplinary joint effort including people from several departments of a company (Figure 3.1) who will carry out research and development (at both the laboratory and pilot-plant levels), synthesis and scale-up processes, market and/or field trials, project implementation, and production on an industrial scale.

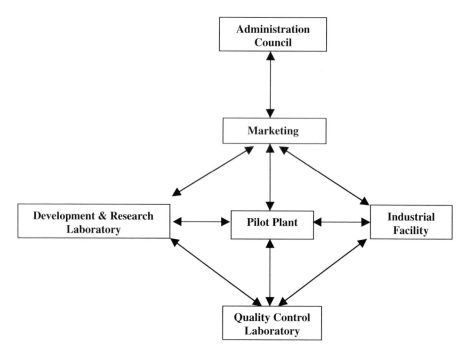

FIGURE 3.1
Relationships among departments to develop a new product.

Product/Process Development Chain

Before starting a process or the development of a product, it is necessary to determine the feasibility of a project. The most relevant questions are (Kossen, 1994):

1. Does the product and its method of production meet the standards for health, safety, and the environment?
2. Are problems expected due to legislation or the attitude of the public?
3. Do we have an acceptable patent position?
4. Do we have the necessary tools to make the product?
5. Do we know the market (potential clients and competitors)?
6. Does the product meet the standards of potential clients?
7. Will the product be on the market in time?
8. Will the product result in an acceptable profit for the company?

   Design involves research of process options and matching these against specific objectives. The first step in process synthesis is to develop a block diagram with the main stages. Then, possible alternative operations are selected for each stage. Bioprocess flowsheets are often synthesized in a sequential fashion, proceeding from one unit to the next until product specifications are known. Individual units are subsequently optimized to improve plant performance. Although this approach may produce economically adequate processes, alternative designs may be more profitable. To avoid this problem, a systematic synthesis and design procedure that considers the overall process rather than the individual units is required (Wheelwright, 1987).

   To achieve high overall process efficiency requires optimization of the complete sequence of upstream processing operations, fermentation, subsequent downstream processing operations, and an integrated approach to design (Samsalatli and Shah, 1996). The interactions between operations can be significant in determining overall process performance (Kelly and Hatton, 1991; Narodoslawsky, 1991; Clarkson et al., 1993; Middelberg, 1995). However, to achieve an overall process performance is not an easy task, because it will depend on the degree of knowledge of the biological system employed and final product specifications. This knowledge comes from the research and development laboratories and pilot plant (experimental data on physiology, kinetics, etc.), while other design data are acquired from the literature or past experience. Then, synthesis and bioprocess design are developed through sequential stages that Kossen (1994) calls the *product/process development chain* (PDC) (Figure 3.2), which includes the following steps.

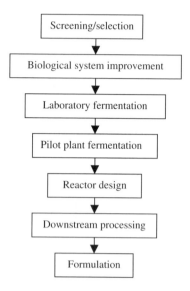

FIGURE 3.2
Product/process development chain (PDC). (Modified from Kossen, N.W.F., in *Advances in Bioprocesses Engineering*, E. Galindo and O. T. Ramírez, Eds., Kluwer, Dordrecht, 1994, pp. 1-11.)

Screening/Selection

During this stage, a screening of the possible procedures available for obtaining the product of interest must be considered in terms of particular chemical characteristics and the costs and availability of raw materials. To illustrate, consider the case of vanillin production. Vanillin is one of the most important aromatic flavor compounds used in foods, beverages, perfumes, and pharmaceuticals and is produced on a scale of more than 10,000 tons per year through chemical synthesis. Alternative biotechnology-based approaches for vanillin production are based on bioconversion of lignin, phenolic stilbenes, isoeugenol, eugenol, ferulic acid, or aromatic amino acids and on *de novo* biosynthesis, applying fungi, bacteria, plant cells, or genetically engineered microorganism from sugars as glucose (Priefert et al., 2001).

Biocatalysis/Bioconversion

Inexpensive, readily available, and renewable natural precursors, such as fatty or amino acids, can be converted to more highly valued products. Biocatalysis competes best with chemical catalysis in the following types of reactions (Krings and Berger, 1998):

- Introduction of chirality
- Functional of chemically inert carbons
- Selective modifications of one functional group in multifunctional molecules
- Resolution of racemates

Biocatalysis is carried out by cultured plant cells or plant callus tissue, prokaryotic or eukaryotic microorganisms, or isolated enzymes as biocatalyst.

De Novo *Synthesis*

A wide range of industrial bioprocesses employ whole cells (plants, animals, or microorganisms that catabolize carbohydrates, fats, and proteins) to produce various metabolites starting from single or complex substrates, through *de novo* synthesis. This property is traditionally used during the production of fermented foods with their amazing number of aroma chemicals.

Biological System Improvement

Once selected, the biological system subsequently will be examined to improve its product yield and performance. Major goals are enhanced production of metabolites (e.g., amino acid and antibiotic production) and improved properties of starter cultures, such as in dairy product production (e.g., reproducible growth characteristics, increased flavor formation and proteolytic activities, or better autolytic properties). In other cases strains are improved to adapt them to industrial operating conditions, such as temperature, pH, mechanical stress, high substrate and/or product concentrations, etc. Methods employed for the biological system improvements are:

- Classical
- r-DNA
- Metabolic engineering

Classical

This method, used for many years, consists of the application of various agents (physical or chemical mutagens), causing mutations in the genes of the microorganism and changing its cellular metabolism. After screening, those strains that better fulfill the objectives sought are selected; however, this process is totally random, and the probability of success in many cases is slim. A screening of strains or cellular lines with improved characteristics is also utilized.

r-DNA

Since the 1970s, advances in genetics and molecular biology have been impressive. Recombinant DNA techniques have allowed the creation of a great quantity of new strains. At the moment, recombinant or genetically modified microorganisms produce therapeutic and diagnostic proteins, diverse enzymes, and other types of products; however, these systems require some particular attention, such as:

- *Strain maintenance.* Any microorganism is always in a constant competition with other microorganisms that share its environment. Recombinant microorganisms compete with wild strains that in time could displace the recombinant one; therefore, in recombinant strains, genes that confer resistance to antibiotics are introduced, and antibiotics are added to broth culture to maintain the strain. In many cases, though, this procedure causes additional problems in purification processes, as the antibiotics are pollutants that must be removed. In some cases, this problem is avoided with recombinant strains that protect themselves from the conditions of their environment, such as the case of yeasts resistant to high ethanol concentrations employed during alcohol production. At high ethanol concentrations, wild yeast cannot survive.

- *Strain stability.* The new genetic material that confers the peculiar characteristics of production and resistance to the recombinant microorganism is generally inserted in plasmid, which could get lost during processing. This phenomenon is most critical in continuous reactors with extended times of operation. It will be very important, then, to verify during the investigation stage the stability of the inserted genetic material.

- *Gene turn on.* With the purpose of avoiding a premature metabolic collapse or to take advantage of a physiologic state of the cell, the promoter of the inserted genes can be "turned on" by means of specific compounds called *inductors* or by a temperature change. The inductor is added or the temperature changed at the appropriate moment.

Metabolic Engineering

Research programs to improve industrial microorganisms used for fermentation were initially focused on strain selection after classical mutagenesis, later followed by more direct approaches using genetic engineering. The main drawbacks of these approaches are that they are time consuming, side effects occurring in the selected or constructed strains are difficult to predict and assess, and the full range of engineering possibilities cannot be exploited, due to lack of knowledge about the interrelated regulatory and metabolic processes going on in a cell (Kuipers, 1999).

In order to enhance the yield and productivity of metabolite production, researchers have focused almost exclusively on enzyme amplification or other modifications of the product pathway, but overproduction of many metabolites at high yields requires significant redirection of flux distribution in the primary metabolism, which may not readily occur following product deregulation because metabolic pathways have evolved to exhibit control architectures that resist flux alterations at branch points. This problem can be addressed through the use of some general concepts of metabolic rigidity, including a means for identifying and removing rigid branch points within an experimental framework (Stephanopoulos and Vallino, 1991).

The recent burst in activity regarding genome sequences of food-related microorganisms has opened the way for functional genomics approaches with novel analysis techniques that include transcriptome, proteome, and metabolome analysis, as well as structural genomics.

DNA micro-arrays (transcriptome) (Graves, 1999), and proteome, which is an improved two-dimensional electrophoresis method, combined with matrix-assisted laser desorption/ionization time of flight mass spectroscopic analysis (MALDI/TOFF) (Blackstock and Weir, 1999), are two powerful tools for genetic analysis. These methods provide information on differential gene expression, global changes in protein production level, and protein–protein interaction through the isolation of protein complexes. The huge amount of experimental data generated is gathered into large databases, the interpretation of which greatly depends on novel bioinformatics methodology that should enable linking of the databases and facilitate the classification and interpretation of results. Eventually, researchers would like to integrate all data on the transcriptome and proteome techniques into a metabolic and regulatory model. The combination of transcriptome with proteome and metabolome research and the elucidation of structure–function relationships of biomolecules will eventually result in a true understanding of whole cell functioning (Kuipers, 1999).

Laboratory Fermentation

During this stage the best conditions for cellular growth and maintenance as well as metabolites production must be found. Three fields must be researched:

- Physiology
- Choice of substrate
- Fermentation regime

Physiology

An understanding of the underlying physiology of the organism can provide insight into the design of a new bioprocess, as well as the base model control of cultivation (Gregory et al., 1996). Metabolites production and microbial growth depend on nutrient availability and physical–chemical factors. Among the most important factors are temperature, pH, light, and sensitivity to mechanical stress. For plant and animal cell cultures, mechanical stress is critical, requiring in some cases immobilization of the cells. It will also be important to know the rheological changes taking place in broth cultures as a consequence of cellular metabolism because such changes can be important at the level of blended and mass transfer, in some cases setting the maximum benchmarks of operation. Usually oxygen availability is very important, and certain substrates and products may inhibit growth and product synthesis.

Choice of Substrate

The essential elements for cellular growth are carbon, nitrogen, and, to a lesser extent, phosphorus and sulfur. Microorganisms can use a variety of carbon sources, from complex ones, such as starch and wood hydrolizates, to refined sources, such as glucose syrup or sucrose. Trace metal requirements must be met with salt solutions. Complex nitrogen and carbon sources are required in some fermentation where the slow release of nutrients may be important in regulating these metabolisms (Blanch and Clark, 1997). In vegetable cell cultures, the use of specific compounds such as growth and differentiation factors may also be important.

The choice of a substrate will depend on the raw material contribution to the production cost and on its influence in downstream processing operations, as well as on final product specifications. For high-value products, the raw materials may not comprise a significant part of the production cost, but selection of raw material may nevertheless be important in maintaining consistent product quality (Blanch and Clark, 1997). For low-value products, the cost of raw materials determines the economical success of a process such as single-cell protein production, for example.

Fermentation Regime

Fermentation can occur under three regimes, the choice of which depends on the particular characteristics of the biological system:

- *Continuous.* This fermentation regime greatly favors the overall process performance because it allows the continuous operation of other equipment, minimizing time-outs and increasing productivity. However, the application of continuous fermentation to biological processes is very limited for a number of reasons, including difficulty maintaining the stability of recombinant strains and keeping the equipment sterile for long periods of time, as well as the high sensibility of biological systems that can be altered by minimum interference, with subsequent loss of the stationary state.
- *Batch.* Batch fermentation is the most used regime in bioprocesses and allows maximum substrate conversion. Given the nature of the involved operations, however, time-outs increase substantially compared to continuous operation. Also, when the production line includes continuous operation at points downstream of the reactor, auxiliary equipment must be considered, such as feeding or storing tanks.

- *Feed-batch.* In systems with substrate and/or product inhibition, as well as agitation and mass transfer constraints due to changes in broth culture rheology, the feed-batch regime is a good alternative. Feed-batch allows the management of diverse dilution rates through feeding patterns (constant, exponential, pulses, etc.), thus controlling both the microbial specific growth rate and metabolic state.

Pilot-Plant Fermentation

Once a new product has been developed in the laboratory, a process to produce and purify the product on a large scale is required. Historically, most processes in biologically based industries have made recourse to existing processes and relied on the use of pilot-plant facilities in which to test proposed new process sequences.

Scale-up at the pilot-plant level is the first step in the scaling process and is of vital importance for the later stages and successful bioprocess design. For each unit operation, besides their technical evaluation, it is important to carry out an economic evaluation in such a way that the selection of operation and equipment represents the best technical and economic alternatives for the overall process. This does not mean that every unitary operation and equipment required in a bioprocess should be assayed, but only those with insufficient knowledge or previous experience. Generally, included in this stage are scale-up bioreactors and equipment for specific operations of separation and purification — for example, different chromatography columns. Here, the experience and criteria of the engineering team are fundamental to decide which unit operation will be investigated at the pilot-plant scale.

During this stage, it is very important to take into account the strength of interactions among unit operations employed in the bioprocesses. Some researchers suggest the use of graphic visualization of those interactions through what they call *windows of operation* (Woodley and Titchener-Hooker, 1996; Zhou and Titchener-Hooker, 1999). A window of operation is defined as being the operational space determined by the system (chemical, physical, and biological) and process (engineering) constraints and correlations governing a particular process or operation under consideration. The constraints may be in the form of economical limitations, physical laws, or biological effects, depending upon the application (Woodley and Titchener-Hooker, 1996). The strength of this approach lies in the ability to visualize process operability in such a clear way. Windows of operation enable the engineer to visualize rapidly the variable region in which it is possible to achieve a specified level of process performance. By changing the desired specifications for the process performance, the windows of operation can be used to indicate the feasibility and ease with which the specification can be met. The window can also be employed to investigate the effects of altering the key operating variables on the likely overall process behavior (Zhou and Titchener-Hooker, 1999).

Reactor Design

Generally, the equipment employed during the up- and downstream processing can be purchased directly from catalogs and adapted to the process requirements; however, specific reactors must be constructed for the particular features of each biological system. The design methodology of reactor systems includes exploration and optimization of the following aspects of process development (Atkinson, 1983):

1. Utilization of the extensive literature on process engineering
2. Selection of an appropriate reactor system after consideration of the process requirement
3. Effective laboratory and pilot-scale experimentation
4. Rational extrapolation of experimental data to the plant on a commercial scale
5. Development of design procedures and cost models
6. Minimization of the delay between initial concept and full-scale production

Based upon the project annual production capacity, the reactor size is calculated and the design is finalized with the development of the following components:

1. Provision of aeration, blending, heating, and cooling capacity
2. Control of substrate concentration, biomass, foam, pH, etc.
3. Facilities for aseptic monitoring operations

Downstream Processing

The desired product of a bioprocess may be either the cells themselves or a specific metabolite of the cell. Metabolic products are either intracellular or extracellular, and the location of the product has a major impact on the purification process. Typically, biological products are present in fermentation broth and cell culture supernatants in low concentrations. Low product concentration coupled with large amounts of interfering species can seriously complicate the task of purification.

When the desired product is an intracellular compound, cells must be mechanically disrupted. These cells are exposed to high liquid shear rates by passing them through an orifice under high pressure (Blanch and Clark, 1997). For this operation homogenizers are used. The most important operating cost is the expenditure in energy.

Disrupted cells release a large quantity of cell debris; therefore, the purification process is often complex and may consist of many unit operations. Multiple separations demand more equipment and labor and generally lower the yield of the final product (Blanch and Clark, 1997). Excessive cell debris breakage and micronization can, for example, increase the load on centrifugal operations, and the passage of fines to packed chromatographic columns can lead to low productivity and high media replacement costs (Siddiqui et al., 1995).

Isolating the product through a series of separation and recovery steps is often referred to as *downstream processing*. The exact number of steps involved will depend on the original material used, the concentration and physico-chemical properties of the product, and the final purity required. The first task in formulating a purification strategy is to define or acknowledge the required purity of the product. The allowable ranges of impurity concentrations and the specific impurities that may be tolerated will be dictated by the end use of the product.

Once the purity criteria have been established, the specific purification procedures can be selected. In general, individual recovery operations can be grouped into different categories, depending on their general purpose (Blanch and Clark, 1997):

1. *Separation of insolubles.* Insoluble material includes whole cells, cell debris, pellets of aggregate protein, and undissolved nutrients. Common operations for this purpose are sedimentation, centrifugation, filtration, and membrane filtration.

2. *Isolation and concentration.* This step generally refers to the isolation of desired product from unrelated impurities. Significant concentration is achieved in the early stages, but concentration accompanies purification as well. This category includes extraction, adsorption, ultrafiltration, and precipitation.

3. *Primary purification.* More selective than isolation, some purification steps can distinguish between species having very similar chemical and physical properties. Primary purification techniques include chromatography, electrophoresis, and fractional precipitation.

4. *Refolding.* Although refolding is not properly a recovery operation, the production of some proteins requires this step. Recombinant bacteria or yeasts produce several pharmaceutical or therapeutic proteins, but not all of them are biologically active, as not all of them are properly folded. Proteins produced naturally by organisms that are not recombinant frequently undergo post-translational modifications (e.g., proteolytic cleavage of precursor protein, macromolecular assembly, glycosylation). Biologically inactive recombinant proteins are typically activated in stirred tanks with buffer solutions and denaturant agents that first unfold the proteins; then process conditions are changed to allow proper protein refolding.

5. *Final purification.* This step is necessitated by the extremely high purity required of many bioproducts, particularly pharmaceuticals

and therapeutics. After primary purification, the product is nearly pure but may not be in proper form. Partially pure solids may still contain discolored material or solvent. Crystallization and drying are typically employed to achieve final purity. Some types of chromatography are also utilized for protein final purification.

Screening Units

The bioprocess unit design procedure consists of a preliminary screening procedure followed by the actual design calculations, which are used to determine effluent stream characteristics and cost information. The downstream processing design methodology described by Steffens et al. (2000), Leser and Asenjo (1992), and Wheelwright (1987), recommends selecting units that exploit the greatest differences in physical properties between components. Two types of tests are used to eliminate units that are not feasible for a particular separation:

1. *Design constraints* refer to physical limits on pieces of equipment. For example, packed-bed chromatography columns cannot process streams that contain solids.

2. *Binary ratio checks* allow screen units using physical properties information for the component that is to be separated. Two numbers are compared to identify candidate separation operations:

   a. *Binary ratios.* The potential driving force for the separation of any of two components is quantified by calculating the ratio of the physical property governing the separation in the unit being considered (binary ratio).

   b. *Feasibility indices.* The binary ratios are compared to feasibility indices η (given in the literature), which are design parameters for each separation technology. The indices define how large the binary ratio must be before a separation is feasible.

Unit Design

When a candidate unit passes the screening tests for a particular separation, design and cost calculations for that unit can be conducted. Two different approaches are used to determine the effluent stream characteristics for a separation unit depending upon the unit type: splitting or fractionating units.

Stream-Splitting Units

These units operate by splitting the feed into two streams with significantly different compositions, hence achieving separation as in distillation operations. In fact, stream-splitting unit operations are designed using a similar concept to that used in shortcut distillation design. Initially, a list of the feed components is generated according to their physical properties that allow the separation. Two key components are selected as an adjacent pair from this list. Prior to any design calculations, the binary ratio of the two key components is calculated and compared with the feasibility index for the unit under consideration. If the binary ratio is greater than the feasibility index, design calculations are performed.

To calculate the effluent stream compositions, the upper key component and any components higher in the list are considered as one. Similar calculations are performed for the lower key component and the components below it. Design assumptions and mass balances are then used to calculate the effluent stream compositions. For each unit, the design procedure is repeated $n - 1$ times at the most, where n is the number of components in the feed to the unit. Each design is performed using a different pair of adjacent components as the keys (Steffens et al., 2000).

Many articles and books have been written in detail about downstream processing unit designs. Design assumptions and cost information for the most common stream-splitting unit operations are summarized below:

- *Settlers and centrifuges.* Both solid and liquid separation units rely upon density differences between insoluble particles and the surrounding fluid to operate. Sedimentation relies on gravity and settling to achieve solid–liquid separation, and is generally performed in rectangular or circular flow tanks. Centrifugation, on the other hand, involves mechanical applications of centrifugal force to obtain a solid concentrate and clarified supernatant. The keys are chosen from the feed-solid components, which are ranked according to settling velocity (Wheelwright, 1991). The upper key component and any components with a higher settling velocity completely distribute into the slurry stream, while the lower key component and slower components move into the less dense stream. Soluble components are assumed to uniformly distribute between the two effluent streams. The slurry solid concentration, c_s, is specified as a design parameter (Steffens et al., 2000). Centrifugal separation is in widespread use in the biotechnology industry, with the disk-stack centrifuge being the machine used most commonly for the separation of biological materials. Depending on the nature of the product and the mode of centrifuge operation, a wide variety of disk-stack separator types are available (Clarkson et al., 1996). Energy expenses are the principal operating costs.

- *Conventional filters.* Conventional filters are probably the most common means for separating solids from liquids on the basis of a single physical parameter: particle size. This parameter serves to choose the key components. The small key is assumed to be completely permeable and the large key impermeable. The wash rate, *w*, and cake-solids concentration, c_k, are both design parameters used to estimate the filter area, assuming constant specific cake resistance. Compared with centrifugation, filtration consumes less energy and requires considerably less capital investment. The operating costs principally include energy and filter aid expenses. Because typical bacteria diameters are about 1 μm, the use of conventional filters for their separation is not recommended.

- *Microfilters.* Microfilters are capable of retaining particles 0.1 to 10 μm in diameter — for example, whole cells or cells debris. The key components for a microfiltration unit are chosen from the particulate components in the incoming stream. For these equipment designs, flux and membrane area must be estimated. Operating costs consist of energy and membrane replacement.

- *Ultrafilters.* Ultrafiltration membranes can retain particles and macromolecules 1000 to 500,000 Da in size. The key components are chosen by molecular size and membrane retention capacity. Membrane area and flux must be estimated, and operating costs are generally dominated by membrane replacement and energy costs.

Fractionating Units

These units generally operate batchwise and produce several fractions or cuts. Chromatography columns, where the various components are sequentially eluted, are a particularly common example of a fractionating unit (Steffens et al., 2000). Chromatography columns, which fractionate rather than split, cannot be designed using the key component technique. A bioprocess stream may contain different types of complex biological compounds that interact with each other and the column, making the system difficult to model (Leser et al., 1996). For this reason, an empirical approach that estimates the effluent composition for a chromatography column using physical properties differences is used:

- *Ion-exchange columns.* Feed components are sorted according to their net charge, which is also used to calculate the binary ratios. Components with a charge opposite that of the resin are assumed to bind completely, while those with the same charge do not bind at all. The column volume is estimated by calculating a binding capacity, B_c, and residence time t_r (Steffens et al., 2000). For any type of column chromatography, the main costs are the package or resin costs and the solvents employed for elution.

- *Gel-filtration columns.* Feed components are sorted by means of their molecular weights, which are also used for calculating the binary ratios (Leser et al., 1996). Column volume is determined by specifying the sample volume, B_{sam} (percent of column volume), and the residence time, t_r.

- *Hydrophobic interaction and reverse phase columns.* Hydrophobicity is the physical property used to design a hydrophobic interaction or reverse-phase chromatography column. Column volume is calculated in the same way as for ion exchange, except that the binding fraction, F_i, for each component is assumed to have a linear relationship with the hydrophobicity, ϕ (Steffens et al., 2000).
- *Affinity columns.* Affinity columns are based on highly specific interactions between the desired compound and the adsorbent. The adsorbents employed have chemical groups called *ligands*, which are capable of binding specifically with the solute. This type of column is often employed in protein purification. The bases of their design are similar to those for ion exchange columns except that only one substance binds to the resin.

Formulation

Once the purity requirements and specifications by food and drug agencies have been achieved, the product manufactured must be formulated in a commercial form. Other compounds, such as excipients (e.g., starch), stabilizers, emulsifiers, antioxidants, and a large quantity of several compounds, can be mixed with the active principle until the commercial formulation is achieved. When the product is an enzyme, it is very important to decide on the final product formulation carefully because the biological activity, measured in international units of activity (IUA), depends directly on the final concentration of enzyme and fixes its selling price.

Economic Analysis

The analysis of the economic feasibility of a chemical or biological process is focused on two points:

- Cost estimates
- Total capital investment

The accuracy of these estimates depends on the extent to which the process is defined; many processes in the research and development phase will undergo changes before they are implemented to production plant scale. Cost estimates and capital investment in the product selling price allow calculation of different economical parameters, such as return on investment (ROI), payback time, present worth, and internal rate of return, with which it is possible to determine whether the proposed process will be profitable.

Cost Estimates

The total cost of a product includes the operating costs associated with manufacturing the product along with general expenses and is referred as the cost estimate. Manufacturing costs include those elements that contribute directly to the costs of production (operating costs, fixed costs, and plant overhead costs). Operating costs are a function of the production volume and include raw materials, labor, utilities, and supplies. Fixed costs relate to the physical plant and do not change with productivity levels; they include depreciation and interest, taxes, and insurance. The category denoted as plant overhead includes charges for services that are not directly attributable to the cost of the product, such as janitorial services, accounting, personnel, etc. Finally, the category denoted as general expenses includes charges for marketing, research and development, and general administration charges (Blanch and Clark, 1997).

Total Capital Investment

Total capital investment is the amount of capital required to construct and equip the plant (fixed capital), plus the capital necessary for its operation (working capital). Fixed capital includes the cost of land, building construction, and engineering design. Working capital refers to the funds used to provide an inventory of raw materials, supplies, and cash to pay salaries, usually during the first 3 months of plant operation.

The starting point for estimating the operating and capital costs of a product is the process flow sheet. The flow sheet and the material and energy balances, based on a desired annual production capacity, allow establishment of a flowchart indicating all liquid, gas, and solid flows in the process. When these flows are set, the equipment can be sized and the raw materials and utilities can be specified.

Single Cell Production: An Example of Bioprocess Design

A process for the production of a food-grade, single cell protein (SCP), developed in the Department of Biotechnology and Bioengineering of the Center for Research and Advanced Studies of the National Polytechnic Institute (CINVESTAV-IPN), is used to evaluate the bioprocess synthesis technique (De la Torre et al., 1994).

A burst of activity appeared in the field of SCP production in the 1950s, 1960s, and early 1970s. Many companies made important improvements in the production processes and in this way reduced the operating and capital costs in order to be competitive with protein supplements of vegetable origin, such as soy. Some of these plants are operating today to produce food-grade yeast autolysates.

From 1985 to 1991, a research team at CINVESTAV-IPN developed a high-cell-density process for food-grade *Candida utilis* production from sugar-cane molasses. A continuous, 10.5-m³, jet-loop fermentor was designed to take advantage of its intrinsic high oxygen transfer rate and energy efficiency. This fermentor was provided with a computer system for online data acquisition and for controlling the molasses flow rate in response to inferred ethanol production rates. The operating costs were minimized because of the efficient conversion of molasses into biomass and of the low water consumption in the fermentation stage. The capital costs were kept down due to the high productivity achieved in the fermentation process. An overview of the most important aspects of the process developed is presented below.

Laboratory Fermentation

Preliminary experiments were conducted on a bench scale for strain screening and to define culture medium and operating conditions in a 0.030-m³ stirred tank reactor (STR). Later, continuous fermentation was carried out on the same STR. It was found, for example, that under the operation conditions, assayed C/N ratios from 5 to 9.5 at dilution rates in the range of 0.14/h to 0.25/h did not affect the biomass yield based on sugars.

Scale-Up

Once the fermentation process conditions were established, work proceeded to the pilot-plant scale and then to the industrial scale. The most important step in this process was the choice of the critical variable that could be used as a scale-up criterion. Under carbon substrate limiting conditions, the most efficient carbon substrate conversion to biomass was achieved. Under these conditions, biomass productivity was related to the oxygen transfer rate (OTR); therefore, OTR was employed as a criterion to scale-up, first from bench to pilot plant (0.03-m³ STR to 1.00-m³ STR) and then from pilot plant to industrial reactor (1.00-m³ STR to 10.50-m³ jet-loop reactor). The biomass volumetric productivity and the biomass yield were similar for both STR reactors, 6.0 kg/m/h and 0.45 ± 0.03 kg cell/kg sugar, respectively, which indicates that OTR choice as a scale-up criterion was correct.

The jet-loop reactor had a higher OTR than the STRs; therefore, the biomass volumetric productivity increased to 11.2 kg/m/h for a biomass concentration of 80 kg/m³ (8% w/v). A significantly higher biomass yield was observed in the jet-loop reactor (0.56 kg cells/kg sugar), and yeast cells were larger than those cultivated in the STRs. Reduced yields in STRs might be a result of ethanol production under oxygen-limiting conditions. However, physiological responses of the cells continuously submitted to transient environmental conditions, such as recycling from a region of high turbulence to one of low turbulence in the jet-loop reactor, might be contributing factors in enhancing biomass yields.

Downstream Processing

Cell Separation

Centrifugation tests were done in a disk-stack centrifuge (FESX 512S, Alfa-Laval). The yeast cream solids concentration after two centrifugation steps was 16% (dry weight). In addition, vacuum leaf filter tests were conducted. Cotton fabric and Decalite 477 were selected as filter medium and filter aid, respectively. Trials with a rotary filter yielded a 24% solids cake. To get a 24% final solids concentration, economic analysis showed that filtration alone is a better alternative than centrifugation followed by filtration of broth solids concentrations higher than 7%.

Thermolysis

In order to increase the availability of nutrients and to develop flavor notes, yeast used as food or feed is usually broken down. Three alternatives were investigated: autolysis, autolysis-thermolysis, and thermolysis. Experiments were carried out to increase protein digestibility while minimizing available lysine and thiamin losses. The independent variables were temperature, pH, and time, and the dependent variables were protein digestibility, thiamin, and available lysine. Results are shown in Table 3.1. A two-step thermolysis was the best alternative to increase protein digestibility with no adverse effects on available lysine content and protein quality. The product has been used as a flavor enhancer additive and to produce chicken-like flavors.

Final Design

All information acquired during the research time was employed to build and operate a demonstration industrial plant with a 10.5-m^3 fermentor. A flowsheet of the process developed is shown in Figure 3.3. Molasses is diluted with tap water 1:1 by volume and heated to 90°C. The settled solids

TABLE 3.1

Results of Autolysis and Thermolysis of *Candida utilis*

Treatment	Available Lysine (Relative to A)	Protein Digestibility (Relative to A)
None (A)	100	1.00
Autolysis (24 h at 50°C)	93	1.53
Autolysis-thermolysis (24 h at 50°C, 2 h at 90°C)	85	1.76
Thermolysis (1 h at 70°C, 2 h at 85°C)	98	1.85

Source: Modified from De la Torre et al., in *Advances in Bioprocess Engineering*, Kluwer, Dordrecht, 1994, 67-74. With permission.

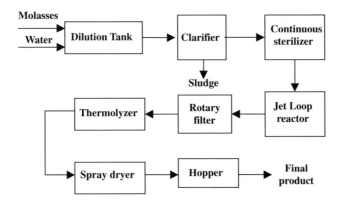

FIGURE 3.3

SCP production from molasses; schematic flowsheet. (Modified from De la Torre et al., in *Advances in Bioprocess Engineering*, Kluwer, Dordrecht, 1994. With permission.)

are separated from the molasses. Clarified molasses is fed to the fermentor through sterilizing equipment. A metering pump supplies phosphoric acid while a pH controller continuously regulates the ammonia supply. The fermentor operates under gas hold-up controlled conditions to optimize oxygen transfer. The yeast is concentrated and washed using a rotary vacuum filter. The yeast cake obtained is fed to thermolyzer tanks to be liquefied and increase its digestibility. Finally, the yeast is spray dried.

Economic Analysis

Table 3.2 shows an economical comparison among different yeast production processes. Speichim and Vogelbusch process data were obtained from industrial facilities operating in Cuba, while CINVESTAV process data were gathered from trials at the industrial plant. The total capital investment for a new facility using the CINVESTAV process is about 80 to 85% that required for conventional processes. Raw material costs are reduced to nearly 20%, mainly as a result of a higher biomass yield on molasses and minimal use of ammonia, which is utilized instead of ammonium sulfate. The specific energy consumption of the fermentors employed is 0.39, 0.55, and 0.53 kW-h kg-cell^{-1} for the airlift fermentor (Speichim), STR (Vogelbush), and jet-loop fermentor (CINVESTAV), respectively, so energy consumption is very similar for the STR and jet-loop fermentor. As a result of culturing at high cell density, water consumption is reduced by two thirds, but utilities cost for the CINVESTAV process is slightly higher, because the Vogelbusch and Speichim processes employ combustion gases for yeast drying, while the CINVESTAV process uses indirect heating to improve product quality.

TABLE 3.2

Comparative Economic Data for Yeast Production[a]

Process	Total Capital Investment (Relative to A)	Capital-Related Cost (Relative to A)[b]	Raw Material Cost (Relative to A)	Utilities Cost (Relative to A)	Total Production Cost (Relative to A)
Vogelbusch	1.250	1.250	1.226	0.939	1.166
Speichim	1.175	1.175	1.229	0.851	1.120
CINVESTAV	1.000	1.000	1.000	1.000	1.000

[a] Basis: 12×10^6 kg/yr.
[b] Annual capital-related cost estimated as 22% of the total capital investment.
Source: Modified from De la Torre et al., in *Advances in Bioprocess Engineering*, Kluwer, Dordrecht, 1994, 67-74. With permission.

Conclusions

Bioprocess design must carefully consider the characteristics of the biological system and the desired final product as well as chemical engineering principles to achieve the best performance of the involved process. It is very important to keep in mind that bioprocess performance depends on the interaction of all unit operations involved. A bioprocess cannot be optimized by taking into consideration each processing unit separately but must consider the whole of the processing line. This is due partially to the several alternatives for each proposed unit operation involved. Several possibilities exist for a specific bioprocess, and several processes for the production of the same bioproduct can be adequate. In this new century, functional genomics will help us understand the entire cell functioning, and this knowledge will allow redirection of cell flux distribution to overproduce metabolites and thus increase bioprocess performance.

References

Atkinson, B. and Mavituna, F. (1983) *Biochemical Engineering and Biotechnology Handbook*, Macmillan Publishers, New York, pp. 593–601.

Blackstock, W.P. and Weir, M.P. (1999) Proteomics: quantitative and physical mapping of cellular proteins, *Trends Biotechnol.*, 17, 121–127.

Blanch, H.W. and Clark, D.S. (1997) *Biochemical Engineering*, Marcel Dekker, New York, pp. 452–682.

Bulmer, M., Clarkson, A.I., Titchener-Hooker, N.J., and Dunnill, P. (1996) Computer-based simulation of the recovery of intracellular enzymes and its pilot-scale verification, *Bioprocess Eng.*, 15, 331–337.

Clarkson, A.I., Lefevre, P., and Titchener-Hooker, N.J. (1993) A study of process interaction between cell disruption and debris clarification stages in the recovery of yeast intracellular products, *Biotechnol. Progr.*, 9, 462–467.

Clarkson, A.I., Bulmer, M., and Titchener-Hooke, N.J. (1996) Pilot-scale verification of a computer-based simulation for the centrifugal recovery of biological particles, *Bioprocess Eng.*, 14, 81–89.

De la Torre, M., Flores, L.B., and Chong, E. (1994) High cell density yeast production: process synthesis and scale-up, in *Advances in Bioprocesses Engineering*, E. Galindo and O.T. Ramírez, Eds., Kluwer, Dordrecht, pp. 67–74.

Fraga, E.S. (1996) Discrete optimization using string encodings for the synthesis of complete chemical processes, in *State of the Art in Global Optimization*, C.A. Floudas and P.M. Pardalos, Eds., Kluwer, Dordrecht, pp. 627–651.

Graves, D.J. (1999) Powerful tools for genetic analysis come of age, *Trends Biotechnol.*, 17, 127–134.

Gregory, M.E., Bulmer, M., Bogle, I.D.L., and Titchener-Hooker, N. (1996) Optimizing enzyme production by baker's yeast in continuous culture: physiological knowledge useful for process design and control, *Bioprocess Eng.*, 15, 239–245.

Gritis, D. and Titchener-Hooker, N. (1989) Biochemical process simulation. I. Chem. E Symposium Series 114, Computer Integrated Process Engineering, 69–78.

Kelly, B.D. and Hatton, T.A. (1991) The fermentation/downstream processing interface, *Bioseparation*, 1, 333–349.

Kossen, N.W.F. (1994) Bioreactor engineering, in *Advances in Bioprocesses Engineering*, E. Galindo and O.T. Ramírez, Eds., Kluwer, Dordrecht, pp. 1–11.

Krings, U. and Berger, R.G. (1998) Biotechnological production of flavours and fragrances, *Appl. Microbiol. Biotechnol.*, 49, 1–8.

Kuipers, O.P. (1999) Genomics for food biotechnology: prospects of the use of high-throughput technologies for the improvement of food microorganisms, *Curr. Opin. Biotechnol.*, 10, 511–516.

Leser, E.W. and Asenjo, J.A. (1992) Rational design of purification processes for recombinant proteins, *J. Chromatogr.-Biomed. Appl.*, 584(1), 43–57.

Leser, E.W., Lienqueo, M.E., and Asenjo, J.A. (1996) Implementation in an expert system of a selection rationale for purification processes for recombinant proteins, *Ann. N.Y. Acad. Sci.*, 782, 441–455.

Lienqueo, M.E., Leser, E.W., and Asenjo, J.A. (1996) An expert-system for the selection and synthesis of multistep protein separation processes, *Comp. Chem. Eng.*, 20(SA): S189–S194.

Middelberg, A.P.J. (1995) The importance of accounting for bioprocesses interaction, *Australian Biotechnol.*, 5, 99–103.

Narodoslawsky, N. (1991) Bioprocess simulation: a system theoretical approach to biotechnology, *Chem. Biochem. Eng. Q.*, 5, 183–187.

Petrides, D.P. (1994) Biopro designer: an advanced computing environment for modeling and design of integrated biochemical process, *Comp. Chem. Eng.*, 18(SS): S621–S625.

Petrides, D.P., Calandranis, J., and Cooney, C. (1995) Computer-aided-design techniques for integrated biochemical processes, *Gen. Eng. News*, 15(16), 10.

Priefert, H., Rabenhorst J., and Steinbüchel, A. (2001) Biotechnological production of vanillin, *Appl. Microbiol. Biotechnol.*, 56, 296–314.

Samsalatli, N.J. and Shah, N. (1996) Optimal integrated design of biochemical processes, *Comput. Chem. Eng.*, S20, 315–320.

Siddiqui, S.F., Bulmer, M., Ayazi Shamlou, P., and Titchener-Hooker, N.J. (1995) The effects of fermentation conditions on yeast cell debris particle size distribution during high pressure homogenization, *Bioprocess Eng.*, 14, 1–8.

Steffens, M.A., Fraga, E.S., and Bogle, I.D.L. (2000) Synthesis of bioprocesses using physical properties data, *Biotechnol. Bioeng.*, 68(2), 218–230.

Stephanopoulus, G. and Vallino, J.J. (1991) Network rigidity and metabolic engineering in metabolite overproduction, *Science*, 252, 1675–1681.

Wheelwright, S.M. (1987) Designing downstream processing for large-scale protein purification, *Bio/Technology*, 5, 789.

Wheelwright, S.M. (1991) *Protein Purification: Design and Scale-Up of Downstream Processing*, Hanser, Munich, pp. 76.

Woodley, J.M. and Titchener-Hooker, N.J. (1996) The use of windows of operation as a bioprocess design tool, *Bioprocess Eng.*, 14, 263–268.

Zhou, H.Y. and Titchener-Hooker, N.J. (1999) Visualizing integrated bioprocess designs through windows of operation, *Biotechnol. Bioeng.*, 65(5), 550–557.

Zhou, H.Y., Holwill, I.L.J., and Titchener-Hooker, N.J. (1997) A study of the use of computer simulations for the design of integrated downstream processes, *Bioprocess Eng.*, 16, 367–374.

4

Gas Hold-Up Structure in Impeller Agitated Aerobic Bioreactors

Ashok Khare and Keshavan Niranjan

CONTENTS

Introduction

Viscous media are frequently encountered in aerobic biotechnological processes, and gas bubbling into such media is an essential feature. An efficient contact between gas and liquid phases is necessary to maintain high reaction rates. The production capacity of aerobic bioreactors is often limited by the rate at which oxygen is made available to microorganisms. High viscosity of certain biological media (e.g., xanthan gum and polysaccharide fermentation) makes it more difficult to transfer oxygen. It is therefore extremely important to understand and decipher the mechanism of oxygen transfer, especially in media that are highly viscous and non-Newtonian. The key variable indicating the efficiency of gas–liquid contacting, and in turn influencing oxygen transfer, is gas hold-up, i.e., fraction of the dispersion constituting the gas phase. This chapter examines the gas hold-up structure in impeller agitated viscous liquids, reviews the link between operating variables and gas hold-up, and presents models that can be used to analyze gas–liquid mass transfer in such systems.

1-56676-892-6/03/$0.00+$1.50

Gas Hold-Up Structure in Viscous Liquids

Gas hold-up structure in highly viscous Newtonian and non-Newtonian impeller agitated liquids (μ, viscosity of Newtonian liquids; μ_a, apparent viscosity of non-Newtonian liquids > 0.4 Pas) is distinctively characterized by a nearly bimodal bubble size distribution (Schugerl, 1981; Nienow, 1990; Khare and Niranjan, 1995). When such liquids are aerated, large bubbles (diameter > 20 mm), some as large as the impeller, are formed along with tiny bubbles (0.1 to 5 mm in diameter). During aeration, large bubbles form rapidly due to coalescence. These bubbles quickly ascend through the liquid due to high buoyancy forces (i.e., they have relatively shorter residence times) and cannot be recirculated back into the impeller region by the liquid flow. In contrast, tiny bubbles slowly ascend through the liquid, and they have comparatively longer residence times. Further, these bubbles are also recirculated back into the impeller region by the circulating liquid. Tiny bubbles are formed in all parts of the reactor. In the immediate vicinity of the rotating impeller blades turbulence is high; the greater the turbulence, the smaller are the bubbles formed. Away from the impeller, tiny bubbles are formed by the break-up of larger bubbles by local turbulence. Finally, tiny bubbles are also formed in the upper regions of the liquid when large bubbles break through the surface and disengage into the headspace. The tiny bubbles thus formed do not merely accumulate in the reactor; they constantly leave it by (1) disengaging at the surface, and (2) coalescing with other small or big bubbles. This ensures their regular turnover, which is significant in mass transfer.

Gas hold-up in impeller agitated viscous liquids can be regarded as consisting of two parts: a hold-up due to larger bubbles and a hold-up due to tiny bubbles. This fact fundamentally distinguishes hold-up structure in viscous liquids from that observed in low-viscosity aqueous solutions and forms the basis of the analysis of gas hold-up in such liquids. The simultaneous formation and disengagement of tiny bubbles also causes the hold-up to vary with time after the commencement of aeration, when the rates of formation exceed the rate of disengagement. With the build-up of tiny bubbles, parity is established between the two rates, and the tiny bubble hold-up attains steady state. The time taken to attain this parity determines the rate of turnover of tiny bubbles in the reactor.

In practice, the gas hold-up at a given impeller speed is determined by the difference between the height of dispersion at any time (H_D) and that of clear liquid height (H_C). At the onset of aeration, the immediate increase in the H_C is due to large bubbles. As mentioned above, with continuous aeration the concentration of tiny bubbles in the liquid progressively increases. This causes the liquid height to increase with time and eventually attain a steady value (H_{Df}), when the rate of generation of tiny bubbles equals the rate of their disappearance. Once a stable dispersion height is reached, the gas flow and agitation can be simultaneously switched off. Large bubbles have been observed to leave the system within approximately 10 sec. At this moment, the dispersion height consisting of tiny bubbles and liquid (H_{tf}) can be noted. The final total gas hold-up (ε_f), the hold-up of tiny bubbles (ε_{ft}), and the hold-up of large bubbles (ε_{fL}) can then be calculated as:

$$\varepsilon_f = (H_{Df} - H_C)/H_C \tag{4.1}$$

$$\varepsilon_{ft} = (H_{tf} - H_C)/H_C \tag{4.2}$$

$$\varepsilon_{fL} = \varepsilon_f - \varepsilon_{ft} \tag{4.3}$$

The typical variation of tiny bubble hold-up with time is shown in Figure 4.1.

Two alternative approaches can be taken to model the time dependency of gas hold-up. Philip et al. (1990) considered the sparged gas to divide itself into two fractions: one flowing through the liquid as tiny bubbles (i.e., F) and the other as large bubbles. Further, they assumed that, initially, only tiny bubbles are formed, and no disengagement occurs.

Considering these assumptions and the variation of the tiny bubble hold-up (ε_t) with time (t) (Figure 4.1), Philip et al. evaluated F, defined as the ratio

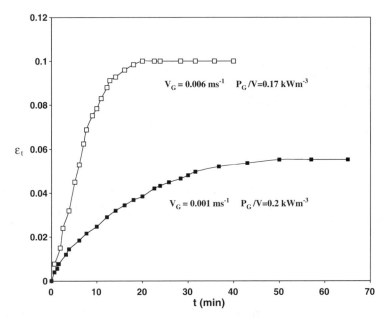

FIGURE 4.1
Typical variation of tiny bubble hold-up with time in castor oil; CBDT impeller, vessel diameter = 0.6 m.

of the volumetric rate of generation of tiny bubbles to the volumetric gas flow rate through the system, as follows:

$$F = \left(\frac{d\varepsilon_t}{dt}\right)_{t=0} \frac{V}{Q_G} \qquad (4.4)$$

where $(d\varepsilon_t/dt)_{t=0}$ is the initial slope of the curve as seen in Figure 4.1; V is the volume of the reactor, and Q_G is the volumetric gas flow rate.

Muller and Davidson (1992) took an alternative approach in which they assumed that the rate of disappearance of tiny bubbles at any instant is dependent on the tiny bubble inventory at that instant, and deduced the following first-order equation:

$$\frac{d\varepsilon_t}{dt} = \frac{1}{\tau}(\varepsilon_{ft} - \varepsilon_t) \qquad (4.5)$$

where ε_t is the hold-up of tiny bubbles at any time t during aeration. Solving Eq. (4.5) with $\varepsilon_t = 0$ at $t = 0$, Muller and Davidson (1992) showed that the dynamic hold-up curve is as follows:

$$\varepsilon_t = \varepsilon_{ft}(1 - e^{-t/\tau}) \qquad (4.6)$$

The τ in Eqs. (4.5) and (4.6) is the characteristic time constant, which can be estimated from the slope of the plot of $\ln(1 - \varepsilon_t/\varepsilon_{ft})$ vs. t. Comparing Eqs. (4.4) and (4.5), it can be deduced that:

$$\frac{FQ_G}{V} = \frac{(\varepsilon_{ft} - \varepsilon_t)}{\tau} \qquad (4.7)$$

as $Q_G = AV_G$ and $V = AH_C$, where A is the cross sectional area of the reactor, V_G is the superficial gas velocity, and H_C is the clear liquid height (i.e., before the commencement of aeration). Noting that $\varepsilon_t = 0$ at $t = 0$, Eq. (4.7) yields:

$$\tau = \frac{H_C \varepsilon_{ft} A}{V_G FA} \qquad (4.8)$$

The numerator and denominator in Eq. (4.8) signify the volume of tiny bubbles at the steady state and the flow rate of gas into the tiny bubbles, respectively; in other words, τ gives an estimate of the mean residence time of tiny bubbles.

Effect of Tiny Bubbles on Mass Transfer

Having considered the hold-up structure in viscous liquids in terms of the coexistence of tiny and large bubbles, we can now discuss its implications on mass transfer rate for the relatively simple case of a continuously stirred tank aerobic bioreactor. As described earlier, Khare and Niranjan (1995) assumed the sparged gas to be divided into two fractions: (1) a completely mixed tiny bubble fraction having lower oxygen partial pressure than the sparged air, and (2) a large bubble fraction flowing through the liquid in a plug flow manner with oxygen partial pressure comparable to that of the inlet air. This difference in oxygen partial pressure between the two fractions is due to the difference between their residence times. Given that the tiny bubble fraction receives a supply of oxygen continuously through the fraction of sparged air flowing through it (FQ_G), an oxygen balance equation covering this fraction can be written as:

$$\frac{FQ_G P^0 Y^0}{RT} - \frac{FQ_G P^0 Y}{RT} - (k_L a)_t (C_2^* - C_L)V = 0 \qquad (4.9)$$

where P^0 is the atmospheric pressure, Y^0 is the mole fraction of oxygen in the inlet, Y is the mole fraction of oxygen in the tiny bubble fraction, R is the universal gas constant, T is absolute temperature, $(k_L a)_t$ is the volumetric mass transfer coefficient due to tiny bubbles, C_2^* is the solubility of oxygen corresponding to the tiny bubble fraction, and C_L is the concentration of dissolved oxygen in the feed to the reactor.

The first two terms in Eq. (4.9) represent the rates at which oxygen flows in and out of the tiny bubble fraction, respectively, and the third term represents the rate of oxygen transfer to the liquid through this fraction. The solubility of oxygen corresponding to its partial pressure in the tiny bubble fraction is given by $C_2^* = P^0 Y / H^*$, H^* being the Henry's constant. Further, also following assumption (2), the solubility of oxygen corresponding to its partial pressure in the large bubble fraction is given by $C_1^* = P^0 Y^0 / H^*$. Considering this, Eq. (4.9) can be rearranged as:

$$C_1^* - C_2^* - N_D (C_2^* - C_L) = 0 \qquad (4.10)$$

where C_1^* is the solubility of oxygen corresponding to large bubble fraction. It follows from Eq. (4.10) that:

$$C_2^* = \frac{C_1^* + N_D C_L}{1 + N_D} \qquad (4.11)$$

N_D in Eqs. (4.10) and (4.11) is the dimensionless number, given by:

$$N_D = \frac{(k_L a)_t \, VRT}{FQ_G H^*} = \frac{\dfrac{V}{FQ_G}}{\dfrac{H^*}{(k_L a)_t \, RT}} \tag{4.12}$$

From Eq. (4.12) the significance of N_D can be developed. The numerator represents the mean residence time for tiny bubbles to traverse through the liquid, whereas the denominator gives a characteristic time for oxygen transfer from these bubbles. Evidently, a low value of N_D indicates that the tiny bubble fraction is actively transferring oxygen. For the extreme case where $N_D \to 0$, Eq. (4.11) gives $C_2^* = C_1^*$, indicating that the solubilities and oxygen partial pressures of the large and tiny bubble fractions are similar.

In contrast, a high value of N_D indicates greater mean residence time of the tiny bubbles in relation to the characteristic time for oxygen transfer, which enables tiny bubbles to equilibrate with the liquid. For the case where $N_D \gg 1$, Eq. (4.11) simplifies to $C_2^* = C_L$.

It is now possible to estimate the ratio of the rates of oxygen transfer between large bubbles and tiny bubbles, which can be given by:

$$\frac{r_t}{r_L} = \frac{(k_L a)_t (C_2^* - C_L)}{(k_L a)_L (C_2^* - C_L)} \tag{4.13}$$

where r_t and r_L are the rate of oxygen transfer through tiny and large bubbles, respectively, and $(k_L a)_L$ is the volumetric mass transfer coefficient due to large bubbles. Substituting for C_2^* from Eq. (4.11), Eq. (4.13) becomes:

$$\frac{r_t}{r_L} = \frac{1}{1 + N_D} \frac{(k_L a)_t}{(k_L a)_L} \tag{4.14}$$

Taking a conservative estimate of $(k_L a)_t / (k_L a)_L = 1$ and $N_D = 1$, Eq. (4.14) will be reduced to:

$$\frac{r_t}{r_L} = \frac{1}{2} \tag{4.15}$$

This implies that in the continuously stirred tank aerobic bioreactor considered above, half of the oxygen transfer might be taking place through tiny bubbles. This analysis reveals that tiny bubbles in viscous liquids cannot be assumed to equilibrate with the liquid phase. In fact, these bubbles can actively contribute to the oxygen transfer, although at a lower driving force.

Experimental Observations on Gas Hold-Up in Impeller Agitated Systems

This section describes the behavior of gas hold-up in low-viscosity liquids (e.g., water), high-viscosity non-Newtonian carboxymethol cellulose(CMC) solution (apparent viscosity $[\mu_a]$ = 0.5 to 0.8 Pas), and Newtonian castor oil (viscosity $[\mu]$ = 0.76 Pas) generated by three disk impellers: the flat-bladed disk turbine (DT) and two of its modified designs (available commercially), concave-bladed disk turbine (CBDT), and Scaba 6SRGT.

The fundamental difference between the gas hold-up structure in low- and high-viscosity liquids lies in the bubble size distribution. In low-viscosity liquids, a statistically normal bubble size distribution is quickly established; that is, the spread of bubble size around a mean value is uniform. Moreover, the gas hold-up does not depend on the aeration time, but varies only with specific power dissipation (P_G/V) and gas velocity (V_G) (Figure 4.2). Further, it is also clear from this figure that gas hold-up in such liquids is independent of impeller design, an observation also reported by Saito et al. (1992) and Bakker et al. (1994).

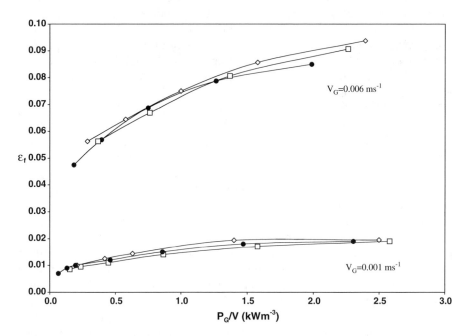

FIGURE 4.2
Variation of final total gas hold-up (ε_f) with specific power dissipation (P_G/V) in water; vessel diameter = 0.6 m. ($\Diamond$) DT; ($\Box$) CBDT; ($\bullet$) Scaba 6SRGT.

In high-viscosity castor oil and CMC solution, the gas hold-up is characterized by time dependency, as well as the existence of tiny and large bubbles (Khare and Niranjan, 1999, 2002). The hold-up must therefore be analyzed by distinguishing between the contributions of tiny and large bubbles (Figures 4.3 and 4.4). It is evident that in both viscous liquids the final steady-state values of the total gas hold-up (ε_f) and the hold-up of tiny bubbles (ε_{ft}) increase with V_G. In contrast, ε_f and ε_{ft} exhibit different trends with respect to P_G/V. For example, while ε_f increases with P_G/V, ε_{ft} either passes through a maxima (e.g., at the lower V_G), or decreases progressively (e.g., at the higher V_G) in the CMC solution (Figure 4.3). On the other hand, ε_f as well as ε_{ft} pass through a maxima in castor oil (Figure 4.4).

It is understood that at a particular gas velocity, increasing P_G/V (i.e., increasing impeller speed) facilitates bubble break-up, which in turn would be expected to enhance the population of tiny bubbles. However, higher $P_G/V/$(or /speed) also induces stronger liquid circulation (hence, intense coalescence), which adversely affects the build-up of tiny bubbles. In the case of shear thinning liquids this effect is more pronounced, as stronger liquid circulation also implies reduction in effective viscosities. The net hold-up of tiny bubbles is therefore a result of the balance between these two conflicting effects. Thus, the plot of ε_{ft} vs. P_G/V can slope either upward or downward depending on which of the two effects predominates. In general, it can be noted that the ε_{ft} dictates the trend of the ε_f, if the fractional contribution of the former is significant to the latter (i.e., $\varepsilon_{ft}/\varepsilon_f$). For example, from Figures 4.3 and 4.4, it is clear that ε_f and ε_{ft} follow the same trend in castor oil, but not in the CMC solution (despite the fact that both liquids have comparable effective viscosities). This is due to the fact that tiny bubbles constitute only 15 to 45% of the total hold-up in the CMC solution, whereas they constitute 60 to 90% in the castor oil.

It is also interesting to note from Figures 4.3 and 4.4 that, unlike water, the gas hold-up in the high-viscosity CMC solution and the castor oil is influenced by the impeller design. For example, in the CMC solution (Figure 4.3), the DT gives higher ε_f and ε_{ft} values than its modified designs (CBDT and Scaba 6SRGT) at the lower gas velocity, whereas at the higher gas velocity these impellers give comparable hold-up values (except for the ε_{ft} values produced by CBDT at the higher gas velocity, which are significantly lower than the DT). This implies: (1) in the CMC solution, the influence of impeller type on gas hold-up is more pronounced at the lower gas velocity; and (2) the DT shows better performance over its modified designs with regard to generating gas hold-up. In the case of castor oil (Figure 4.4), the modified disk impellers (Scaba 6SRGT and CBDT) generally give higher hold-up than the DT. Further, all impellers produce comparable maximum ε_f and ε_{ft} values, but modified disk impellers generate these values at significantly lower P_G/V levels compared to DT. In short, Scaba 6SRGT and CBDT appear to be superior to DT in castor oil. This implies that the influence of impeller design on gas hold-up is markedly

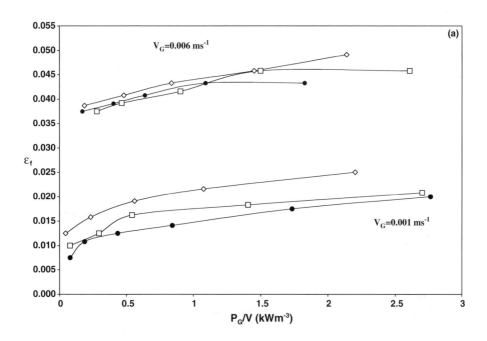

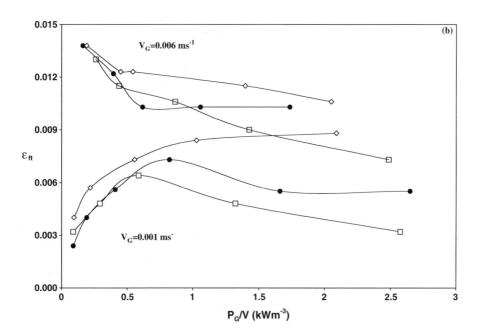

FIGURE 4.3
Variation of final gas hold-up with specific power dissipation (P_G/V) in the CMC solution; vessel diameter = 0.6 m. (a) Total gas hold-up (ε_f); (b) tiny bubble hold-up (ε_{ft}). Keys same as in Figure 4.2.

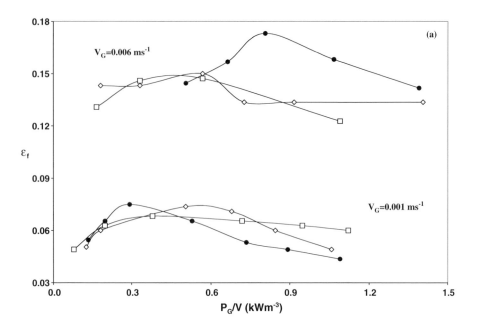

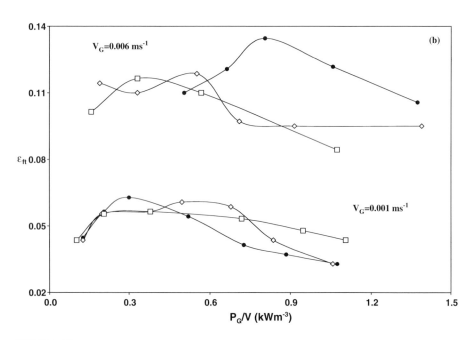

FIGURE 4.4
Variation of final gas hold-up with specific power dissipation (P_G/V) in castor oil; vessel
diameter = 0.6 m. (a) Total gas hold-up (ε_f); (b) tiny bubble hold-up (ε_{ft}). Keys same as in
Figure 4.2.

different in high-viscosity non-Newtonian and Newtonian liquids having comparable viscosities.

As mentioned, the gas hold-up in CMC solution and castor oil is time dependent (Figure 4.1). In CMC solution, the time taken for gas hold-up to reach steady state varies from 30 to 120 sec, whereas it varies between 600 and 6000 sec in castor oil (Khare and Niranjan, 1999, 2002). This implies that τ in non-Newtonian liquids is much lower than in Newtonian liquids having comparable viscosities; therefore, it appears that, between the two liquids, tiny bubbles in castor oil would be expected to be relatively less effective with respect to their contribution to mass transfer. Figures 4.5 and 4.6 show the variation of τ values with P_G/V in the CMC solution and castor oil, respectively. It can be seen that at a given gas velocity τ decreases with P_G/V in both liquids. This implies that increasing P_G/V effectively enhances the contribution of tiny bubbles to the mass transfer. It is, however, essential to note that higher P_G/V also lowers the hold-up of tiny bubbles (hence, the interfacial area) in both CMC solution (Figure 4.3b) and castor oil (Figure 4.4b). Under such circumstances, caution must be exercised, as decreasing area may offset the benefits of lower τ values. Figures 4.5 and 4.6 also

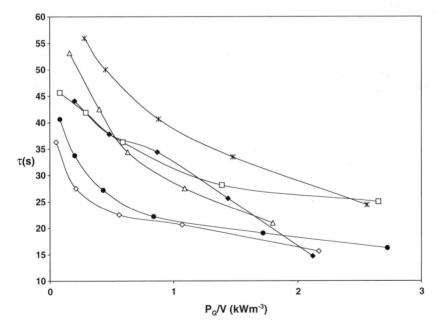

FIGURE 4.5
Variation of characteristic time constant (τ) with specific power dissipation (P_G/V) in the CMC solution; vessel diameter = 0.6 m. For $V_G = 0.001/ms^{-1}$: ($\Diamond$) DT; ($\square$) CBDT; ($\bullet$) Scaba 6SRGT. For $V_G = 0.006/ms^{-1}$: ($\blacklozenge$) DT; (*) CBDT; ($\triangle$) Scaba 6SRGT.

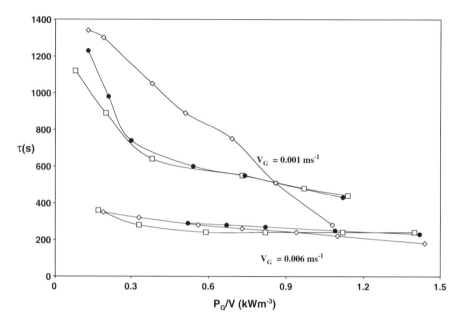

FIGURE 4.6
Variation of characteristic time constant (τ) with specific power dissipation (P_G/V) in castor oil; vessel diameter = 0.6 m. Keys same as in Figure 4.2.

demonstrate that impeller type influences τ values. For instance, DT gives the lowest τ values in CMC solutions (Figure 4.5), whereas DT gives the highest τ values in castor oil (Figure 4.6). Khare and Niranjan (1995) suggested that an efficient impeller must not only dissipate low power, but also give high hold-up (ε_{ft} and ε_f) and low τ values. Considering this, it is clear from Figures 4.3 to 4.6 that DT shows potential for enhanced performance over its modified designs in the CMC solutions, whereas the latter impellers perform better in castor oil.

   Finally, the effect of scale of operation on hold-up values and trends has been found to be significant; this is evident from Figures 4.7a and b, where hold-up in castor oil has been compared in two vessels measuring 0.3 and 0.6 m in diameter. The gas hold-up (ε_{ft} and ε_f) passes through a maxima in the 0.6-m vessel, whereas it progressively increases with P_G/V in the 0.3-m vessel. Further, these figures show that at a given P_G/V, gas hold-up (ε_f and ε_{ft}) in the 0.3-m vessel is significantly higher than in the 0.6-m vessel. This clearly shows that gas hold-up cannot be correlated with power dissipation levels alone; geometric size and related factors must also be taken into account.

Conclusions

This article discusses gas hold-up structure in impeller agitated high-viscosity liquids, which are often encountered in aerobic bioreactors. The hold-up

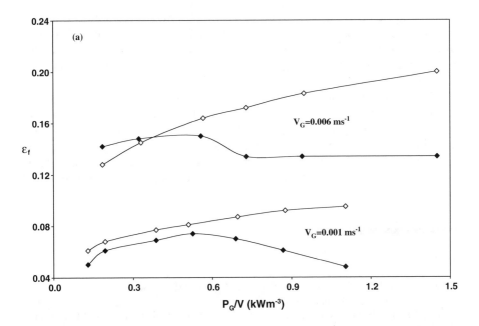

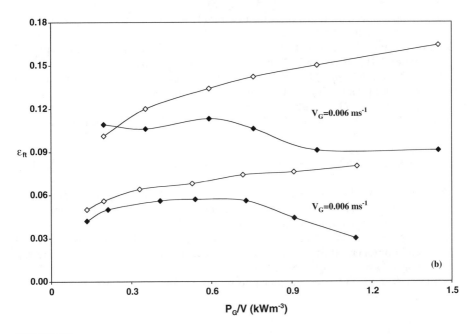

FIGURE 4.7
Variation of final gas hold-up with specific power dissipation (P_G/V) in castor oil; DT impeller; vessel diameter = 0.6 m ($\blacklozenge$) and 0.3 m ($\lozenge$). (a) Total gas hold-up (ε_f); (b) tiny bubble hold-up (ε_{ft}).

in such liquids consists of two parts: (1) due to the large bubbles (diameter > 20 mm) and (2) due to tiny bubbles (diameter = 0.1 to 5 mm). It has been assumed, often erroneously, that the tiny bubble constituent of the gas hold-up is not actively transferring oxygen, as these bubbles tend to attain equilibrium with the liquid phase due to high residence times. This chapter challenges this assumption and suggests that tiny bubbles can play a significant role in mass transfer. In addition, it has been shown that, unlike in low-viscosity liquids, gas hold-up in high-viscosity liquids is time dependent and, interestingly, need not always increase with increasing specific power dissipation levels. Impeller design and scale also play a critical role in determining hold-up, as well as its variation with operating parameters (i.e., speed and gas velocity).

References

Bakker, A., Smith, J.M., and Dyers, K.J. (1994) How to disperse gases in liquids, *Chem. Eng.*, 101(12), 98–104.

Khare, A.S. and Niranjan, K. (1995) Impeller agitated aerobic bioreactor: the influence of tiny bubbles on gas hold-up and mass transfer in highly viscous liquids, *Chem. Eng. Sci.*, 50(7), 1091–1105.

Khare, A.S. and Niranjan, K. (1999) An experimental investigation into the effect of impeller design on gas hold-up in a highly viscous Newtonian liquid, *Chem. Eng. Sci.*, 54(8), 1093–1100.

Khare, A.S. and Niranjan, K. (2000) Comparison of gas hold-up in impeller agitated water and high viscosity liquids, *J. Chem. Eng. Jpn.*, 33(5), 815–817.

Khare, A.S. and Niranjan, K. (2002) The effect of impeller design on gas hold-up in surfactant containing highly viscous non-Newtonian agitated liquids, *Chem. Eng. Process.*, 41, 239-249.

Muller, F.L. and Davidson, J.F. (1992) On the contribution of small bubbles to mass transfer in bubble columns containing highly viscous liquids, *Chem. Eng. Sci.*, 47, 3525–3532.

Nienow, A.W. (1990) Gas dispersion performance in fermenter operation, *Chem. Eng. Progr.*, 8, 61–71.

Philip, J., Proctor, J.M., Niranjan, K., and Davidson, J.F. (1990) Gas hold-up and liquid circulation in internal loop reactors containing highly viscous Newtonian and non-Newtonian liquids, *Chem. Eng. Sci.*, 45, 651–664.

Saito, F., Nienow, A.W., Chatwin, S., and Moore, I.P.T. (1992) Power, gas dispersion and homogenisation characteristics of Scaba and Rushton turbine impellers, *J. Chem. Eng. Jpn.*, 25(3), 281–287.

Schugerl, K. (1981) Oxygen transfer into highly viscous media, *Adv. Biochem. Eng.*, 19, 71–174.

5

Production and Partial Purification of Glycosidases Obtained by Solid-State Fermentation of Grape Pomace Using Aspergillus niger 10

Sergio Huerta-Ochoa, María Soledad de Nicolás-Santiago, Wendy Dayanara Acosta-Hernández, Lilia Arely Prado-Barragán, Gustavo Fidel Gutiérrez-López, Blanca E. García-Almendárez, and Carlos Regalado-González

CONTENTS

1-56676-892-6/03/$0.00+$1.50
© 2003 by CRC Press LLC

Introduction

Grape pomace is a waste material from winemaking comprised of residual sugars, gums, stalks, seeds, and a large number of polyphenolic and terpenoid compounds. Grape pomace can be divided in two main parts: grape pulp (rich in gums and sugars) and grape pips (rich in fibers and tannins). The use of grape pomace has varied from country to country. For example, in France, current legislation makes it compulsory to transform grape pomace from cooperative wineries that send the waste to large distilleries where residual sugars are converted into a distilled liquor, which is sold as brandy. The remaining fractions are used to produce natural dyes (antocyanins), grapeseed oil, compost, and fuel for the distillery. The net result is an environmentally friendly industry with almost zero discharge, because the spillage is usually disposed of in agricultural fields or is treated to comply with environmental regulations.

A review of the U.S. patent list indicates that grape pomace has been used to obtain anthocyanin pigments (Philip, 1976; Crosby et al., 1984; Shrikhande, 1984; Langston, 1985) and aroma compounds (Brillouet et al., 1989; Dupin et al., 1992; Günata et al., 1988, 1996). Hang (1988) used a technique of solid-state fermentation (SSF) in order to obtain significant microbial production of citric acid from grape pomace. Additionally, two U.S. patents use this material for enzyme production or as part of a composting technology (Graefe, 1982a,b). The use of grape pomace has recently focused on animal feed through protein enrichment by SSF using cultures of *Trichoderma viride* (Bensoussani and Serrano-Carreon, 1997). The economic appraisal of microbial protein production, however, has remained uncertain because of the low cost of soybean protein.

Work during the last decades has shown that grapes accumulate monoterpens, C13-norisoprenoids, and shikimate-derived metabolites mainly as odorless mono- and di-glycosidic conjugates (Voirin et al., 1990; López-Tamames et al., 1997; Charles et al., 2000; Ferreira et al., 2000). Flavor enhancement of grape juice and wine through hydrolysis of the glycosides has therefore attracted much attention (Günata et al., 1990a, 1997). Disaccharide glycosides can be hydrolyzed by a sequential reaction requiring, first, α-L-rhamnopyranosidase (EC 3.2.1.40), α-L-arabinofuranosidase (EC 3.2.1.55), and β-apiofuranosidase and, second, β-glucopyranosidase (EC 3.2.1.21) (Günata et al., 1990a; Spagna et al., 1998). Such terpens are value-added materials of nearly U.S.$1000/kg in a pure form due to their high flavor potency. Additionally, there are important amounts of polyphenols, namely tannins, in grape seeds. These can be transformed by fermentation or enzyme reactions by tannase into gallic acid, which is in high demand for the production of a variety of antioxidants in food and pharmaceutical industries. Gallic acid presently has a value of U.S.$200 to U.S.$500/kg.

Glycosidases obtained from grapes, filamentous fungi, and yeast have been considered for research due to their involvement in the hydrolysis of glycosidic flavor precursors. Thus, both grape pulp and grape pips can be used as substrates for production of microbial enzymes or value-added chemicals (terpenoids or polyphenols).

In this work, one of the above-mentioned alternatives is explored for upgrading grape pomace by the production of glycosidases using SSF of grape pulp. Several reports on β-glucopyranosidase, α-L-rhamnopyranosidase, α-L-arabinofuranosidase, and β-apiosidase are available (Dupin et al., 1992; Günata et al., 1997; Guo et al., 1999); however, culture conditions for glycosidase production have only been studied using submerged cultures (Dupin et al., 1992) or purified from commercial crude enzyme preparations (Günata et al., 1997; Spagna et al., 1998, 2000). Their use in winemaking has allowed hydrolysis of monoterpenyl, arabinosyl, rhamnosyl, and apiosylglucosides. Thus, SSF can be used to produce this type of enzymes and could be of practical significance because it may simplify materials handling. Simplifications include the use of solid substrates with minimal upstream (pasteurization and inoculation) and downstream operations (leaching by screw pressing). On the other hand, liquid fermentations require more complex unit operations (extraction of enzyme inducers, mash sterilization, stirred vat fermentation, ultrafiltration of spent wort, and liquid waste treatment).

Materials and Methods

Microorganism

Aspergillus niger, strain number 10 (ORSTOM, France) (Raimbault, 1981), was used in this study. The strain was kept in porcelain pearls at 4°C.

Solid Substrate

Grape (*Vitis vinifera* cv. Chenin Blanc) pomace obtained from the Freixenet Company (Cadereyta, Querétaro, México) was dried and separated into three parts (branches, seeds, and skins) by sieving. The skins were used as substrate for glycosidase production and were kept at room temperature until used.

Solid-State Culture Media

Solid substrate (dry skins) was supplemented by addition of other nutrients (nitrogen source and mineral salts) and an inducer (red wine, Cabernet Sauvignon 1998, Santa Rita, Chile) for glycosidase production. Initial media compositions tested are shown in Table 5.1.

Effect of C/N Ratio and Water Activity on Glycosidase Production

To study the effect of C/N ratio and water activity of the culture medium on glycosidase production, a 2^2 factorial experiment was employed, based on the medium selected in the previous step and data reported in the literature (Table 5.2). Casein peptone was used as a nitrogen source.

Inoculum Production

A porcelain pearl was inoculated into 250-mL conical flasks containing 50 mL of potato dextrose agar (PDA) (Bioxon, México) before solidification and mixed. The inoculated conical flasks were incubated at 30°C for 5 days. Spores were harvested using 50 mL of sterile water added with 0.2% Tween 80 solution. Spore suspension was filtered, and the concentration of spores (spores/mL) was quantified using a Newbauer chamber. The inoculum size used in the experiments was 2×10^7 spores per gram of dry matter.

Fermentation Columns

Small glass columns (20 g of wet material) were used for solid-state fermentation experiments, according to the methodology developed by Raimbault

TABLE 5.1

Culture Media Composition Used for Glycosidase Production by Solid-State Fermentation with *Aspergillus niger* 10

	Media Compound					
Culture Media	$(NH_4)_2SO_4$ (mg/g dry substrate)	$(NH_4)_2HPO_4$ (mg/g dry substrate)	KH_2PO_4 (mg/g dry substrate)	$MgCl_26H_2O$ (mg/g dry substrate)	Wine[a] (mL/g dry substrate)	Glucose (mg/g dry substrate)
Control	3.0	1.50	0.038	0.038	—	—
Washed[b]	3.0	1.50	0.038	0.038	—	—
Plus wine	1.5	0.56	0.018	0.018	0.75	—
Plus glucose	2.4	0.90	0.030	0.030	—	1.50
C/N 15.9	25.8	9.53	0.038	0.038	—	—

[a]1998 Cabernet Sauvignon (Santa Rita) made in Chile.
[b]Washed three times with hot distilled water and dried at 60°C.

TABLE 5.2

Culture Media for the 2^2 Factorial Design Experiment

	Media Compounds			
Treatment	KH_2PO_4 (mg/g dry substrate)	$MgCl_2 6H_2O$ (mg/g dry substrate)	Casein Peptone (mg/g dry substrate)	Ethylenglycol (mL/g dry substrate)
1	8.73	8.73	203.9	—
2	8.73	8.73	203.9	0.05
3	8.73	8.73	47.26	—
4	8.73	8.73	47.26	0.05

and Alazard (1980). Experiments were carried out at 30°C and a saturated airflow rate of 60 cm^3/min.

Enzyme Crude Extract

Enzyme crude extract was obtained using a modification of the method of Roussos et al. (1992). Fermented wet samples were weighed and mixed with the same weight of 0.1 M citrate–phosphate buffer (pH 5.6). The crude extract was obtained by pressing the mixture at 2000 psi in a hydraulic press (Erkco model PH-51T, Aeroquip). The extract was filtered and kept at 4°C for further purification steps.

Physicochemical Analyses: Carbon Dioxide (CO_2) Evolution

Metabolic activity during the fermentation process was followed by outlet gas evolution (Saucedo-Castañeda et al., 1994) using a gas chromatograph fitted on line (Cadena-Méndez et al., 1993).

Proximal Analyses

Protein, ash, ether extract, moisture, C/N ratio, and crude fiber content were determined according to the methods of AOAC (Helrich, 1990).

Glycosidase Activity

α-L-arabinofuranosidase and α-L-rhamnopyranosidase activity was determined using the method reported by Günata et al. (1997). 150 μL of enzyme extract was added to 150 μL of the proper substrate (p-nitrophenol-α-L-arabinofuranoside, 4 mM; p-nitrophenol-α-L-rhamnopyranoside, 4 mM), mixed and incubated for 10 min at 40°C. Enzyme reaction was stopped by adding 900 μL of 0.1-M Na_2CO_3 solution and cooling at room temperature. Absorbance was read at 400 nm using a Shimadzu UV-160 spectrophotometer and compared against a standard curve of *p*-nitrophenol (0.0 to 0.1 μ*M*). An activity unit (U) was defined as the amount of enzyme necessary to release 1 nmol of *p*-nitrophenol per minute at reaction conditions.

Protease Activity

Protease activity was determined using the method reported by Perlman and Lorand (1970). Crude extract (0.5 mL) was added to 0.5 mL of 2% casein solution (pH 5.6) and incubated at 30°C for 10 min. Then 1 mL of 0.4-*M* TCA was added, mixed, and centrifuged at 16,800× *g*. Next, 2.5 mL of 0.1-*N* Na_2CO_3 and 0.5 mL of Folin–Ciocalteu reagent were added to 0.5 mL of the supernatant for color development. Sample absorbance was read at 660 nm and compared against a standard curve of tyrosine.

Protein

Protein was quantified using the Bradford microassay (Bio-Rad) based on the Bradford (1976) method. Sample absorbance was read at 595 nm and compared against a standard curve of bovine serum albumin (BSA) in the range of 1 to 10 μg/mL.

Partial Characterization of the Crude Extract

Optimum pH and pH Stability

For optimum pH determination, glycosidase activity was assayed in a pH range of 1 to 10 using different buffer solutions (Table 5.3) at 40°C for 10 min. For enzyme stability, the same pH range was tested, keeping the enzyme solutions at 4°C for 14 h, after which glycosidase activity was assayed at 40°C for 10 min. Residual enzyme activity was reported as relative activity (%).

TABLE 5.3

Buffer Solutions Used for Optimum pH and pH Stability Determination

Buffer Solution	pH Range
HCl, 0.1 *N*	1–2
Citrate, 0.3 *M*	3–4
Phosphate	5–8
Carbonate–bicarbonate	9.2–10

Optimum Temperature and Temperature Stability

Optimum temperature for glycosidase activity was determined by assaying enzyme activity between 15°C and 70°C for 10 min. For temperature stability, enzyme extract samples of 150 μL were kept at the temperature range tested (15 to 70°C) for 1 h, and glycosidase activity was assayed at 40°C for 10 min. Residual enzyme activity was reported as relative activity (%).

Prepurification

Precipitation by salting out was carried out using 80% saturation of ammonium sulfate under continuous agitation at 4°C and kept for 2 h. The precipitate was centrifuged at 12,000× g at 4°C for 15 min. The precipitate was resuspended in 10 mM citrate–phosphate buffer (pH 5.6). The resuspended precipitate was dialyzed against the same buffer at 4°C for 24 h, with the buffer solution being changed every 6 h. Glycosidase activity and protein were determined, followed by freeze drying (LabConCo equipment).

Glycosidase Purification

The lyophilized sample was resuspended in 0.3-M citrate buffer (pH 2.6). An Econo-Column (Bio-Rad) filled with DEAE-cellulose (Whatman) was equilibrated using this buffer. Two-milliliter aliquots were filtered in membranes of 0.45 μm and injected onto the anion exchange column, then 4-mL fractions were collected at a flow rate of 4 mL/min. From fraction 5, elution was conducted using the same buffer added with 1-M NaCl, and a linear gradient from 0 to 50% of this buffer was used up to fraction 23. From fractions 23 to 35, the gradient varied from 50 to 100%. Protein and glycosidase activity were determined for each fraction, and those having activity were mixed and dialyzed against deionized water at 4°C for 36 h.

Electrophoretic Studies (Sodium Dodecyl Sulfate–Polyacrylamide Gel Electrophoresis; SDS-PAGE)

Electrophoresis under denatured conditions was carried out according to Laemmli (1970) using a Mighty Small SE 250 (Hoeffer) vertical electrophoresis chamber. Gels of 10% T (percent of acrylamide in the solution acrylamide plus *bis*-acrylamide) and 8% C (percent of *bis*-acrylamide in the monomers solution) were used for the separating gel, while the stacking gel had 4% C. Low-molecular-weight protein markers (Sigma) were used for SDS-PAGE: aprotinine (6500 Da), α-lactalbumin (14,200 Da), trypsin inhibitor (20,000 Da), trypsinogen (24,000 Da), carbonic anhydrase (29,000 Da), glyceraldehyde 3-phosphate dehydrogenase (36,000 Da), egg albumin (45,000 Da), and bovine serum albumin (66,000 Da). Injection volume was 12 μL. Protein alignment was conducted at 10 mA, while protein band separation was conducted at 15 mA. The rapid silver staining method (Ausubel et al., 1995) was used to identify the protein bands. Proteins were fixed with a formaldehyde solution (methanol 40%, 0.5 mL of 37% [w/w] formaldehyde per liter of solution) for 10 min with constant stirring.

After 5 min of washing, the gels were submerged in $Na_2S_2O_3$ solution (0.2 g/L). After washing with deionized water, a 0.1% silver nitrate solution was used for protein dying for 10 min. Silver was removed by washing, and the gel was submerged in a revealing solution (3% sodium carbonate, 0.0004% $Na_2S_2O_3$, and 0.5 mL of formaldehyde per liter of solution) for 1 min until bands were revealed. Five milliliters of 2.3-M citric acid per 100 mL of revealing solution were added. Finally, gels were submerged in 50 mL of drying solution (10% ethanol and 4% glycerol).

Results and Discussion

Solid Substrate Characterization

Grape pomace was sundried until it reached 11% moisture, and separated into three parts (branches, seeds, and skins) by sieving (Table 5.4). Due to grape pomace compositional changes according to season, type of grapes, weather conditions, etc., no comparison with similar wastes was attempted. Skins were taken as solid substrate for solid-state fermentation experiments. Besides the presence of expected aroma precursors in the skins, solid waste had high protein and crude fiber content (Table 5.5).This fact allowed us to use the skins as substrate of *Aspergillus niger* 10 for enzyme production. This fungus has a broad enzyme synthesis spectrum and can be induced to produce glycosidases as well as proteases, hemicellulases, and lipases.

Culture Media Selection

Five culture media (Table 5.1) were selected for preliminary fermentation experiments. These media were selected in order to find a proper initial culture medium:

TABLE 5.4

Results after Grape Pomace Sieving

Part	Percent of Dry Matter
Seeds	58.46
Skins	32.01
Branches	9.53

TABLE 5.5

Proximal Analysis of Skins Separated from Grape Pomace and Results from Other Reports

	Percent Wet Basis	
Parameter	This Work	Literature[a]
Protein	14.6	12–17
Ash	6.31	3–13
Ether extract	2.25	5–10
Moisture	11.0	nr
C/N ratio	20.6	nr
Crude fiber	31.5	22–35

[a]Bensousani and Serrano-Carreón (1988).
Note: nr = not reported.

1. *Control*: Solid waste without any treatment
2. *Washed*: Solid waste washed three times with hot distilled water to remove possible unwanted compounds
3. *Plus wine*: Solid waste supplemented with commercial wine to add aroma precursors that may induce glycosidase production
4. *Plus glucose*: Solid waste supplemented with glucose to induce high microbial growth
5. *C/N ratio, 15.9*: Solid waste supplemented with ammonium sulfate and ammonium phosphate to modify the initial C/N ratio of 20

Microbial growth and enzyme production were quantified by measuring CO_2 evolution during fermentation and glycosidase activity, respectively, in the crude extract (α-L-rhamnopyranosidase and α-L-arabinofuranosidase). Results of enzyme activity after 65 h of fermentation are shown in Figure 5.1.

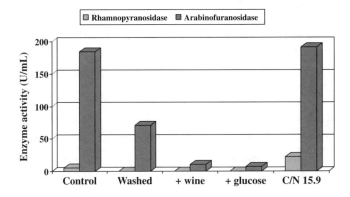

FIGURE 5.1
Effect of media supplementation on enzyme production by *Aspergillus niger* 10 growth on grape skins as solid substrate.

Under the fermentation conditions tested, the main activity produced was arabinofuranosidase. Control and C/N 15.9 culture media showed important enzyme production. The highest arabinofuranosidase and rhamnopyranosidase activities were produced when the fungus grew in the C/N 15.9 culture medium. Washed culture media presented some arabinofuranosidase activity and little rhamnopyranosidase activity, probably due to nutrient removal during the washings. On the other hand, the plus-wine and plus-glucose culture media were the poorest for enzyme production, probably due to the presence of simple carbon sources, which inhibited enzyme synthesis (Solis-Pereira et al., 1996). Thus, the C/N 15.9 medium was chosen for further enzyme production. Moreover, the maximum specific respiration rate, $\mu(CO_2)$, for this media was also the largest: 0.312/h (Table 5.6), indicating that *A. niger* invaded the solid substrate much faster with this culture media. Although enzyme production is not completely associated with fungal growth, the magnitude of the $\mu(CO_2)$ for this culture media is in agreement with the glycosidase production obtained.

Kinetic studies of microbial growth (CO_2 evolution) and enzyme production using a culture medium with a C/N ratio of 15.9 were conducted over 34 h. From kinetic data (Figure 5.2) it was observed that enzyme production started after 22 h of fermentation, when microbial growth was at the end of the exponential phase. During the beginning of the stationary phase, arabinofuranosidase activity increased up to 170 U/mL in 4 h (between 22 and 26 h of fermentation time). During the next 8 h, the arabinofuranosidase production rate slowed to about 75%; however, the activity after 32 h of fermentation increased to a value of 250 U/mL.

From data reported in the literature (Dupin et al., 1992) and preliminary experimental results (data not shown), we decided to change the nitrogen source from inorganic to organic (peptone) to improve enzyme production. The effect of water activity (a_w) and C/N ratio on glycosidase production was also studied. It was observed that the highest arabinofuranosidase activity was obtained using a C/N ratio of 10 and a_w value of 0.979 (Table 5.7) after 29 h of fermentation.

Figure 5.3 shows the kinetic data of microbial growth (CO_2 evolution) and enzyme production using the above-mentioned conditions for 44 h of

TABLE 5.6

Maximum Specific Respiration Rate Based on CO_2 Evolution Kinetics

Culture Media	$\mu\ (CO_2)\ (hr^1)$
Control	0.254
Washed	0.165
Plus wine	0.117
Plus glucose	0.134
C/N ratio 15.9	0.312

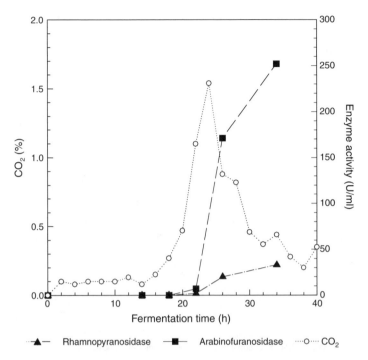

FIGURE 5.2

Kinetics of CO_2 evolution and enzyme production during solid-state fermentation (SSF) of grape skins using *Aspergillus niger* 10 and supplementing with C/N 15.9 culture media.

TABLE 5.7

Effect of Water Activity (a_w) and C/N Ratio on Enzyme Production

	Arabinofuranosidase Activity (U/mL)			
C/N Ratio	$a_w = 0.985$	$a_w = 0.979$	$a_w = 0.975$	$a_w = 0.968$
10	—	243	—	110.1
15	149.7	—	45	—

fermentation time. An enzyme production pattern similar to that found in Figure 5.2 was observed until 34 h of fermentation. After this fermentation time, arabinofuranosidase production increased almost three times in the next 6 h, up to 670 U/mL in 44 h. However, rhamnopyranosidase production was not improved by the changes tested in the culture medium.

Partial Characterization of Arabinofuranosidase and Rhamnopyranosidase

This study was conducted primarily to monitor the effect of pH and temperature on arabinofuranosidase and rhamnopyranosidase stability and activity in the crude extract. The crude extract was obtained from SSF experiments using a culture medium with C/N = 10 after 30 h of fermentation.

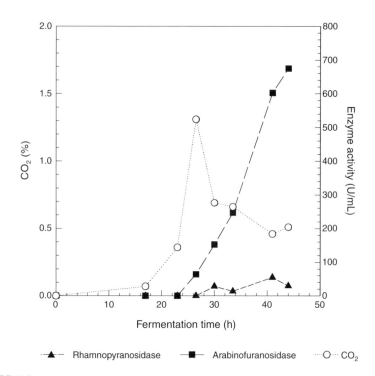

FIGURE 5.3

Kinetics of CO_2 evolution and enzyme production during solid-state fermentation (SSF) of grape skins using *Aspergillus niger* 10 and supplementing with C/N 10 culture media with a_w of 0.979.

Arabinofuranosidase activity was the highest at pH 3.0 (Figure 5.4a). This optimum pH is close to the pH of wine (3.8) used in this work.

Günata et al. (1997) purified from a commercial enzyme preparation an arabinofuranosidase with an optimum pH of 4.0 (Klerzyme 200, Gist-Brocades, France). Spagna et al. (1998) also reported an optimum pH of 4.0 for α-L-arabinofuranosidase purified from another commercial preparation (AR 2000, Gist-Brocades, France). This enzyme showed only a relative activity of 40% at pH 3.3.

Arabinofuranosidase activity showed high stability at pH 3.0 (Figure 5.4b) where no relative activity was lost after 14 h of incubation at 4°C; however, arabinofuranosidase activity had poor stability at higher pH values. At pH 3.5, the activity decreased to about 50% that shown without 14 h incubation, and it further decreased at higher pH values (Figure 5.4b). At pH 3.0 the stability was good (100% activity retention), but the activity relative to that at pH 9.0 was insignificant.

Results on pH for optimum activity and stability for rhamnopyranosidase are shown in Figures 5.4a and b. The activity was low (<30 U/mL) and was nearly constant in the pH range of 3.0 to 4.5, while in the pH range of 5.0 to 9.2 it was insignificant; therefore, stability was meaningless under these conditions. Spagna et al. (2000) reported an optimum pH of

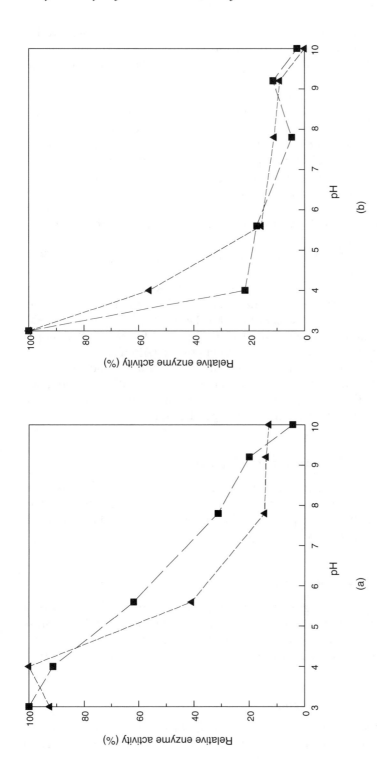

FIGURE 5.4

(a) Effect of pH on enzyme activity in the crude extract incubated for 10 min at 40°C. (b) Effect of pH on enzyme stability in the crude extract incubated for 14 h at 4°C. Arabinofuranosidase, ■; rhamnopyranosidase, ▲.

4.0 for an α-L-rhamnopyranosidase purified from a commercial preparation (AR 2000). This result agrees with the optimum pH, 4.5 to 5.0, obtained by Günata et al. (1997) for a rhamnopyranosidase from a commercial enzyme preparation (Klerzyme 200).

The temperature for optimum activity of arabinofuranosidase was 50°C (Figure 5.5a) at pH 5.6. This is lower than the optimum temperature obtained by Günata et al. (1997) with arabinofuranosidase purified from a commercial enzyme preparation (Klerzyme 200). Spagna et al. (1998) reported a higher optimum temperature (65°C) at pH 3.3 for α-L-arabinofuranosidase also purified from a commercial preparation (AR 2000).

Arabinofuranosidase was highly stable at 40°C (Figure 5.5b); however, at higher temperatures, the stability decreased sharply (50% at 60°C), while at lower temperatures (≤30°C) the initial activity was insignificant (Figure 5.5a) and stability was irrelevant.

Rhamnopyranosidase showed a temperature of optimum activity of 50°C while the stability was best at 60°C; however, at higher or lower temperatures the stability was adversely affected (Figure 5.5b). A similar optimum temperature value was obtained by Günata et al. (1997) for a rhamnosidase from a commercial preparation (Klerzyme 200). However, Spagna et al. (2000) reported an optimum temperature of 70°C at pH 3.4 for an α-L-rhamnopyranosidase purified from a commercial preparation (AR 2000).

Data obtained give an idea about pH and temperature values where the maximum arabinofuranosidase activity can be found. Stability data obtained are useful in finding appropriate storage conditions of this enzyme and to set the limits of pH for chromatographic purification.

Partial α–L-Arabinofuranosidase Purification

Different purification techniques have been used to purify α-L-arabinofuranosidase from commercial enzyme preparations (Günata et al., 1990b; Spagna et al., 1998). In this work, precipitation with ammonium sulfate and anion exchange chromatography (DEAE-cellulose) were tested.

Ammonium Sulfate Precipitation

The crude extract resulting from ammonium sulfate precipitation showed a purification factor of 42.7, with a yield of 8.3% of the initial activity. Because 90% of the original activity was lost, 0.5-M dithiothreitol (DTT; Bio-Rad) was added to the crude extract. This increased enzyme activity yield to 29.1%.

Anion Exchange Chromatography

The resuspended precipitate was freeze-dried before ion exchange experiments. This resulted in 64% of activity lost. This was probably associated

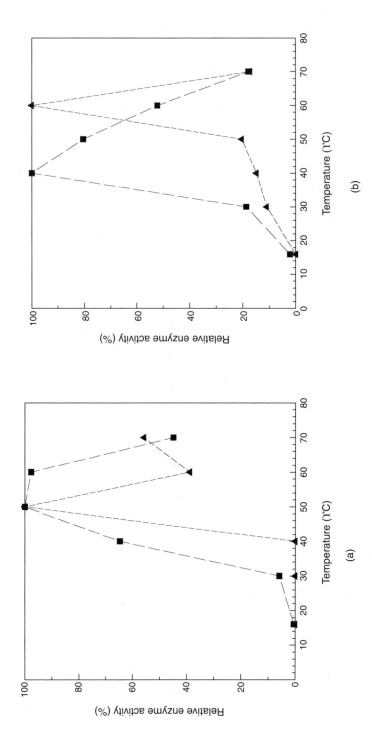

FIGURE 5.5

(a) Effect of temperature on enzyme activity in the crude extract incubated for 10 min at 40°C and pH 5.6. (b) Effect of temperature on enzyme stability in the crude extract incubated for 1 h at 40°C. Arabinofuranosidase, ■; rhamnopyranosidase, ▲.

with the poor enzyme stability at pH 5.6, which was used to handle the extract. In addition, perhaps a cryo-protective agent was required for this sample. The best results obtained from chromatographic purification experiments were conducted using an anionic exchange column (DEAE-cellulose) equilibrated with 0.1-M citrate buffer (pH 4.5). Elution profiles showed two peaks (Figure 5.6).

One peak eluted at the beginning of the salt gradient (probably acid proteins) without arabinofuranosidase activity, while the second peak eluted at 50% of NaCl concentration used. This peak had nearly all of the injected arabinofuranosidase activity. Two peaks with arabinofuranosidase activity were observed by Günata et al. (1997) from an enzyme preparation (Klerzyme 200) when a DEAE-Sepharose column was used (Imidazole buffer, 25 mM, pH 7.5).

After two purification steps, a yield of 2.4% and a purification factor of 177 were achieved (Table 5.8). The purification steps used by Günata et al. (1990b) were gel permeation, ion exchange, and affinity chromatography.

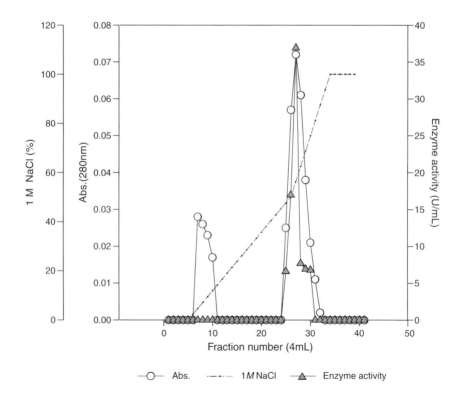

FIGURE 5.6

Anion exchange chromatography (DEAE-cellulose) of an arabinofuranosidase extract produced by solid-state fermentation of grape skin and ammonium sulfate precipated. (A) Citrate buffer (0.1 M, pH 4.5); (B) citrate buffer (0.1 M, pH 4.5) plus NaCl (1 M).

TABLE 5.8

Results of Purification Experiments of Arabinofuranosidase Produced by Solid-State Fermentation of Winemaking Waste (Skins) by *Aspergillus niger* 10

Purification Step	Volume (mL)	Activity (U/mL)	Protein (mg/mL)	Specific Activity (U/mg)	Purification Factor	Yield (%)
Crude extract	80	283	0.72	393	1.0	100
$(NH_4)_2SO_4$ precipitation	20	93.8	5.6×10^{-3}	1680	42.7	8.3
Anion exchange[a]	15	36.9	5.3×10^{-4}	69,500	177	2.4

[a] Resin used was DEAE-cellulose; the resuspended precipitate was lyophilized before anion exchange (64% of the activity was lost).

Using these purification steps, yield and purification factors of 3% and 27.4, respectively, were obtained. Spagna et al. (1998), using precipitation (ethanol, KCl), ultrafiltration, and adsorption (Bentonite, chitosan), obtained activity yield and purification factors of 70% and 10.4, respectively.

SDS-PAGE Electrophoresis

The SDS-PAGE electrophoresis of the fraction having arabinofuranosidase activity after anion exchange chromatography showed two protein bands with molecular weights of 74 and 130 kDa (Figure 5.7). The activity of each of these bands was not tested, so it is uncertain if one or both bands had

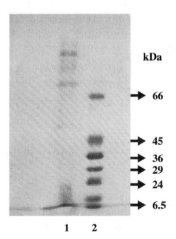

FIGURE 5.7

SDS-PAGE electrophoresis of the fraction with arabinofuranosidase activity after anion exchange chromatography. Lane 1, purified fraction; lane 2, low-molecular-weight markers (Sigma).

arabinofuranosidase activity. Günata et al. (1990b) reported an α-L-arabinofuranosidase with molecular weight of 61 kDa.

This research showed an alternative for upgrading grape pomace by
producing glycosidase enzymes from the solid-state fermentation of the
skins (32% of dry grape pomace) with *Aspergillus niger* 10. Different protocols have been reported in the literature to measure the effect of pH and
temperature on enzymatic activity, which did not allow comparison to our
results. It was observed that the arabinofuranosidase produced here had
a lower optimum pH than those reported previously. Enzyme stability was
also remarkable at pH 3.0.

Conclusions

Using the skins from grape pomace as substrate for solid-state culture (C/
N = 10, a_w = 0.979) with *Aspergillus niger* 10, a crude extract with glycosidase
activity (arabinofuranosidase, 674.5 U/mL; rhamnopyranosidase, 29.5 U/
mL) was obtained after 44 h of incubation at 30°C. The optimum temperature
for both arabinofuranosidase and rhamnopyranosidase activities in the
crude extract was 50°C. The pH for optimum arabinofuranosidase activity
in the crude extract was 3.0. On the other hand, rhamnopyranosidase activity
showed poor activity in the pH range of 3.0 to 10.0; therefore, it was not
possible to observe a clear optimum pH. An α-L-arabinofuranosidase
obtained from the solid-state fermentation experiments was partly purified
using ammonium sulfate precipitation and anion exchange chromatography.
The activity yield (2%) was low, while a high purification factor, 177, was
obtained. The purified sample showed two protein bands with molecular
weights of 74 and 130 kDa. Further work is required to identify which bands
correspond to arabinofuranosidase. The properties shown by this partly
purified enzyme represent an important advantage, because arabinofura-
nosidases have been used to improve aroma in wines that have low pH
(about 3.0).

Acknowledgment

Thanks are given to CONACYT for financial support through the project
"Solid State Fermentation for Upgrading Grape Pomace" (UC MEXUS-CON-
ACYT, 1999).

References

Ausubel, F., Brent, R., Kingston, R., Moore, D., Seidman, J.G., Smith, J.A., and Struhl, K. (1995) *Short Protocols in Molecular Biology*, 3rd ed., John Wiley & Sons, New York, pp. 10–38.

Bensoussani, M. and Serrano-Carreon, L. (1997) Upgrading agro-industrial products/wastes, in *Advances in Solid State Fermentation*, S. Roussoss, B.K. Lonsane, M. Raimbault, and G. Viniegra-González, Eds., Kluwer, Dordrecht, pp. 223–233.

Bradford, M.M. (1976) A rapid and sensitive method for the quantitation of microgram quantities of protein utilizing the principle of protein-dye binding, *Anal. Biochem.*, 72, 248–254.

Brillouet, J.M., Günata, Z., Bitteur, J., Cordonnier, R., and Bosso, C. (1989) Terminal apiose, a new sugar constituent of grape juice glycosides, *J. Agric. Food Chem.*, 37, 910–912.

Cadena-Méndez, M., Cornejo-Cruz, J.M., Prieto, N.J.M., Gaitán-González, J., Carrasco-Sosa, S., González-Camarena, F.E., Gutiérrez-Rojas, M., and Saucedo-Castañeda, G. (1993) Características de Medición de un Metabolímetro para fermentadores de sustrato sólido, *Rev. Mex. Ing. Biomed.*, 14, 311–319.

Charles, M., Martin, B., Ginies, C., Etievant, P., Coste, G., and Guichard, E. (2000) Potent aroma compounds of two red wine vinegars, *J. Agric. Food Chem.*, 48, 70–77.

Crosby, W.H., Fulger, C.V., Haas, G.J., and Nesheiwat, D.M. (1984) Stabilized Anthocyanin Food Colorant, U.S. Patent No. 4,481,226.

Dupin, I., Günata, Z., Sapis, J.C., Bayonove, C., M'Bairaroua, O., and Tapiero, C. (1992) Production of β-apiosidase by *Aspergillus niger*: partial purification, properties, and effect on terpenyl apiosylglucosides from grape, *J. Agric. Food Chem.*, 40, 1886–1891.

Ferreira, V., López, R., and Cacho, J.F. (2000) Quantitative determination of the odorants of young red wines from different grape varieties, *J. Sci. Food Agric.*, 80, 1659–1667.

Graefe, G. (1982a) Method for Producing High-Grade Fertilizer, U.S. Patent No. 4,311,510.

Graefe, G. (1982b) Method for Producing High-Grade Fertilizer, U.S. Patent No. 4,311,511.

Günata, Z., Bitteur, S., Brillouet, J.M., Bayonove, C., and Cordonnier, R. (1988) Sequential enzymic hydrolysis of potentially aromatic glycosides from grape, *Carbohydrate Res.*, 184, 139–149.

Günata, Y.Z., Bayonove, C.L., Tapiero, C., and Cordonnier, R.E. (1990a) Hydrolysis of grape monoterpenyl β-D-glucosides by various β-glucosidases, *J. Agric. Food Chem.*, 38, 1232–1236.

Günata, Z., Brillouet, J.M., Voirin, S., Baumes, R., and Cordonnier, R. (1990b) Purification and some properties of an α-L-arabinofuranosidase from *Aspergillus niger*. Action on grape monoterpenyl arabinofuranosylglucosides, *J. Agric. Food Chem.*, 38, 772–776.

Günata, Z., Bitteur, S., Baumes, R., Brillouet, J.M., Tapiero, C., Bayonove, C., and Cordonnier, R. (1996) A Method Is Disclosed for Obtaining Aroma Components and Aromas from Their Glycosidic Precursors Containing a β-Apioside with the Use of β-Apiosidase, U.S. Patent No. 5,573,926.

Günata, Z., Dugelay, I., Vallier, M.J., Sapis, J.C., and Bayonove, C. (1997) Multiple forms of glycosidases in an enzyme preparation from *Aspergillus niger*: partial characterization of β-apiosidase, *Enzyme Microbiol. Technol.*, 21, 39–44.

Guo, W., Salmon, J.M., Baumes, R., Tapiero, C., and Günata, Z. (1999) Purification and some properties of an *Aspergillus niger* β-apiosidase from an enzyme preparation hydrolyzing aroma precursors, *J. Agric. Food Chem.*, 47, 2589–2593.

Hang, Y.D. (1988) Grape Pomace as Substrate for Microbial Production of Citric Acid, U.S. Patent No. 4,791,058.

Helrich, K., Ed. (1990) *Official Methods of Analysis*, Vol. 1, 15th ed., AOAC, Arlington, VA, pp. 910–928.

Laemmli, U.K. (1970) Cleavage of structural proteins during assembly of head of bacteriophage T4, *Nature*, 227, 680–685.

Langston, M.S.K. (1985) Anthocyanin Colorant from Grape Pomace, U.S. Patent No. 4,500,556.

López-Tamames, E.N., Carro-Mariño, Y.Z., Günata, C., Sapis, R., Baumes, C., and Bayonove, C.L. (1997) Potential aroma in several varieties of Spanish grapes, *J. Agric. Food Chem.*, 45, 1729–1735.

Perlman, G.E. and Lorand, L. (1970) Proteolytic enzymes, *Meth. Enzymol.*, 19, 397–399.

Philip, T. (1976) Recovery of Anthocyanin from Plant Sources, U.S. Patent No. 3,963,700.

Raimbault, M. (1981) Croissance de Champignons Filamenteux sur Substrat Amylacé, Ph.D. thesis, ORSTOM-Paris, 291 pp.

Raimbault, M. and Alazard, D. (1980) Culture method to study fungal growth in solid fermentation, *Eur. J. Appl. Microbiol. Biotechnol.*, 9, 199–209.

Roussos, S., Raimbault, M., Saucedo-Castañeda, G., and Lonsane, B.K. (1992) Efficient leaching of cellulases produced by *Trichoderma harzianum* in solid state fermentation, *Biotechnol. Techniques*, 6, 429–432.

Saucedo-Castañeda, G., Trejo-Hernández, M.T., Lonsane, B.K., Navarro, J.M., Roussos S., Dufour, D., and Raimbault, M. (1994) On-line automated monitoring and control systems for CO_2 and O_2 in aerobic and anaerobic solid-state fermentation, *Process Biochem.*, 29, 13–24.

Shrikhande, A.J. (1984) Extraction and Intensification of Anthocyanins from Grape Pomace and Other Material, U.S. Patent No. 4,452,822.

Solis-Pereyra, S., Favela-Torres, E., Gutiérrez-Rojas, M., Roussos, S., Saucedo-Castañeda, G., Gunasekaran, P., and Viniegra-González, G. (1996) Production of pectinases by *Aspergillus niger* in solid state fermentation at high initial glucose concentrations, *World J. Microbiol. Biotechnol.*, 12, 257–260.

Spagna, G., Romagnoli, D., Angela, M., Bianchi, G., and Pifferi, P.G. (1998) A simple method for purifying glycosidases: α-L-arabinofuranosidase and β-D-glucopyranosidase from *Aspergillus niger* to increase the aroma of wine, part I, *Enzyme Microb. Technol.*, 22, 298–304.

Spagna, G., Barbagallo, R.N., Martino, A., and Pifferi, P.G. (2000) A simple method for purifying glycosidases: α-L-rhamnopyranosidase from *Aspergillus niger* to increase the aroma of moscato wine, *Enzyme Microb. Technol.*, 27, 522–530.

Voirin, S.G., Baumes, R.L., Bitteur, S.M., Günata, Z.Y., and Bayonove, C.L. (1990) Novel monoterpene disaccharide glycosides of *Vitis vinifera* grapes, *J. Agric. Food Chem.*, 38, 1373–1378.

6

Protein Crystallography Impact on Biotechnology

Manuel Soriano-García

CONTENTS

Introduction

It is well known that the bulk of our knowledge about the three-dimensional structures of biological macromolecules comes from x-ray crystallographic and nuclear magnetic resonance (NMR) investigations. Both methods produce extremely large amounts of data and are entirely dependent upon the availability of powerful computers and sophisticated processing algorithms for the interpretation of these data. Combining structural information from several experimental techniques can, in many cases, provide the basis for a structural solution where only partial data are available from any single technique. This allows solutions to be obtained for structural problems that would otherwise be intractable.

By defining the three-dimensional structures of biological macromolecules, scientists can create a map of life, a guide for exploring the biological and chemical interactions of the vast variety of molecules found in living organisms. Using the x-ray crystallography and NMR techniques, we can solve the structure of a molecule (small molecule or a macromolecule) — that is, determine the exact location of each of its atoms and produce relevant information on its function. Studies in structural biology may lead to new insights into how biological systems are formed and nourished, how they survive and grow, and how they are damaged and die.

Protein crystallography has had a profound impact on advancing the fields of biology, biochemistry, biotechnology, and pharmacology. Developments over the past 26 years in biochemistry, recombinant-DNA technology, and synchrotron radiation are combined in modern-day protein crystallography. More complicated molecules and assemblages are being worked out in atomic detail.

In organisms, proteins are present in smaller quantities in the cell; however, these biological macromolecules can now be cloned, expressed, purified, sequenced, and characterized to produce sufficient protein to carry out all spectroscopic, biochemical, and pharmacological studies. Site-directed mutagenesis can introduce specific changes into such proteins in order to adapt them for clinical and industrial applications. Alternatively, DNA encoding of a mutant protein, or even the protein itself, can be synthesized directly. These developments open up the field to protein engineering.

Enzymes, antibodies, membrane receptors, and most other proteins have well-defined three-dimensional structures and bind their ligands in specific conformations. The topographies of the complementary surfaces of the ligand and protein that determine affinity and specificity must be defined if the specificity, the thermal stability, or other properties are to be modified in a rational way (Blundell, 1994). For example, the application of enzymes in industrial processes often calls upon properties not found in enzymes isolated from their natural sources. That is, if enzyme technologists wish to have high overall stability under specific process conditions (high temperatures, organic solvents, detergents, optimum pH for maximum enzymatic activity, and oxidants) as well as maintain its high catalytic activity to non-natural substrates, then this problem is complex and multifactorial. The answers must come from our knowledge of the structure and function of those proteins under these experimental conditions.

Macromolecular crystallography is a relatively new branch of classical x-ray crystallography. The theory and practice of x-ray diffraction have already been well worked out; however, significant differences can be noted. For example, all protein crystals are labile and contain high solvent content, contributing to poor quality of the diffraction data.

While it is far beyond the scope of this chapter to develop in detail the theory of x-ray diffraction or its rigorous mathematical derivation, a brief outline is necessary for the nonspecialist who uses the atomic coordinates determined from a crystallographic investigation to judge the quality of the results and to understand the limitations of the method. For a detailed description of x-ray crystallography in general, *Fundamentals of Crystallography* (Giacovazzo et al., 1992) is an excellent textbook. The standard texts for macromolecular crystallography are *Protein Crystallography* (Blundell and Johnson, 1976), *Principles of Protein X-Ray Crystallography* (Drenth, 1999), and *Crystallization of Biological Macromolecules* (McPherson, 1999), among others. In this chapter, the fundamentals of diffraction and the advances made by macromolecular crystallography in biotechnology are highlighted, focusing on biologically significant structures and the papers describing the key technical achievements of each.

Fundamentals of Diffraction

Crystals

A crystal is the heart of any x-ray crystallographic experiment directed toward obtaining knowledge about the three-dimensional structure of a molecule. A crystal is a three-dimensional object with very well-defined faces and edges (Figure 6.1) which contains an array of atoms, ions, molecules, or groups of these. The basic unit of a crystal is called the *unit cell*, which is defined by three nonparallel axes (a, b, and c) which, when repeated in three dimensions, build the crystal.

The intersection of these axes defines the origin of the unit cell. The angles between axes are the *unit cell angles* (α, β, and γ). The angles α, β, and γ are defined between the b/c, a/c, and a/b axes, respectively. The unit cell itself may be subdivided by the internal crystallographic symmetry. This fact means that only a part of the unit cell (asymmetric unit) must be investigated. Once locating all atoms in the asymmetric unit has been accomplished, application of the symmetry generates the full unit cell.

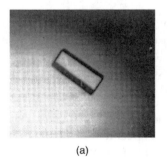 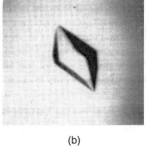

(a)   (b)

FIGURE 6.1
Crystals of triose phosphate isomerase (TIM) from *Trypanosoma cruzi*. (From Maldonado, E. et al., *Protein & Peotide Lett.*, 4, 139-144, 1997. With permission.)

Symmetry elements are composed of rotation axes, mirror planes, inversion centers, or proper combinations of these with a translation, in addition to various crystal lattices (P, primitive; A-, B-, C-centered; F (face)-centered; I (body)-centered). The proper combination of these lattice types in three dimensions and the symmetry operators give rise to 240 space groups. Fortunately, in the case of proteins or nucleic acids, chiral centers exist; thus, symmetry operators that invert the hand of the molecule, such as mirror planes and inversion centers, cannot exist. Due to this fact, the number of space groups for chiral molecules is only 65.

Diffraction

The fundamental principles of x-ray diffraction are well established. X-ray scattering is due to the electrons and, apart from the resonance effects of anomalous scattering, is classical Thomson scattering. Because the atoms, and hence their electrons, are ordered in a crystal at certain angles in three-dimensional space, their scattered radiation will have a constructive interference that gives rise to a diffraction spot or reflection, while elsewhere they will be a destructive interference. Each spot has three characteristics associated with it: its three-dimensional indices (h, k, l, or Miller indices), its intensity (I_{hkl}), and its phase angle (ϕ_{hkl}). The intensity depends on how the scattering from different atoms in the same unit cell interfere with each other; the phase angle relates the various diffraction points to each other and to the unit cell origin. For the scattering in the wavevector direction u from an element of electron density $\rho(\mathbf{r})$ positioned at $\mathbf{r}$, the phase shift relative to scattering from a point at the origin is $2\pi\, \mathbf{r} \cdot \mathbf{s}$, where $\mathbf{s} = (\mathbf{u} - \mathbf{u_o})/\lambda$ when the incident x-rays are directed along $\mathbf{u_o}$ and have a wavelength of λ. On integrating over all volume V of the unit cell and normalizing to the scattering from a free electron, the total scattered wave $F(s)$ is expressed as:

$$F(s) \;=\; \int_v \rho(r)\, exp\,(2\pi\, i\, r \cdot s)\, dv \qquad\qquad (6.1)$$

From this equation, the x-ray scattering is calculated if we know the electron-density distribution, but what we actually wish to do is the inverse. Given the scattering, what is the structure? In an x-ray experiment, the intensities of the diffraction pattern are measured but not their relative phases. Unfortunately, the calculation of an electron density map requires both intensities and phases. This problem is known as the *crystallographic phase problem*.

In Eq. (6.1), the diffraction from a molecular crystal takes a discrete form; however, the diffraction occurs only at specific directions according to the Laue conditions: $\mathbf{a} \cdot \mathbf{s} = h$, $\mathbf{b} \cdot \mathbf{s} = k$, $\mathbf{c} \cdot \mathbf{s} = l$, where h, k, and l are integers. This defines a lattice dictated by the crystal parameters $\mathbf{a}$, $\mathbf{b}$, and $\mathbf{c}$ in a manner reciprocal to the unit-cell dimensions; in other words, spots are closer together as cell lengths get longer. As electronic overlaps between atoms are small enough to be negligible, the integral in Eq. (6.1) can be replaced by a sum over the atomic position $\mathbf{r}_j$ for each atom j in the unit cell. In this sum, the integration for each atom gives its atomic scattering factor f_j taking into account the atomic mobility parameter defined by $B = 8\pi^2 <u^2>$, where u is an atomic displacement along $\mathbf{s}$ and the symbol $<>$ indicates averaging. Furthermore, $|\mathbf{s}| = 2s$ and $s = \sin\theta/\lambda$, where θ is the Bragg reflection angle. Finally, we have the working equation:

$$F(hkl) = \sum_{j=1}^{n} f_j \exp-(B_j \sin^2\theta / \lambda^2) \exp 2\pi i(hx_j + ky_j + lz_j) \qquad (6.2)$$

where (x_j, y_j, z_j) are the atomic positions in fractions of unit-cell dimensions; h, k, and l are the Miller indices; and F(hkl) is known as the structure factor. By Fourier analysis, the electron density distribution is then

$$\rho(x_j, y_j, z_j) = 1/V \sum_{h=-\infty}^{\infty} \sum_{k=-\infty}^{\infty} \sum_{l=-\infty}^{\infty} F(hkl) - (B_j \sin^2\theta / \lambda^2) \exp(-2\pi (hx_j + ky_j + lz_j)) \quad (6.3)$$

where V is the unit cell volume and the sum is over all Miller indices. Equations (6.2) and (6.3) together provide the practical foundation for crystallographic structure analysis. All atoms contribute to each structure factor, and all reflections contribute to each point of electron density.

Crystallization

It seems that the most critical step in a macromolecular structure determination is crystallization of the molecule; it requires the greatest labor and ingenuity to produce a single crystal of a biological macromolecule or complexes. The general approach to crystallization of proteins is to start with a solution of protein and to slowly change the experimental conditions so that saturation is exceeded. These experimental conditions are pH, temperature of the solution, ionic strength, and the addition of ions, organic molecules, detergents, oils, cofactors, and inhibitors to the solution, among many others. The objective is to approach supersaturation slowly so that a few nucleation sites are created which, in time, will grow larger. The goal is to have a few crystals with a reasonable size of approximately 0.3 to 0.4 mm in all three dimensions that diffract the x-ray radiation at high atomic resolution. In the last 10 years, recombinant techniques have provided recombinant proteins with mutations and deletions of part of the protein molecule that make the protein molecule more soluble and easier to crystallize.

Figure 6.2 illustrates one of the popular ways of crystallizing a protein, a procedure known as the "hanging drop" crystallization method. McPherson's book, *Crystallization of Biological Macromolecules* (1999), and *Crystallization of Nucleic Acids and Proteins: A Practical Approach* (1998) by Ducruix

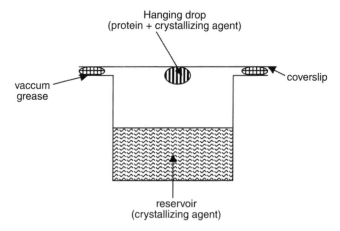

FIGURE 6.2

"Hanging drop" crystallization experiment. The crystallizing agent is placed inside the reservoir. A drop of protein solution is hanging from the coverslip in a closed system seal with vacuum grease. Through vapor diffusion, the two solutions slowly reach equilibrium. Later the hanging drop will be concentrated beyond the protein saturation point, nucleation will occur, and finally we will obtain crystals.

and Geigé are excellent publications dealing with the crystallization of proteins and nucleic acids, respectively. Crystallization methods have advanced little in recent years. Of significance, however, is the range and quality of surfactants used for membrane protein crystallography (Adir, 1999). For example, the structure determination of bacteriorhodopsin (Pebay-Peyroula et al., 1997) with crystals grown from lipidic cubic phases (Rosenbusch et al., 1998) has shown new variations in the way detergents can be used for crystallization.

Data Collection

Crystals of proteins and nucleic acids are quite different from small organic or inorganic crystals, in that they contain a relatively high solvent content and poorly diffract x-ray radiation. Usually, around 50% of the volume of a crystal is solvent, but sometimes this figure is higher. As a result, these crystals are fragile and must be kept in a moist environment (in a capillary) to prevent them from drying out. Recently, the lifetime of biological macromolecular crystals exposed to x-ray radiation has increased by freezing the crystals using liquid nitrogen.

The advantage of having crystals with high solvent content is that the proteins or nucleic acids molecules in the crystal are still in an environment similar to the physiological one. This is good news to biochemists but not to the crystallographer, as the crystals are not well ordered and cause subsequent loss of diffraction and atomic resolution.

The crystals are mounted on a thin-walled glass capillary and set in an automatic diffractometer equipped with a rotating anode and an electronic area detector or a film-like image plate device for collecting diffraction data from biological macromolecule crystals. Recent developments include the use of charge-coupled devices (CCDs) (Westbrook and Naday, 1997). Stable, high-brilliance beams provided by the third-generation synchrotron source at the Advanced Photon Source, Advanced Light Source, Spring, and European Synchrotron Radiation Facility (Wakatsuki et al., 1998) have enabled the use of smaller and weakly diffracting crystals. Microfocus beamlines improve further the radiation intensity available and the size of crystal that can be accommodated (Cusack et al., 1998). These improvements in beam intensity and new detectors have led not only to faster data collection, but also to high-quality data at higher resolution. For example, the collection time of 80 seconds for a lysozyme dataset at 2-Å resolution has been reported at the Advanced Photon Source, and complete 4-wavelength multiwavelength anomalous dispersion (MAD) datasets can be collected routinely in about an hour (Bränden, 1999).

The collected data must be corrected for known systematic errors such as background radiation, decay of the crystal due to overtime exposure to x-rays, and absorption of the x-ray radiation as it passes through the crystal. Reflections that should have identical values in intensities because of symmetry are averaged together to get the best possible estimation of their true value. The agreement of these intensities measurements is called R_{merge} and gives an indication of the accuracy of the data. For biological macromolecules crystals, R_{merge} is usually less than 10%. For data processing, the HKL suite (Otwinowski and Minor, 1996; AGW Leslie, 1996), DPS (Steller et al., 1997), and D*Trek (Kabsch, 1988) are used, among others.

Structure Solution

In order to calculate an electron density map, both the structure factor amplitudes $|F_{hkl}|$ and phase angle ϕ_{hkl} of each reflection are required. The structure factor amplitudes are calculated from the experimental intensities [$|F_{hkl}| = (I_{hkl})^{1/2}$], but the phase angles are lost. In the case of macromolecules, the phase problem is solved either by using a closely related structure as a search fragment (molecular replacement method) or preparing isomorphous heavy-atom derivatives (multiple isomorphous replacement method, MIR). The latter method requires considerable work and skill. A recent alternative approach that is gaining rapidly in popularity is the multiwavelength anomalous dispersion (MAD) method, at the cost of extra time spent in the wet laboratory expressing and crystallizing a protein with its methionine residues replaced by selenomethionine; with synchrotron measurements of the reflections intensities at three or more different x-ray wavelengths, an electron density map can be obtained. Even MAD phasing involves a problem in finding the selenium atoms when many crystallographic independent sites are present; without these atoms to provide the reference phases, no map can be calculated. The software packages for MIR and MAD phasing are CCP4 (Collaborative Computer Project Number 4, suite-programs for protein crystallography), CNS (Brünger et al., 1998), and SOLVE (Terwilliger and Berendzen, 1999), among others. AMoRE (Navaza and Saludjian, 1997) and XPLOR (Brünger, 1992) continue to be the most popular molecular replacement packages.

Refinement and Validation

Once the electron density (Figure 6.3) has been fitted with a structure, it can be used to calculate what the diffraction of the model should look like. The structure factor amplitudes of this calculated diffraction can then be compared to those actually measured. This is expressed as the crystallographic reliability factor (R_f) and is defined as:

$$\mathrm{Rf} = \frac{\sum_{hkl} |Fo| - |Fc|}{\sum_{hkl} |Fo|} \qquad (6.4)$$

where F_o are the observed or measured structure factor amplitudes, and F_c are the ones calculated from the model, summed over all measured reflections. That is, the perfect fit of the calculated structure factors to the observed will give an R_f of zero, whereas a completely random structure will give an R_f of 0.60. Usually, a fairly good starting model built into an MIR electron density map will have an R_f around 0.4 to 0.5, whereas a good starting structure from molecular replacement is 0.35 to 0.45, depending on how similar the search structure is to the unknown structure.

One of the biggest problems of macromolecular crystallographic refinement is the low number of observations per parameter refined. The solution to this problem comes from the fact that the atomic positions are not totally independent; that is, the atoms are bonded together, some groups of atoms are known to be coplanar, and some of them are chiral centers having specific torsion angles, among other stereochemical considerations. This *a priori* stereochemical information can be incorporated as observations and added to the quantity to be minimized. Molecular dynamics can be used to sample a larger section of conformational space, thereby increasing the radius of convergence for refinement. During the refinement procedure, water molecules that are hydrogen bonded to the protein in the crystal lattice will also appear in the electron density map. These water molecules are included as a part of the model and refined along with the protein molecule. The electron density map is visualized on a computer graphic and represents the average structure of all the molecules in the crystal averaged over the data collection time. Both the static and dynamic disorders make a smearing effect on the electron density of an atom. This is taken into account by the thermal vibra-

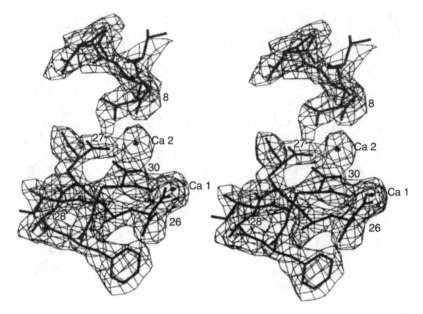

FIGURE 6.3
Stereo view of the 2Fo–Fc map contoured at 1σ showing several amino acid residues and calcium ions within the electron density. (From Soriano-García, M. et al., *Biochemistry*, 31, 2554-2562, 1992. With permission.)

tion factor or isotropic temperature factor, B. Each atom in the crystal will be described by four parameters to be refined: x, y, z, and B.

The information content of a protein crystal structure is determined by the correctness and completeness of the structure refinement procedure. Measures of the accuracy of the structure are provided by validation tools (Wilson et al., 1998) and recently by the program SFCHECK (Vaguine et al., 1999). The program PROCHECK (Laskowski et al., 1993) is still the program used most.

Analysis of Results

The end of an x-ray crystallographic analysis (Figures 6.4 and 6.5) of a particular biological macromolecule has a list of atomic coordinates, occupancies, and isotropic temperature factors that describe the three-dimensional structure of the molecule under investigation. The overall quality of the atomic coordinates is usually measured by four quantities: the resolution, the crystallographic R_f factor, the cross-validation or free R value (R_{free}), and its molecular stereochemistry. In general, the higher the resolution, the better the structure obtained. For example, at resolutions lower than 3.5 Å, it is very difficult to trace the main chain unambiguously; certain structural elements such as α-helices and β-sheets are clear, but loops are sometimes difficult to assign; and aromatic rings of tryptophan (Trp, W), phenylalanine

148 Food Science and Food Biotechnology

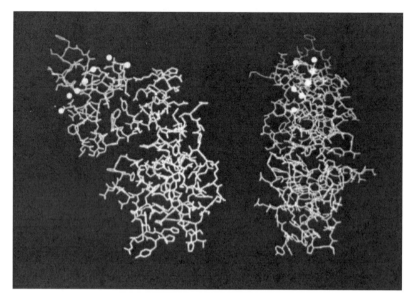

FIGURE 6.4
Complete three-dimensional structure of a protein. (From Soriano-García, M. et al., *Biochemistry,* 31, 2554–2562. With permission.)

(Phe, F), and tyrosine (Tyr, Y) are well placed, but other side chains are difficult to place. Between resolutions of 2.5 and 2.0 Å, the orientations of all side chains become less ambiguous, solvent molecules are found, and individual isotropic temperature factors are refined. At resolutions lower than 2.0 Å, the coordinates and temperature factors become more precise and discrete alternative conformations for side chains may become visible. A well-refined structure has a R_f factor of 0.15 to 0.20. Low-resolution structures (3.5 to 2.8 Å) seldom get below 0.20 due to the lack of high-resolution data that would allow very accurate positioning of the atoms. For crystallographic refinement, the introduction of cross-validation (the free R value) (Brünger, 1992) has significantly reduced the danger of over-fitting the diffraction data. A correct structure will have good stereochemistry as well as both R_f and R_{free} factors. This structure will have the following root mean square deviations from ideal stereochemical restraints: <0.015 Å for bond distances, <3.0 for bond angles, <0.015 Å for planar atoms out of the plane, no incorrect chiral centers, and close van der Waals contacts.

Impact on Biotechnology

Very exciting and often surprising new structures of increasingly complex molecules and assemblages continue to appear rapidly (Hendrickson,

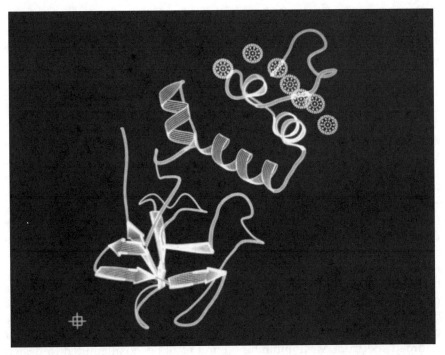

FIGURE 6.5
Schematic representation of the secondary structure elements of a protein. The α-helices, β-sheets, and random secondary structures are shown as spirals, arrows, and thick lines, respectively. (From Soriano-García, M. et al., *Biochemistry*, 31, 2554–2562. With permission.)

1995). Any description of the tremendous achievements of x-ray structural studies in biology has to include the impact of this fundamental research on the fields of medicine and biotechnology. As a result of this, nearly all pharmaceutical and biotechnology companies employ crystallographers in structure-based drug design efforts and use x-ray structures in the engineering of macromolecules with designed properties. From a recent survey of the pharmaceutical/biotechnology sector, it was estimated that 100 drugs derived from biotechnology have been approved so far (Annual report, 1999). These represent the investment of at least 10 or more years of drug discovery and development work. The movement to use these products form a better understanding of why drugs have failed in the past. Many of the early protein drugs derived from biotechnology failed because they were primary molecules with suboptimal affinity or poor half-lives *in vivo*, leading to poor efficacy. In other cases, many of the original protein drug molecules were non-human and caused immune responses against the drug itself. Affinity, half-life, and dosing regime are all interrelated and play a part in determining the clinical efficacy and financial viability of protein-based drugs. Our understanding of the issues that affect success in drug development has been paralleled by the increased capabilities in protein engineering and selection/screening technologies. These have been used to improve the effectiveness of a number of protein drug candidates (McCafferty and Glover, 2000).

The need for structural information about drug–protein complexes is now fundamental for drug design. The two approaches to structure-based drug design of docking of known compounds into a target protein and molecular assembly *in situ* are seen to be merging technologies (Gane and Dean, 2000). A small fraction of papers have been selected here as excellent examples of the quality and pace of work that has occurred in the past year (Gane and Dean, 2000). Matthews et al. (1999) have developed a strategy of structure-led drug discovery for inhibitors of rhinovirus 3C protease. Guided by the crystal structures of protein–peptide complexes, they studied each binding pocket in turn and modified their inhibitors to maximize binding, which they confirmed by activity studies and by solving the structures of many of their protein inhibitor co-crystals. A potent antiviral is now in clinical trials. A more traditional approach dependent on chemistry and the structure–activity relationship (SAR) was used in the development of a matrix metalloproteinase (MMP) inhibitors by Almstead et al. (1999). A series of thiazines and thiazepines was synthesized and proved to be active in many of the MMPs. SARs and crystal data correlated well and resulted in a number of broad spectrum MMP inhibitors. Furet et al. (1999) attempted to design a peptide mimic with agreeable pharmacokinetic properties that is capable of antagonizing the Grb Src homology (SH)2 domain for possible use as an antitumor agent. The x-ray crystal structure of the protein with a bound peptide provided the vectors of essential hydrogen bonds that they maintained in their mimic. Subsequent co-crystallization confirmed their molecular design of a high-affinity ligand.

Mihelich and Schevitz (1999) describe how structure determination revealed an alternative ligand-binding mode, resulting in the modification and final synthesis of a phospholipase A_2 (PLA_2) inhibitor that is currently in clinical assessment as an antiinflammatory. NAD+ is a necessary cofactor of the enzyme glyceraldehyde-3-phosphate dehydrogenase; by exploiting the small differences between its binding pockets in *Trypanosoma brucei* and *Leishmania mexicana* and those in humans, Aranov et al. (1999) synthesized a number of adenosine analogs. Activity studies, co-crystallization with the enzyme, and docking resulted in more than 50-fold selective inhibition of the enzyme by a number of these analogs. Recently, Moult and Melamud (2000) briefly reviewed methods for identifying protein function by analysis of the structure, which has direct relevance to rational drug design and to biotechnology. Nienaber et al. (2000) exploited molecular biological and biophysical techniques to provide very high-resolution data about urokinase, revealing an additional binding site for possible drug design.

The hydrocarbon-dihydroxylating dioxygenases have been cloned in an extremely radiation-resistant bacterium, *Deinococcus radiodurans*, and studied for 20 years. Interest is increasing in how to exploit the enzyme-generated chiral centers for use in synthesizing complex chiral molecules (Hudlicky et al., 1999). Only recently, however, has a model of the active site been available based on the x-ray structure derived for naphthalene dioxygenase (Kauppi et al., 1998). This should ignite interest in protein and metabolic engineering using this broad class of dioxygenases.

Plants are potentially marvelous expression systems for the production of desirable gene products, such as natural plant products and foreign proteins. Our laboratory is currently working on two projects dealing in the field of plant biotechnology. The first project deals with biochemical and structural studies of seed storage proteins of *Amaranthus hypochondriacus*. Within the great genetic diversity existing in Mesoamerica, which is the center of origin and dispersion of numerous species, amaranths occupy a leading position. They constituted one of the essential plants in the basic diet of pre-Hispanic Mesoamerican civilizations and were an essential part of Aztec tributes. Amaranth plant can be used both as a cereal and as a vegetable. The primary interest is in the seeds, which contain carbohydrates and 12 to 16% proteins with a high lysine content. We isolated, purified, and crystallized a 36-kDa amaranth globulin from *Amaranthus hypochondriacus*, and its molecular structure determination is now underway (Vasco-Méndez et al., 1999).

The second project deals with the biochemical characterization and structural study of cysteine proteases isolated from Mexican plants. Cysteine proteases are another class of proteolytic enzymes that use nucleophilic catalysis in hydrolyzing peptide bonds, with the use of a cysteine residue. For example, papain is a cysteine protease isolated from green papaya fruit which is used in meat tenderizers and in laundry detergents. We chose the papain family of cysteine proteases as the focus of our drug design efforts, because these proteases play a key role in the life cycle of parasites and in the pathogenesis of diseases caused by these parasites. Two new cysteine proteases isolated from *Pileus mexicanus* have been biochemically characterized and we are now in the process of crystallization of these enzymes (Oliver-Salvador et al., 1998).

Acknowledgments

I wish to thank Prof. Ignacio Fita, Instituto de Biología Molecular de Barcelona (IBMB), CSIC of Spain, for inviting me to work in his laboratory. And, I also wish to thank IBERDROLA, S.A., for their financial support. This chapter was written during my stay in Barcelona, Spain.

References

Adir, N. (1999) Crystallization of the oxygen-evolving reaction centre of photosystem II in nine different detergent mixtures, *Acta Crystallogr.*, D55, 891–894.

AGW Leslie (1996) *MOSFLN Program for Integration of Molecular Diffraction Data*, user's guide at ftp://ftp.mrc-lmb.cam.ac.uk/pub/mosfln/ver450/.

Almstead, N.G., Bradley, R.S., Pikul, S. de B., Natchus, M.G., Taiwo, Y.O., Gu, F., Williams, L.E., Hynd, B.A., Januez, M.J., Dunaway, M., and Mieling, G.E. (1999) Design, synthesis, and biological evaluation of potent thiazine- and thiazepine-based matrix metalloproteinase inhibitors, *J. Med. Chem.*, 42, 4547–4562.

Annual report (1999) The world's biotechnology companies, *Pharma Business*, 30–52.

Aronov, A.M., Suresh, S., Buckner, F.S., Van Voorhis, W.C., Verlinde, C.L.M.J., Opperdoes, F.R., Hol, W.G.J., and Gelb, M.H. (1999) Structure-based design of sub-micromolar, biologically active inhibitors of trypanosomatid glyceraldehyde-3-phosphate dehydrogenase, *Proc. Natl. Acad. Sci. USA*, 96, 4273–4278.

Blundell, T.L. (1994) Protein engineering: towards rational design, *Trends Biotechnol.*, 12, 145–148.

Blundell, T.L. and Johnson, L.N. (1976) *Protein Crystallography*, Academic Press, San Diego, CA.

Bränden, C.-I. (1999) Fifth European Workshop on Crystallography of Biological Macromolecules, Carl-Ivar Bränden, Ed., May 16–20, 1999, Como, Italy.

Brünger, A.T. (1992) *XPLOR (Version 3.1): A System for X-Ray Crystallography and NMR*, Yale University Press, New Haven, CT.

Brünger, A.T., Adams, P.D., Clore, G.M., DeLano, W.L., Gros, P., Grosse, Kunstleve, R.W., Jiang, J.S., Kuszewski, J., Nilges, M. Pannu, N.S., Read, R.J., Rice, M., Simonson, T., and Warran, G.L. (1998) Crystallographic and NMR system: a new software suite for macromolecular structure determination, *Acta Crystallogr.*, D54, 905–921.

CCP4 (1994) Collaborative Computer Project Number 4 — The CCP4 suite: programs for protein crystallography, *Acta Crystallogr.*, D50, 760–763.

Cusack, S., Belrhali, H., Bram, A., Burghammer, M., Perrakis, A., and Riekel, C. (1998) Small is beautiful: protein micro-crystallography, *Nat. Struc. Biol.*, 5, 634–637.

Drenth, J. (1999) *Principles of Protein X-Ray Crystallography*, 2nd ed., Springer-Verlag, New York.

Ducruix, A. and Geigé, R. (1998) *Crystallization of Nucleic Acids and Proteins: A Practical Approach*, Oxford University Press, London. (1999).

Furet, P., García-Echeverría, C., Gay, B., Schoepfer, J., Zeller, M., and Rahuel, J. (1999) Structure-based design, synthesis, and x-ray crystallography of a high-affinity antagonist of the Grb2-SH2 domain containing an asparagine mimetic, *J. Med. Chem.*, 42, 2358–2363.

Gane, P.J. and Dean P.M. (2000) Recent advances in structure-based rational drug design, *Curr. Opin. Struc. Biol.*, 10, 401–404.

Giacovazzo, C., Monaco, H.L., Viterbo, D., Scordari, F., Gilli, G., Zanotti, G., and Catti, M., (1992) *Fundamentals of Crystallography*, C. Giacovazzo, Ed., International Union of Crystallography (IUCr), Oxford Science Publishing, London.

Hendrickson, W.A. (1995) X-rays in molecular biophysics, *Physics Today*, 48(11) 42–48.

Hudlicky, T., Gonzalez, D., and Gibson, D.T. (1999) Enzymatic dihydroxylation of aromatics in enantioselective synthesis: expanding asymmetric methodology, *Aldrichimica Acta*, 32, 35–62.

Kabsch, W. (1988) Evaluation of single-crystal x-ray diffraction data from a position-sensitive detector, *J. Appl. Crystallog.*, 21, 916–924.

Kauppi, B., Lee, K., Carredano, E., Parales, R.E., Gibson, D.T., Eklund, H., and Ramaswamy, S. (1998) Structure of an aromatic-ring-hydroxylating dioxygenase-naphthalene 1,2-dioxygenase, *Structure Folding Design*, 6, 571–586.

Laskowski, R.A., MacArthur, M.W., Moss, D.S., and Thornton, J.M. (1993) PROCHECK: a program to check the stereochemical quality of protein structures, *J. Appl. Cryst.*, 26, 283–291.

Maldonado, E., Moreno, A., Panneerselvam, K., Ostoa-Saloma, P., Garza-Ramos, G., Soriano-García, M., Pérez-Montfort, R., Tuena de Gómez-Puyou, M., and Gómez-Puyou, A., (1997) Crystallization and preliminary x-ray analysis of triosephosphate isomerase from trypanosoma cruzi, *Protein & Peotide Lett.*, 4, 139-144.

Matthews, D.A., Dragovich, P.S., Webber, S.E., Fuhrman, S.A., Patrick, A.K., Zalman, L.S., Hendrickson, T.F., Love, R.A., Prins, T.J., Marakovits, J.T., Zhou, R., Tikhe, J., Ford, C.E., Meador, J.W., Ferre, R.A., Brown, E.L., Binford, S.L., Brothers, M.A., Delisle, D.M., and Worland, S.T. (1999) Structure-assisted design of mechanism-based irreversible inhibitors of human rhinovirus 3C protease with potent antiviral activity against multiple rhinovirus serotypes, *Proc. Natl. Acad. Sci. USA*, 96, 11000–11007.

McCafferty, J. and Glover, D.R. (2000) Engineering therapeutic proteins, *Curr. Opin. Struc. Biol.*, 10, 417–420.

McPherson, A. (1999) *Crystallization of Biological Macromolecules*, 2nd ed., Cold Spring Harbor Laboratory Press, Cold Spring Harbor, New York.

Mihelich, E.D. and Schevitz, R.W. (1999) Structure-based design of a new class of anti-inflammatory drugs: secretory phospholipase A(2) inhibitors, SPI, *Biochem. Biophys. Acta Mol. Cell. Biol. Lipids*, 1441, 223–228.

Moult, J. and Melamud, E. (2000) From fold to function, *Curr. Opin. Struc. Biol.*, 10, 384–389.

Navaza, J. and Saludjian, P. (1997) AMoRE: an automated molecular replacement program package, *Methods Enzymol.*, 276, 581–594.

Nienaber, V., Wang, J., Davidson, D., and Henkin, J. (2000) Re-engineering of human urokinase provides a system for structure-based drug design at high resolution and reveals a novel structural subsite, *J. Biol. Chem.*, 275, 7239–7248.

Oliver-Salvador, M.C., Laursen, R., Moreno, A., and Soriano-García, M. (1998) Isolation and characterization of two cysteine proteinases from the *Pileus mexicanus* Latex, *Protein Sci.*, 7(suppl. 1), 165.

Otwinowski, Z. and Minor, W. (1996) Processing of x-ray diffraction data collected in oscillation mode, *Methods Enzymol.*, 276, 307–326.

Pebay-Peyroula, E., Rummel, G., Rosenbusch, J.P., and Landau, E.M. (1997) X-ray structure of bacteriorhodopsin at 2.5 Å from microcrystals grown in lipidic phases, *Science*, 277, 1676–1681.

Rosenbusch, J.P. (1998) The structure of membrane proteins: Fact and fancy, *J. Mol. Graphics Modelling*, 16, 287-288.

Rummel, G.J., Hardmeyer, A., Widmer, C., Chiu, M.L., Nollert, P., Locher, K.P., Pe-druzzi, I., Landau, E.M., and Rosenbusch, J.P. (1998) Lipidic cubic phases: new matrices for the three-dimensional crystallization of membrane proteins, *Struct. Biol.*, 121, 82–91.

Soriano-García, M., Padmanabham, K., de Vos, A.M., and Tulinsky, A. (1992) *Biochemistry*, 31, 2554-2562.

Steller, I., Bolotovsky, R., and Rossmann M.G. (1997) An algorithm for automatic indexing of oscillation images using Fourier analysis, *J. Appl. Crystallogr.*, 30, 1036–1040.

Terwilliger, T.C. and Berendzen, J. (1999) Automated MAD and MIR structure solution, *Acta Crystallogr.*, D55, 849–861.

Vaguine, A.A., Richelle, J., and Wodak, S.J. (1999) SFCHECK: a unified set of procedures for evaluating the quality of macromolecular structure–factor data and their agreement with the atomic model, *Acta Crystallogr.*, D55, 191–205.

Vasco-Méndez, N.L., Soriano-García, M., Moreno, A., Castellanos-Molina, R., and Paredes-López, O. (1999) Purification, crystallization, and preliminary x-ray characterization of a 36 kDa amaranth globulin, *J. Agric. Food Chem.*, 47, 862–866.

Wakatsuki, S., Belrhali, H., Mitchell, E.P., Burmeister, W.P., McSweeney, S.M., Kahn, R., Bourgeois, Y., Yao, M., Tomizaki, T., and Theveneau, P. (1998) ID14 "Quadriga," a beamline for protein crystallography at the ESRF, *J. Synchrotron Radiation*, 5, 215–221.

Westbrook, E.M. and Naday, I. (1997) Charged-coupled device-based area detectors, *Methods Enzymol.*, 276, 244–268.

Wilson, K.S., Butterworth, S., Dauter, Z., Lamzin, V.S., Walsh, M., Wodak, S., Pontius, J., Richelle, J., Vaguine, A., Sander, C., Hooft, R.W.W., Vriend, G., Thornton, J.M., Laskowski, R.A., MacArthur, M.W., Dodson, E.J., Murshudov, G., Oldfield, T.J., Kaptein, R., and Rullmann, J.A.C. (1998) Who checks the checkers? Four validation tools applied to eight atomic resolution structures, *J. Mol. Biol.*, 276, 417–436.

7

Stability of Dry Enzymes

Mauricio R. Terebiznik, Viviana Taragano, Vanessa Zylberman, and
Ana M.R. Pilosof

CONTENTS

Introduction

The last several years have witnessed considerable increase in the use of
enzymes as industrial catalysts, for clinical diagnostics, and for pharmaceutical purposes, as well as in a variety of analytical applications and in molecular biology as DNA restriction and modifying enzymes. The problem of
long-term stability of enzymes, however, places an important limitation on
their use. Enzymes are obtained from vegetables and animals but mainly
from microorganisms as a result of fermentation processes (Poutanen, 1997).
Because most enzymes are not stable in solution, spray-drying or freeze-
drying is often used to improve stability during storage (Pilosof and Terebiznik, 2000). Although freeze-drying is usually favored for drying highly
valued and thermally labile enzymes in small quantities, spray-drying is
largely applied in large-scale production of enzymes because of its much
lower cost. Some enzymes, however, are known to be partially or totally

1-56676-892-6/03/$0.00+$1.50
© 2003 by CRC Press LLC

inactivated during these processes or are still unstable at ambient temperature. Thus, drying enzymes in the presence of carbohydrates or other additives is a common method to protect proteins from the stresses of drying
and storage inactivation (Aguilera et al., 1997; Suzuki et al., 1997; Carpenter
and Crowe, 1998). Generally, the stability of dry enzymes is a function of
the composition of the initial liquid to be dehydrated, storage conditions,
and physicochemical characteristics of the enzyme, so the stability of dry
enzyme products should be considered on an individual basis. This chapter
discusses some of the factors that influence the stability of several dry
enzymes obtained by microbial fermentation: the glass–rubbery transition,
addition of protective carbohydrates, and the role of components of the
enzymatic fermentation extracts.

The Influence of the Glass–Rubbery Transition on Stability of Dry Pectinlyase

Pectinases are enzymes increasingly used in the juice, wine, and oil industries
(Sakai et al., 1993; Dominguez et al., 1994). Commercially available pectinases are a mixture of pectinesterases (PEs), polygalacturonases (PGs), pect-
inlyases (PLs), and hemicellulases. To degrade a pectin molecule completely,
PE and PG must act together to liberate methanol as a byproduct of PE
action. Pectinlyase, however, is able to cleave pectin by itself without the
action of the two other enzymes (Sakai et al., 1993). Commercial preparations
rich in PL and poor in PE are desirable because they avoid the production
of methanol, precipitation of partially de-esterified pectin with endogenous
calcium, and damage to the volatile ester content responsible for the specific
aroma of various fruits (Delgado et al., 1992).

   Taragano and Pilosof (1999) found fermentation conditions for a local isolated strain of *Aspergillus niger* that yielded a pectinolytic extract extremely
rich in PL (65% of the total proteins in the fermentation extract). The enzymatic extract was freeze-dried to 8.1% moisture (dry basis), after having been
filtered stepwise by Millipore membranes of 0.45 μm and 0.22 μm and ultrafiltered through a cellulose regenerated membrane (pore size: 10,000 Da).

   The DSC thermogram in Figure 7.1 shows that the glass transition temperature (Tg) of the PL preparation was 45°C, and the peak denaturation
temperature (Tp) was 116°C. The arrows indicate the temperatures selected
for the study of the kinetics of PL activity loss. Two temperatures were
selected in the glassy state (5 and 28°C), the glass transition temperature was
45°C, one temperature was selected between Tg and the onset of denaturation (58°C), the onset temperature for denaturation was 75°C, and two
temperatures were selected within the denaturation endotherm (100 and
130°C). In the glassy state (T ≤ 45°C), a lag period of approximately 7 days
with no PL activity loss was observed. After that, PL activity was lost

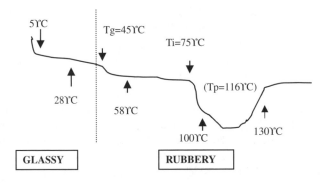

FIGURE 7.1

DSC thermogram for a pectinlyase (PL) preparation at 8.1% moisture (scan rate = 10°C/min). Arrows show the temperatures selected for PL inactivation kinetics studies.

according to a first-order kinetic model (Taragano and Pilosof, 2001). Above Tg, no lag time was observed.

Even under conditions in which the system was in the glassy state (5 or 28°C), a decay in the activity with time was observed (Table 7.1). At 5°C (40°C below Tg), the half-life time was 35 days. Besides the occurrence of PL inactivation in the glassy state, the glass transition had a great impact on rates of PL activity loss. Table 7.1 shows that above Tg the half-lives for PL inactivation greatly decreased (i.e., the time scale).

Denaturation kinetics of proteins in the solid PL preparation were determined by differential scanning calorimetry. The dynamic method of Osawa (1970) based on the increase of the peak denaturation temperature (Tp) as a function of scan rate (β) was used to obtain the activation energy (Ea) of the reaction and the pre-exponential factor of the Arrhenius equation (z).

By plotting $-\ln(\beta/Tp^2)$ vs. $1/Tp$, a straight line should be obtained with a slope proportional to Ea and the x-axis intercept to the pre-exponential factor (z) according to the equation:

$$-\ln(\beta/Tp^2) = \ln(z.R/Ea) - Ea/R.Tp \qquad (7.1)$$

where R is a gas constant (1.98 cal/K.mol).

The application of Eq. (7.1) allows calculation of Ea = 485.3 ± 50.6 kJ/mol and z = (2.72 ± 0.37) × 10^{65} min^{-1}. De Cordt et al. (1994) estimated the Ea for denaturation of solid α-amylase equilibrated with different polyols to be

TABLE 7.1

Half-Lives ($t_{1/2}$) for Dry Pectinlyase Inactivation at Various Temperatures

T	5°C	28°C	45°C	58°C	75°C	100°C	130°C
(T – Tg)	–40	–17	0	13	30	55	85
$t_{1/2}$	35 days	12.9 days	8.4 days	6.9 h	41.3 min	4.5 min	2.6 min

between 313 and 522 kJ/mol. Vrábel et al. (1997) obtained an Ea of 502 kJ/mol for dry yeast invertase, and Bell et al. (1995) determined an Ea of 238 kJ/mol for lyophilized bovine somatotrophin.

The great difference between the activation energies for solid PL denaturation and inactivation above Tg (80.3 kJ/mol) (Taragano and Pilosof, 2001b) indicates that denaturation is not the mechanism responsible for the loss of PL activity. The low activation energy for the inactivation of rubbery PL primarily reflects undergoing reactions (not involving denaturation) causing the loss of enzymatic activity. In fact, activation energies for chemical reactions are several times lower (50 to 100 kJ/mol) than those involving conformational changes (200 to 420 kJ/mol) (Mozhaev, 1993). Recently, the true activation energy of the irreversible step of thermal denaturation of cutinase in a liquid system was determined (Baptista et al., 2000). The corresponding Ea of 62.7 kJ/mol is consistent with the low Ea values obtained for the PL preparation.

Some reports indicate chemical inactivation of enzymes without preliminary unfolding (Manning, et al., 1989; Mozhaev, 1993), which is opposite of the usual behavior of proteins where they are more likely to undergo a chemical reaction when unfolded.

The comparison of rates of protein denaturation and inactivation in solid rubbery PL (Taragano and Pilosof, 2001b) showed that:

- At the onset of denaturation (75°C), denaturation was insignificant ($t_{1/2}$: 4927 days), but PL lost 50% of its activity in 41 minutes.
- At 100°C, the rate of denaturation was approximately 2 orders lower than the rate of inactivation; in 4.5 min, 50% of PL activity was lost but only 3% of proteins were denatured.
- At 130°C, the protein denaturation rate of the solid PL preparation was 4 orders higher than the inactivation rate, reflecting the higher temperature dependence of protein denaturation, and the Arrhenius plots crossover at 105°C (Figure 7.2).

Thus, when raising the temperature above 105°C, the conformational change of the protein would be the dominant reaction. Below that temperature, the dominant reaction would be an irreversible inactivation reaction involving the active site of the enzyme, without denaturation. This inactivation mechanism also occurs below Tg, but at a lower rate.

Different inactivation mechanisms could be responsible for PL activity loss. Some authors (Mozhaev, 1993; Costantino et al., 1994) have proposed that some reactions such as hydrophobic aggregation, deamidation, cross-linking, racemization, and β-elimination, could easily occur in low-moisture systems because of the proximity of the reactants. In particular, it has been shown that due to the relatively high effective protein concentration of proteins in the solid state, intermolecular processes such as aggregation are more prevalent than intramolecular processes (Bell et al., 1995; Duddu and

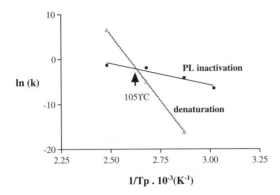

FIGURE 7.2
Superimposed Arrhenius plots for inactivation and denaturation of solid PL above the glass transition temperature (Tg).

Dal Monte, 1997). However hydrophobic or covalent aggregation through disulfide bonds was not observed in solid PL (polyacrylamide gel electrophoresis, or PAGE). Nevertheless, other chemical reactions involving the active site of enzyme, such as oxidation, might take place.

The results suggest that reactions causing the loss of PL activity do not involve denaturation. Strategies for preventing inactivation of PL in the solid state should focus on the arrest of the irreversible inactivation reaction involving the active site of the enzyme. The glassy state does not guarantee the stability of dry pectinlyase; however, the glass transition of the solid enzyme accelerates the inactivation reaction.

Carbohydrate-Induced Protection of Dry Enzymes

Addition of stabilizing solutes to labile enzymes is a common way of protecting them during preparation, drying, and storage. A wide variety of solutes, including sugars, polyols, aminoacids, methylamines, and salts, are effective when minimizing thermal denaturation in aqueous systems (De Cordt et al., 1994; Lozano et al., 1994; Salahas, et al. 1997).

A single mechanism for the solute-induced protection of proteins in aqueous systems against the denaturing effect of heat as well as in the frozen state has been addressed by Timasheff and co-workers (Lee and Timasheff, 1981; Arakawa and Timasheff, 1982, 1985; Timasheff, 1982; Timasheff and Arakawa, 1988). Denaturation of proteins in the presence of stabilizing solutes is thermodynamically less favorable than in water because these solutes are preferentially excluded from contact with the surface of the proteins. Thermodynamically, the positive free-energy change and the magnitude of this unfavorable free-energy shift are proportional to the surface area of the protein. Consequently, the native structures of monomers or the structure of oligomeric proteins are stabilized because denaturation or dissociation would lead to a greater contact surface between the protein and the solvent and therefore augment this thermodynamically unfavorable effect (Carpenter and Crowe, 1988).

In contrast to the behavior shown in solutions, only certain solutes (i.e., disaccharides) can preserve enzyme activity during air-drying and subsequent storage (Carpenter and Crowe, 1988). During the final stages of air-drying, the major constraint that must be overcome is removal of the hydration shell of the enzyme, which, for at least some labile enzymes, can result in irreversible inactivation upon rehydration (Carpenter and Crowe, 1989). The mechanism of this level of protection is different from that occurring in solution (Crowe et al., 1998). It has been suggested that sugars can continue to protect the dried protein by hydrogen bonding to the protein at some critical point during dehydration (Carpenter and Crowe, 1988; Carpenter et al., 1990), thus serving as water substitutes when the hydration shell of the enzyme is removed. This is generally referred to as the *water replacement hypothesis.*

The processes of freezing and drying should be considered as separate stresses during freeze-drying, which often call for multiple stabilizing excipients in a single formulation (Carpenter et al., 1997). Excipients that have a stabilizing effect during freezing (i.e., cryoprotectans) are generally solutes that tend to be excluded from the surface of the protein which has the thermodynamic consequence of stabilizing the native conformation of the protein (Timasheff, 1982; Timasheff and Arakawa, 1988). The water replacement hypothesis is the primary hypothesis advanced to rationalize the role of stabilizers during the drying phase of lyophilization.

Disaccharides such as trehalose, maltose, or sucrose protect extremely fragile biomolecules such as DNA restriction and modifying enzymes during drying (Colaço et al. 1992; Uritani et al., 1995; Rossi et al., 1997). Interestingly, trehalose did not show any improved stabilization of these enzymes in solution when compared to that observed with glycerol, supporting the concept that the mechanism by which solutes stabilize the enzyme structure in solution is different from that occurring during the terminal stages of drying (Carpenter et al., 1990).

The *vitrification hypothesis* states that a good stabilizer is an additive that readily forms a glass with a high glass transition temperature and has been proposed to explain the protective role of sugars on dry enzymes. The rationale for this hypothesis is based upon the fact that "glassy" means extremely high viscosity. It is assumed that all molecular motions relevant to instability are correlated with the reciprocal of viscosity (Slade and Levine, 1991); however, the stabilizing effect of sugars seems to be related more to the amorphous state of the matrix, rather than to their glassy or rubbery nature. Several studies support the conclusion that the importance of the amorphous properties of protein and additive is the allowance for effective hydrogen bonding between these two components. A glassy additive that does not have the interaction will not protect the protein against dehydration damage (Crowe et al., 1998).

The formation of an amorphous phase with the protein, mechanically immobilized in the glassy solid carbohydrate matrix, stabilizes protein structure during dehydration and subsequent storage. The restriction of translational and relaxation processes is thought to prevent protein unfolding, and spatial separation between protein molecules (i.e., dilution of protein molecules within the glassy matrix) is proposed to prevent aggregation (Crowe et al., 1998).

It has been demonstrated that vitrification (glass formation) of a pectin-lyase preparation does not assure enzyme stability, and similar results have been observed for dry enzymes in carbohydrate matrices. Significant invertase, lactase, and restriction enzyme *Eco*RI inactivation occurred in sugar, maltodextrins, or polyvinylpyrrolidone (PVP) systems kept well below their glass-transition temperatures (Schebor et al., 1996; Mazzobre et al., 1997; Rossi et al., 1997). More recently, it was confirmed that vitrification was not necessary for protection of the enzyme *Eco*RI (Buera et al., 1999); in fact, the water replacement hypothesis seems to be more consistent with experimental data.

The glassy state of the matrix was not a sufficient condition to ensure fungal α-amylase stability. In fact, while glassy trehalose and lactose matrices (15 and 30°C below Tg) allowed the retention of approximately 80% of α-amylase activity (after 96 h at 70°C), remaining activity in raffinose or PVP matrices (30 and 60°C below Tg) was 50% or less (Figure 7.3). At zero water content, where no plasticization of the matrices can take place, the protective effect of the matrices in the glassy state was inversely related to their molecular weight. As molecular weight decreases, molecules can be more densely packed when concentrated in a glassy state, with smaller holes. Thus, free volume effects, arising mainly from differences in molecular packing of matrices, could play an important role in determining thermal stability of enzymes (Terebiznik et al., 1998a).

Figure 7.4 shows that at a constant glassy condition (T – Tg = –20°C), which reflects different water contents of the matrices, the same relationship between molecular weight of matrices and α-amylase protection was found. Matrices that allowed higher enzyme stability had a lower combined water content. The plasticization of the enzyme by water contained in the glassy matrix could be another factor that determines the degree of protection provided by the matrix (Terebiznik et al., 1998a).

In addition to the possible effect on enzyme stabilization of the physical characteristics of the matrices discussed above (free volume and plasticization), the interaction between matrices and enzyme, which could result in specific protective effects, must also be considered. The enzymes α-glucosidase and α-amylase present in the same dry formulation were protected at different degrees by trehalose (Figures 7.5 and 7.6). The protective effect of trehalose was observed to be related to the amount of sugar added to the enzyme fermentation extract. The remaining activity of α-amylase was greatly enhanced when trehalose was added up to 10% w/v. Beyond this amount of trehalose, no further effect on enzyme stability was observed.

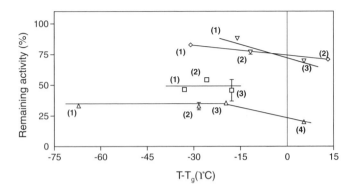

FIGURE 7.3
Remaining activity of α-amylase freeze-dried in lactose (◊), PVP (Δ), raffinose (o), and trehalose (∇) matrices after 96 h of heat treatment at 70°C as a function of (T − Tg) of the matrices. The numbers in parentheses indicate the RH% of the matrices: (1), 0; (2), 11; (3), 20; (4), 42.

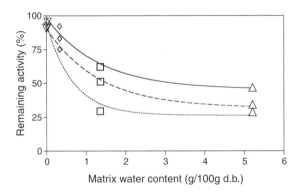

FIGURE 7.4
Remaining activity of α-amylase freeze-dried in lactose (◊), PVP (Δ), raffinose (□), and trehalose (∇) matrices after 48 h (——), 96 h (– – –), and 360 h (------) of heat treatment at 70°C as a function of the water content of the matrices corresponding to a Tg of 90°C (T − Tg = −20°C).

Only in matrices at 0% RH could α-glucosidase be effectively stabilized by 10% w/v of trehalose. For 11% RH, a 20% w/v trehalose was necessary to give around 50% α-glucosidase remaining activity. A low protective effect of α-glucosidase was observed for all the trehalose concentrations tested when matrices were equilibrated above 22% RH.

Crystal formation of additive during storage may be an important factor affecting loss of enzyme activity. Crystallization of amorphous solids is a consequence of holding the system above its Tg and is time-dependent phenomena (Roos and Karel, 1991). Once amorphous components (e.g., carbohydrates, polyethylene glycol) crystallize, they essentially form a separated phase (Karmas, 1994), and the protective effect on the enzyme may be

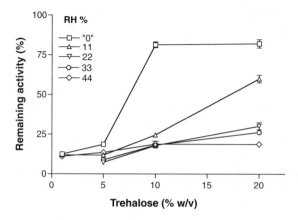

FIGURE 7.5
Remaining activity of α-glucosidase in freeze-dried crude enzymatic fermentation extracts as affected by RH and trehalose concentration. 100% α-glucosidase activity corresponds to 1.5 IU/mL.

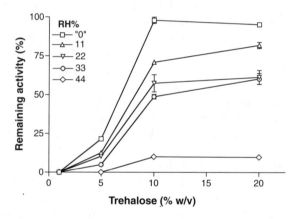

FIGURE 7.6
Remaining activity of α-amylase in freeze-dried crude enzymatic fermentation extracts as affected by RH and trehalose concentration. 100% α-amylase activity corresponds to 270 IU/mL.

lost because crystalline compounds are no longer associated with the enzyme and reactions leading to inactivation are free to occur.

The protective effect of trehalose and sucrose on lactase and enzyme *Eco*RI inactivation during storage has been shown to be related to the extent of sugar crystallization (Mazzobre et al., 1997; Rossi et al., 1997). The kinetics of glucose-6-phosphate dehydrogenase inactivation in two amorphous carbohydrate systems with similar Tg values has been examined (Sun and Davidson, 1998). The superior stability of the glucose/trehalose system over the glucose/sucrose one was associated with a number of closely related dynamic properties of the glucose/trehalose glass, including the ability to resist phase separation and crystallization during storage.

Crystallization of amorphous sugars occurs above Tg and below Tcr (crystallization temperature) as a function of time (Roos and Karel, 1991). The reported Tcr for trehalose is 102°C (Cardona et al., 1997), and instantaneous crystallization will occur at (Tcr − Tg) = 57°C. A trehalose/amylase dry preparation equilibrated at 22% relative humidity (RH), had (T − Tg) values between 35 and 55°C for heating temperatures between 80 and 100°C. Thus, approaching 100°C, crystallization would occur instantaneously. Figure 7.7 shows that above (T − Tg) = 35°C the α-amylase deactivation rate was greatly enhanced according to increasing trehalose crystallization rates.

The effect of crystallization on water balance in a system containing crystalline and amorphous components depends on the type of crystals. Formation of anhydrous crystals (i.e., sucrose and lactose) releases water but during formation of hydrated crystals water is absorbed (Karmas, 1994). In the last case, partial crystallization of the matrix could remove water from the amorphous phase, thus increasing its glass transition temperature and leading to a delay in enzyme inactivation. A first phase of rapid loss of α-amylase activity until 50 to 60% was observed in crystallizing trehalose matrices (40 to 45°C above Tg). It was followed by a second phase in which activity changed little over time or reached a defined plateau. This behavior was attributed to dihydrate crystal formation (Terebiznik et al., 1997).

Trehalose has been described to act as the best stabilizer of several enzymes (Colaço et al., 1992; Schebor et al., 1996; Cardona et al., 1997; Suzuki et al., 1997; Terebiznik et al., 1998a). The extraordinary effect of trehalose in dried systems has been attributed to several of its properties such as the glassy forming properties, the ability to make hydrogen bonds, reduce large-scale fluctuations of protein specific motions, and give a dense molecular packaging, the high size exclusion effect, the direct interaction on the molecular structure of proteins, and the water sequestering action during dihydrate crystal formation in rubbery systems (Carpenter and Crowe, 1988, 1989; Colaço et al., 1994; Donnamaría et al., 1994; Rossi et al., 1997; Terebiznik et al., 1997, 1998a; Cordone et al., 1998; Sola-Penna and Meyer-Fernandes, 1998). Despite the proposed stabilization mechanisms shared by trehalose, the question of why trehalose is more effective than other sugars is still not fully answered.

The trisaccharide raffinose offers good protection of enzyme activity, and its role as a novel excipient matrix for labile enzyme stabilization deserves further investigation. Raffinose protected enzymes lactase, invertase, and restriction enzymes (Cardona et al., 1997; Mazzobre et al., 1997; Rossi et al., 1997) at the same level as trehalose in various conditions of temperature and water content. However, it did not protect α-amylase at the same level as trehalose or lactose (Figure 7.3).

The non-reducing sugar/alcohol lactitol has been reported to be a good protector for freeze-dried enzymes (Gibson et al., 1992). Table 7.2 shows that lactitol performed better than trehalose in protecting a dry pectinlyase preparation (Taragano and Pilosof, 1999) during storage at 58°C, and this system was rubbery. At 130°C, both trehalose and lactitol crystallized, leading to

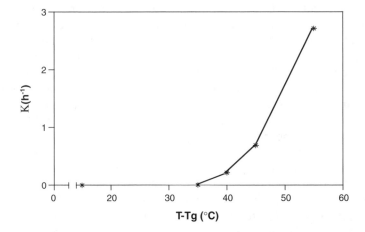

FIGURE 7.7
Deactivation rates of α-amylase freeze-dried in trehalose matrices (20 g/100 mL fermentation extract) as a function of (T − Tg).

inactivation of the enzyme, but the pectinlyase preparation without added carbohydrates showed a superior stability. Terebiznik et al. (1998b) pointed out the existence of stabilizing solids in enzymatic extracts obtained by solid-state fermentation of fungi. Trehalose and lactitol displace the stabilizing solids from their interactions with the enzyme taking over the stabilizing effect that predominates while the sugars are amorphous. When the sugars crystallize, the enzyme is no more protected and rapidly inactivates.

The Impact of Components of Enzymatic Fermentation Extracts on the Stability of Dry Enzymes

Crude enzymatic fermentation extracts (especially those obtained by solid-state fermentation processes) are usually heterogeneous systems containing components from the fermentation media, metabolites generated by the action of microorganisms over the substrate, and cellular components. The presence of low-molecular-weight compounds (e.g., reducing sugars, amino acids, acids), biopolymers (e.g., cellulose, starch, dextrin, pectin), or their hydrolysis products would produce a plasticization or vitrification of the dry enzyme with a consequent effect over enzyme stability. In addition, some solutes such as reducing sugars could lead to deleterious non-enzymatic browning reactions with enzymes in dried preparations. Nevertheless, these solutes could display specific protective effects if they can act as substrates or substrate analogs for the enzyme. Therefore, the selective depletion of solutes that might contribute to enzyme inactivation could be a more rational approach for improving the stability of dry enzyme preparations.

Szczodrak and Wiater (1998) pointed out the importance of the methods applied for enzyme concentration and purification with regard to the final activity of dried enzyme preparations. They showed that freeze-drying crude lactase preparations of *Penicillium notatum* was followed by 28 to 36% losses of specific and overall enzyme activities, respectively. Enzyme recovery of 89% was reached after lyophilization of culture supernatant precipitated with ethanol in the volume proportion of 1:3. Fractionation with ammonium sulfate or concentration by ultrafiltration was followed by only 4 to 8% losses of activity during the freeze-drying process.

A crude fermentation extract of *Aspergillus oryzae* α-amylase grown over wheat bran contained 2.8 g of solids per 100 mL (Terebiznik et al., 1998b). Only 1.14% of total solids showed a molecular weight (MW) above 300 kDa, while 30.40% of the solids had MWs below 0.5 kDa. The thermal stability of freeze-dried α-amylase was greatly affected by removing different molecular weight solutes from the crude fermentation extract by ultrafiltration (Figure 7.8).

Freeze-drying of the crude fermentation extract resulted in a collapsed dry system characterized by a high degree of stickiness and brown color. Collapse causes the loss of quality of dehydrated, especially freeze-dried, foods and occurs because of the decreased viscosity of the system above Tg so that the system can no longer support its own weight. The depletion of solutes of MW below 10 kDa also avoided the collapse of the enzyme preparation (Terebiznik et al., 1998b). Low-molecular-weight compounds may contribute to enzyme inactivation and to collapse by depressing Tg, which reduces the dry system viscosity, allowing protein to undergo conformational changes or chemical reactions that lead to inactivation, (e.g., Maillard reaction, aggregation, oxidation).

A single Tg (–20°C) was apparent in DSC thermograms of the dry crude fermentation extract equilibrated at 11% RH (Table 7.3). Nevertheless, after removal of components below 10 or 50 kDa by ultrafiltration, three glass transitions within –40 and 60°C were apparent in the dry extract equilibrated at 11% RH. Blends of incompatible polymers usually exhibit many Tg values, which are typical of the same polymer taken individually (Matveev et al., 1997). The appearance of several glass transitions after removal of low-molecular-weight components suggests that components with MWs between 10 and 300 kDa are co-soluble in the presence of low-molecular-weight components. Ultrafiltration greatly reduced the water adsorbed at 11% RH (Table 7.3), affecting the degree of plasticization of the dry extract; however, the Tg of the ultrafiltered extract at 9% moisture was still higher than the Tg of the crude fermentation extract. Thus, the appearence of higher Tg values after ultrafiltration could explain the absence of collapse during freeze-drying.

All Tg values were well below the temperature for residual activity determinations (90°C), so the dry enzyme preparations were rubbery during thermal treatment. It could be expected that the higher (T – Tg) value for the crude extract would be related to a lower residual activity; however, the increase of (T – Tg) of the ultrafiltered extract by increasing plasticization

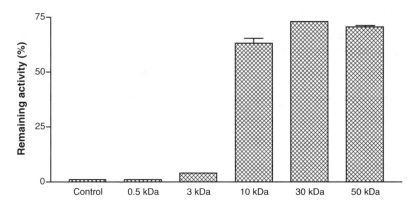

FIGURE 7.8

Remaining activity of α-amylase in crude (control) and ultrafiltered freeze-dried fermentation extracts equilibrated at 11% RH, after heat treatment at 90°C for 2 h. 100% of α-amylase activity corresponds to 200 U/mL. The numbers under the bars indicate the cutoff of the membrane used in each case.

TABLE 7.2

Residual Activity of Pectinlyase Freeze-Dried[a] without Added Carbohydrates and with Trehalose and Lactitol

Enzymatic preparation	Tg[b] (°C)	Residual Activity after 11.5 Hours at 58°C (%)	Residual Activity after 24 Minutes at 130°C (%)
Without added carbohydrates	45.8	85.8	79.6
5% Trehalose	57.6	79.5	30.0
15% Trehalose	55.0	64.2	20.6
5% Lactitol	37.1	98.3	26.7
15% Lactitol	32.9	97.8	33.8

[a] Equilibrated at 11% relative humidity.

[b] Glass transition temperature (Tg) was determined as the midpoint transition in DSC thermograms (scan rate = 10°C/min).

by water (4 to 9% moisture) did not affect the residual activity (Table 7.3). Therefore, the changes in Tg or (T – Tg) caused by removal of low-molecular-weight components would not be the primary factor determining the great stability of α-amylase after ultrafiltration.

Some low-molecular-weight solutes (MW ≤ 10 kDa) are able to take part in non-enzymatic browning reactions. During thermal inactivation at 90°C, various degrees of non-enzymatic browning were observed in the crude or ultrafiltered fermentation extracts, further contributing to enzyme inactivation (Terebiznik et al., 1998b). Izutsu et al. (1991) and Suzuki et al. (1997) observed that treatment conditions leading to rapid inactivation of enzymes in reducing mono- and disaccharide dry matrices were associated with the browning of the samples.

Despite the deleterious effects of low-molecular-weight compounds in the fermentation extract on collapse, Tg, and browning of the dry enzyme

TABLE 7.3

Glass Transition Temperature (Tg)[a] and Residual α-Amylase Activity of Crude and Ultrafiltered Dry Enzymatic Extracts

Enzyme Preparation	Tg (°C)	Equilibrium Moisture Content (%db)	Residual Activity[b] (%)
Dry crude fermentation extract	–20[c]	9 (RH = 11%)	5
Ultrafiltered dry fermentation extract (300–50 kDa)	– 40/–10/60[d]	4 (RH = 11%)	72
Ultrafiltered dry fermentation extract (300–50 kDa)	–10/40[e]	9 (RH = 53%)	72

[a] DSC midpoint transitions at scan rate = 10° C/min.
[b] Residual α-amylase activity after heating 2 h at 90°C at constant moisture.
[c] One Tg value was apparent in DSC thermograms.
[d] Three Tg values were apparent.
[e] Two Tg values were apparent.

preparation, the existence of stabilizing components was demonstrated (Terebiznik et al., 1998b). Purification of the α-amylase enzymatic extract by gel filtration chromatography indicated that in the absence of components between 50 and 300 kDa, α-amylase activity was rapidly lost, suggesting a stabilization mechanism in which amorphous dextrines, pectines, and cellulose or its fungal metabolized products are involved. It was found that only 1.5% of the total solids in the crude fermentation extract had MWs between 50 and 300 kDa. Despite the extraordinary stability given to the enzyme by these molecules, its action could be easily disabled by restitution of low-molecular-weight solutes (<0.5kDa) to an ultrafiltered system.

The following model attempts to explain the interaction between the stabilizers and the enzyme. The α-amylase is stabilized in the dry ultrafiltrated fermentation extract by a group of carbohydrate molecules heterogeneous in MW and nature, composed by dextrines, pectin, or cellulose or their hydrolysis products, giving a degree of protection similar to amorphous trehalose. It is likely that the protecting molecules were able to interact with the enzyme by a mechanism similar to that displayed by trehalose. Burin et al. (2001) observed that dry β-galactosidase was stabilized by its substrate, lactose, to the same extent as trehalose. However, the polymeric nature of the protective compounds implies an advantage over this sugar, because it does not crystallize. The retention of about 70% of α-amylase activity in dry ultrafiltered fermentation extracts (through 50- and 10-kDa membranes) at a relative humidity as high as 75% is additional evidence of the amorphous state of the intrinsic stabilizing compounds. Crystallization of additives may cause rearrangement of molecules resulting in the loss of favorable interactions and significant changes in the stabilizing effect as shown in Table 7.2 for a pectinlyase preparation ultrafiltered through a 10-kDa membrane and dried in trehalose or lactitol matrices.

Conclusions

Over the past several years, evidence has been provided that the glassy state is not sufficient to ensure the stability of dry enzyme preparations. Nevertheless, the glass–rubbery transition generally accelerates rates of enzyme inactivation. Aside from the experimental data supporting the water replacement hypothesis as being the more consistent mechanism involved in sugar stabilization of dry enzymes, evidence suggests a more complex behavior. Some questions must still be addressed:

1. Do specific interaction effects depend on the enzyme structure?
2. What is the relationship between molecular packaging of the sugar matrix and enzyme inactivation?
3. Is the stability of an enzyme included in a glassy matrix affected by its own physical state (glassy/rubbery)?

New strategies for stabilizing enzymes based on more rationale approaches are still worth trying, particularly those based on knowledge of the influence that methods applied for enzyme concentration/purification have on the final activity of dried enzyme preparations. This is certainly an area of future intensive research that will provide better control of enzyme activity during drying and storage as well as the development of rationale downstream processes based on the knowledge of the influence of components of microbial fermentation extracts on the stability of dried enzymes. Finally, strategies for preventing enzyme inactivation in the solid state should focus on the knowledge of the mechanisms involved in the loss of activity that could be different from those occurring in liquid systems.

Acknowledgments

The authors acknowledge financial support from the University of Buenos Aires, Consejo Nacional de Investigaciones Científicas y Técnicas de la República Argentina, and Agencia Nacional de Promoción Científica y Tecnológica de la República Argentina.

References

Aguilera, J.M. and Karel, M. (1997) Preservation of biological materials under desiccation, *Crit. Rev. Food Sci. Nutrit.*, 37, 287–309.

Arakawa, T. and Timasheff, S.N. (1982) Preferential interactions of proteins with salts in concentrated solutions, *Biochemistry*, 21, 6545–6552.

Arakawa, T. and Timasheff, S.N. (1985) The stabilization of proteins by osmolytes, *Biophys. J.*, 47, 411–414.

Baptista, R.P., Cabral, J.S., and Melo, E.D. (2000) Trehalose delays the reversible but not the irreversible step of thermal denaturation of cutinase, in *Proc. 2nd Int. Conf. on Protein Stabilisation*, Centre for Biological and Chemical Engineering, Instituto Superior Técnico, Lisbon.

Bell, N.L., Hagerman, M.J., and Bauer, J.M. (1995) Impact of moisture on thermally induced denaturation and decomposition of lyophilized bovine somathotropin, *Biopolymers*, 35, 201–209.

Buera, M.P., Rossi, S., Moreno, S., and Chirife, J. (1999) DSC confirmation that vitrification is not necessary for stabilization of the restriction enzyme *Eco*RI dried with saccharides, *Biotechnol. Progress*, 15, 577–580.

Burin, L., Mazzobre, M.F., and Buera, M.P. (2001) The performance of different de-hydro-protective agents on thermal stability of β-galactosidase from different sources, in *Engineering and Food at ICEF*, Welti-Chanes, J., Barbosa-Canovas, G.V., and Aguilera, J.M., Eds., Technomic Publishing, Lancaster, PA, 729–733.

Cardona, S., Schebor, C., Buera, M.P., Karel, M., and Chirife, J. (1997) Thermal stability of invertase in reduced-moisture amorphous matrices in relation to glassy state and trehalose crystallization, *J. Food Sci.*, 62, 105–112.

Carpenter, J.F. and Crowe, J.H. (1988) Modes of stabilization of a protein by organic solutes during desiccation, *Cryobiology*, 25, 459–470.

Carpenter, J.F. and Crowe, J.H. (1989) An infrared spectroscopic study of the interactions of carbohydrates with dried proteins, *Biochemistry*, 28, 3916–3922.

Carpenter, J.F. and Crowe, J.H. (1998) Modes of stabilization of a protein by organic solutes during desiccation, *Cryobiology*, 25, 459–470.

Carpenter, J.F., Crowe, J.H., and Arakawa, T. (1990) Comparison of solute-induced protein stabilization on aqueous solution and in the frozen and dried states, *J. Dairy Sci.*, 73, 3627–3636.

Carpenter, J.F., Pikal, M.J., Chang, B.S., and Randolph, T.W. (1997) Rational design of stable liophilyzed protein formulations: some practical advice, *Pharm. Res.*, 14, 969–975.

Colaço, C.A.L.S., Sen, S., Thangavelu, M., Pinder, S., and Roser, B. (1992) Extraordinary stability of enzymes dried in trehalose: simplified molecular biology, *Research*, 10, 1007–1011.

Colaço, C.A.L.S., Smith, C.J.S., Sen, S., Roser, D.H., Newman, Y., Ring, S., and Roser, B.J. (1994) Chemistry of protein stabilization by trehalose, in *Formulation on Delivery of Proteins and Peptides*, American Chemical Society, Washington, D.C., pp. 223–240.

Cordone, L., Galajda, P., Vitrano, E., Gassmam, A., Ostermann, A., and Parak, F. (1998) A reduction of protein specific motions in co-ligated myoglobin embedded in a trehalose glass, *Eur. Biophys. J.*, 27, 173–176.

Costantino, H.R., Langer, R., and Klibanov, A.M. (1994) Solid-phase aggregation of proteins under pharmaceutically relevant conditions, *J. Pharm. Sci.*, 83, 1662–1669.

Crowe, J.H., Carpenter, J.F., and Crowe, L.M. (1998) The role of vitrificaton in anhydrobiosis, *Annu. Rev. Physiol.*, 60, 73–103.

De Cordt, S., Avila, I., Hendrickx, M., and Tobback, P. (1994) DSC and protein-based time–temperature integrators: case study of α-amylase stabilized by polyols and/or sugars, *Biotechnol. Bioeng.*, 44, 859–865.

Delgado, L., Trejo, B., Huitrón, C., and Aguilar, G. (1992) Pectinlyase from *Aspergillus niger* sp. CH-Y-1043, *Appl. Microbiol. Biotechnol.*, 39, 515–519.

Domínguez, H., Núñez, M.J., and Lema, J.M. (1994) Enzymatic pretreatment to enhance oil extraction from fruits and oilseeds: a review, *Food Chem.*, 49, 271–286.

Donnamaría, M.C., Howard, E.I., and Grigera, J.R. (1994) Interaction of water with α, α-trehalose in solution: molecular dynamics simulations approach, *J. Chem. Soc. Faraday. Trans.*, 90, 2731–2735.

Duddu, S.P. and Dal Monte, P.R. (1997) Effect of glass transition temperature on the stability of lyophilized formulations containing a chimeric therapeutic monoclonal antibody, *Pharm. Res.*, 14, 591–595.

Gibson, T.D., Higgins, I.J., and Woodward, J.R. (1992) Stabilization of analytical enzymes using a novel polymer–carbohydrate system and the production of a stabilized single reagent for alcohol analysis, *Analyst*, 117, 1293–1297.

Izutsu, K., Yoshioka, S., and Takeda, Y. (1991) The effects of additives on the stability of freeze-dried β-galactosidase stored at elevated temperature, *Int.J. Pharm.*, 71, 137–146.

Karmas, R. (1994) The Effect of Glass Transition on Nonenzymatic Browning Dehydrated Food Systems, Ph.D. thesis, The State University of New Jersey, Camden.

Lee, J.C. and Timasheff, S.N. (1981) The stabilization of proteins by sucrose, *J. Biol. Chem.*, 256, 7913–7201.

Lozano, P., Combes, D., and Iborra, J.L. (1994) Effects of polyols on α-chymotrypsin thermostability: a mechanistic analysis of the enzyme stabilization, *J. Biotechnol.*, 35, 9–18.

Manning, M.C., Patel, K., and Borchardt, R.T. (1989) Stability of protein pharmaceuticals, *Pharm. Res.*, 6, 903–918.

Matveev, Y., Grimberg, V., Sochava, I., and Tolstoguzov, V. (1997) Glass transition temperature of proteins. Calculation based on the additive contribution method and experimental data, *Food Hydrocolloids*, 11, 125–133.

Mazzobre, M.F., Buera, M.P., and Chirife, J. (1997) Protective role of trehalose on thermal stability of lactase in relation to its glass and crystal forming properties and effects of delaying crystallization, *Lebensm.-Wiss.u.-Technol.*, 30, 324–329.

Mozhaev, V.V. (1993) Mechanism-based strategies for protein thermostabilization, *TIBTECH*, 11, 88–95.

Osawa, T. (1970) Kinetic analysis of derivative curves in thermal analysis, *Thermal. Anal.*, 2, 301–324.

Pilosof, A.M.R. and Terebiznik, M.R. (2000) Spray and freeze-drying of enzymes in *Drying Technology in Agriculture and Food Sciences*, Mujumdar, A.S., Ed., Science Publishers, Enfield, NH, 167–190.

Poutanen, K. (1997) Enzymes: an important tool in the improvement of the quality of cereal foods, *Trends Food Sci. Technol.*, 8, 285–320.

Roos, Y. and Karel, M. (1991) Plasticizing effect of water on thermal behaviour and crystallization of amorphous food models, *J. Food Sci.*, 56, 38–43.

Rossi, S., Buera, M.P., Moreno, S., and Chirife, J. (1997) Stabilization of the restriction enzyme *Eco*RI dried with trehalose and other selected glass-forming solutes, *Biotechnol. Prog.*, 13, 609–616.

Sakai, T., Sakamoto, T., Haellert, J., and Vandamme, E. (1993) Pectin, pectinase and protopectinase: production, properties and applications, *Adv. Appl. Microbiol.*, 25, 213–294.

Salahas, G., Peslis, B., Georgiou, C.D., and Gavalas, N.A. (1997) Trehalose, an extreme temperature protector of phosphoenolpyruvate carboxylase from the C_4 plant *Cynodon* Dactylon, *Phytochemistry*, 46, 1331–1334.

Schebor, C., Buera, M.P., and Chirife, J. (1996) Glassy state in relation to the thermal inactivation of the enzyme invertase in amorphous dried matrices of tehalose, maltodextrin and PVP, *J. Food Eng.*, 30, 269–282.

Slade, L. and Levine, H. (1991) Beyond water activity: recent advances based on an alternative approach to the assessment of food quality and safety, *Crit. Rev. Food Sci. Nutr.*, 30, 115–360.

Sola-Penna, M. and Meyer-Fernandes, J.R. (1998) Stabilization against thermal inactivation promoted by sugars on enzyme structure and function: why is trehalose more effective than others sugars?, *Arch. Biochem. Biophys.*, 360, 10–14.

Sun, W.Q. and Davidson, P. (1998) Protein inactivation in amorphous sucrose and tehalose matrices: effects of phase separation and crystallization, *Biochimica et Biophysica Acta*, 1425, 235–244.

Suzuki, T., Imamura, K., Yamamoto, K., Satoh, T., and Okazaki, M. (1997) Thermal stabilization of freeze-dried enzymes by sugars, *J. Chem. Eng.*, 30, 609–613.

Szczodrak, J. and Wiater, A. (1998) Selection of method for obtaining an active lactase preparation from *Penicillium notatum*, *J. Basic Microbiol.*, 38, 71–75.

Taragano, V.M. and Pilosof, A.M.R. (1999) Application of Doehlert designs for water activity, pH and fermentation time optimization for *Aspergillus niger* pectinolytic activities production in solid state and submerged fermentation, *Enzyme Microbiol. Technol.*, 25, 411–419.

Taragano, V.M. and Pilosof, A.M.R. (2001) Calorimetric studies on dry pectinlyase preparations: impact of glass transition on inactivation kinetics, *Biotechnol. Progress*, 17, 775–777.

Terebiznik, M.R., Buera, M.P., and Pilosof, A.M.R. (1997) Thermal stability of dehydrated α-amylase in trehalose matrices in relation to its phase transitions, *Lebensm.-Wiss.u.-Technol.*, 30, 513–518.

Terebiznik, M.R., Buera, M.P., and Pilosof, A.M.R. (1998a) Thermostability and browning development of fungal α-amylase freeze-dried in carbohydrated and PVP matrices, *Lebensm.-Wiss.u.-Technol.*, 31, 143–149.

Terebiznik, M.R., Zylberman, V., Buera, M.P., and Pilosof, A.M.R. (1998b) Stabilizing crude and ultrafiltrated α-amylase and α-glucosidase from *Aspergillus oryzae* by trehalose, *Biotechnol. Techniques*, 12, 683–687.

Timasheff, S.N. (1982) Preferential interaction in protein–water–cosolvent systems, in *Biophysics of Water*, F. Franks and S. Mathias, Eds., Wiley, New York, pp. 70–72.

Timasheff, S.N. and Arakawa, T. (1988) Mechanisms of protein precipitation and stabilization by co-solvents, *J. Crystal Growth*, 90, 39–44.

Uritani, M., Takai, M., and Yoshinaga, K. (1995) Protective effect of dissacharides on restriction endonucleases during drying under vacuum, *J. Biochem.*, 117, 774–779.

Vrábel, P., Polakovio, M., Stefuca, V. and Báles, V. (1997) Analysis of mechanism and kinetics of thermal inactivation of enzymes: evaluation of multitemperature data applied to inactivation of yeast invertase, *Enzyme Microbiol. Technol.*, 20, 348–354.

8

Trends in Carotenoids Biotechnology

María Eugenia Jaramillo-Flores and Rosalva Mora-Escobedo

CONTENTS

Introduction

Carotenoids, C_{40}-isoprenoids consisting of eight isoprene units, are natural colorants present in a number of fruits and vegetables such as carrot, tomato, spinach, orange, watermelon, melon, papaya, mango, broccoli, lettuce, chili, and sweet potato (Stahl and Sies, 1998), as well as in microorganisms and microalgae. In vegetables and green leaves, the orange-like color of carotenoids is masked by chlorophylls. Carotenoids are widely distributed in nature, and over 600 different types have been identified in several organisms (Olson and Krinsky, 1995).

In photosynthesis, carotenoids act as accessory pigments; in protein complexes they work in photoreception as well as in photoprotection (Havaux, 1998). In human nutrition, they serve various important biological functions and roles as provitamin A, antioxidants, cell communication aids, immunological function improvers, protectors of skin against ultraviolet light, and protectors against macular degeneration. The function of some carotenoids as provitamin A (the most important one in human nutrition) is well documented and established, primarily for β- and α-carotenes. β-cryptoxanthin also has provitamin A activity, which has been reported to be lower than that for β-carotene (Van den Berg et al., 2000).

Regarding the cancer-prevention effects of carotenoids, the suggested roles of these compounds include antioxidant properties and the capacity to induce gap junctional communication (GJC). Carotenoids efficiently quench singlet oxygen and are generally capable of inhibiting free-radical reactions, with an activity that is largely dependent on the number of conjugated double bonds in the molecule (Stahl and Sies, 1998). The antioxidant activity has been demonstrated both *in vitro* and *in vivo*, but the behavior of the *in vivo* antioxidant activity depends on the location and concentration of target cells, tissues, or cell compartments, in addition to various other factors (Mathews-Roth, 1993).

Until relatively recently, an important number of genes for plant and microbial biosynthesis were cloned. The application of carotenoid-biosynthesis engineering in non-carotenogenic organisms such as *Escherichia coli* (for instance, where isoprenoid-route engineering was used to improve production of astaxanthin) and yeasts (including the transformation of *Candida utilis* for lycopene production) has allowed us to achieve the functional heterologous expression of the majority of genes. Lately, highly valuable new carotenoids have been obtained, some of them having improved antioxidant properties resulting from genetic combination and molecular combination of biochemical pathways (Schmidt-Dannert, 2000).

The main pathways involved in synthesis of carotenoids have been described. Nevertheless, little is known about the biosynthesis of various replacing agents, such as methyl groups, as well as regarding additional modifications or rearrangements of structures that give rise to a great diversity of compounds that to date continue being isolated, studied, and characterized.

All carotenoids are synthesized within plastids of plants from the central isoprenoid or terpenoid pathway, which also serves to synthesize many other compounds including terpenes, gibberellins, sterols, etc. The condensation of one dimethyl-allyl diphosphate molecule (DMADP) plus three isopenthyl diphosphate molecules (IDP) in a sequential order gives the diterpene geranyl-geranyl diphosphate (GGDP), which makes up half of all C_{40} carotenoids. The colorless phytoene, the first carotenoid in the pathway, is produced by the head-to-head condensation of two GGDP molecules. Afterwards, unsaturation reactions lengthen the conjugated double-bond system to produce neurosporene or lycopene. The latter is produced by introducing four double bonds into phytoene. Following with the unsaturation reactions, the carotenoid biosynthesis pathway branches to produce cyclic and acyclic carotenoids (Hirschberg, 1999).

Spheroidene, spheroidenone, and spyriloxanthin are acyclic xanthophylls synthesized by phototropic bacteria. Spheroidene and spheroidenone are produced by additional C3 and C4 unsaturations and by the introduction of hydroxyl and methoxyl groups into neurosporene and lycopene C1(C1′) under anaerobic conditions, and the introduction of a keto group in the C2 of neurosporene produces spirilloxanthin. The synthesis of cyclic carotenoids involves cycling of one (as in β-zeacarotene) or both (as in β-carotene) final groups in lycopene or neurosporene.

Normally, β-rings are introduced (e.g., β-carotene), but in higher plants ε-rings (as in α-carotene) and carotenoids with γ-rings (e.g., γ-carotene, found in some fungi) are commonly formed. Most cyclic carotenoids include at least one functional oxygen in one C atom of one of the rings.

Cyclic carotenoids with keto groups in C4(C4′) or hydroxyl groups in C3(C3′) (e.g., zeaxanthin, astaxanthin, echinenone, and lutein) are widely distributed in plants and microorganisms, while, for example, lycopene-hydroxilated derivatives such as lycoxanthin and lycophyll are rarely found. Leaf xanthophylls, violaxanthin, and neoxanthin, as well as algae carotenoids such as fucoxanthin, contain β-rings with epoxy groups in C(5,6). The β-terminal ring with epoxy in C(5,6) is a key intermediary in the formation of a variety of other unusual final groups, such as the cyclopentyl terminal group of capsanthin and capsorubin found in *Capsicum annum* (red pepper) (Schmidt-Dannert, 2000).

Use of Biotechnology for Carotenoid Production

Despite the wide variety of compounds available in nature and their potential for use as food colorants, many of these are found in relatively low concentrations, so that production through biotechnology may represent an excellent option for large-scale production. Thus, obtaining carotenoids through plant-cell-tissue culture, microbial fermentations, and genetic manipulation has been investigated. However, given that many of the producing organisms are considered as non-GRAS (generally recognized as safe), the products obtained are subjected to a battery of food-safety testing prior to their release for use as food additives (Downham and Collins, 2000).

Plant-cell-tissue culture is regarded as a suitable alternative for the production of natural pigments (Cormier, 1997). Carotenoids are found among the pigments produced through plant cell cultures, as, for example, in the case of production of β-carotene by the *Blakeslea trispora* fungus through reactor fermentation. The pigment so obtained is currently available commercially from Gist Brocades (DSM Gist Brocades Delft, Heerlen, The Netherlands) (Downham and Collins, 2000).

Biotechnological Production of Astaxanthin

New, advanced technologies have allowed successful production of astaxanthin from *Haematococcus pluvialis* (Chlorophyceae) at a commercial level, and some companies produce astaxanthin from the yeast *Xanthophyllomyces dendrorhous*. *H. pluvialis* is a freshwater, unicellular biflagellate that accumulates astaxanthin into its aplanospores. A highly-valued carotenoid, astaxanthin has a high added value when used as a food additive in aquaculture. This carotenoid is produced by a wide variety of microorganisms (Margalith, 1999). Astaxanthin is innocuous, especially in marine ecosystems, where it is biosynthesized and accumulated by zooplankton, which is consumed by many marine animals. Astaxanthin is the main pigment that produces the pinkish-red shade characteristic of salmonids, as well as shrimp, lobsters, and crabs (Lorenz and Cysewski, 2000).

Heterotrophic and mixotrophic cultures have been studied in the laboratory (Chen, 1996). Yields up to 40 mg/L or 43 mg/g of dry weight of cells can be achieved in the laboratory with a mixotrophic culture and adequate illumination. If production of natural astaxanthin is to be competitive as compared to synthetic astaxanthin, scaling-up is required; however, large-scale production in open ponds has been unsuccessful given problems caused by contamination. This situation could be resolved by using a more selective medium (Margalith, 1999).

New techniques have been developed for the production of natural astaxanthin from *Haematococcus*, including photo-bioreactor technology, with efforts being made by several firms throughout the world (in the U.S., Japan, France, U.K., Australia, Israel, Sweden, and Australia, among others) to develop commercial photo-bioreactors. Companies such as Aquasearch in the U.S. offer photo-bioreactors for large-scale production of astaxanthin. In Sweden, closed photo-bioreactors illuminated with artificial light have been used for astaxanthin production, and in Hawaii a combination of these techniques with open-pond cultivation has been successful. (Lorenz and Cysewski, 2000).

In a large-scale open system, the production of astaxanthin-rich *Haematococcus* involves a two-phase process. In the first phase, vegetative cells must be produced under nearly optimum growth conditions, carefully controlling pH, temperature, and nutrient level. Temperature and pH can be regulated automatically. In Hawaii, deep water with a temperature of 14°C is used as the cooling liquid for culture systems, along with heat-exchange systems. After a sufficient biomass has been produced, the culture is subjected to nutritional and environmental stresses to induce astaxanthin production. In commercial systems, the production of this carotenoid is induced by limiting nitrates and phosphates, increasing light or temperature levels, or adding sodium chloride to the culture medium. Between 2 and 3 days after the culture is subjected to stress, hematocysts are formed and astaxanthin accumulation starts; 3 to 5 days following hematocyst formation, their astaxanthin content ranges from 1.5 to 3% and they are ready for harvesting.

Because haematocysts are heavier than water, they are harvested by decantation followed by centrifugation, drying, and grinding to ensure maximum astaxanthin bioavailability. For food-grade applications, antioxidants such as ethoxiquin must be added to the harvested paste prior to drying to minimize carotenoid oxidation. Essentially, the production of *Haematococcus* alga meal as a natural source of astaxanthin is relatively simple. Cultures grow rapidly in a simple nutritive medium. Because the pH of the required medium is neutral, however, contamination by other microalgae, amoeba, and protozoa may represent problems that are magnified in the scaling-up stages; thus, advanced control systems are required. Although upstream technology is necessary for the production of natural astaxanthin from *Haematococcus*, this is becoming a commercial reality with great potential for the poultry and salmonid industries, as well as for the dietary supplement industry as a nutraceutic agent (Lorenz and Cysewki, 2000).

Carotenoids are intracellular agents located in membranes of mitochondria, chloroplasts, chromoplasts, and endoplasmic reticulum. While in *Phaffia rhodozyma* or *Haematococcus pluvialis*, carotenoids are located within cytoplasmic lipid globules; in *Dunaliella bardawii* β-carotene is located in the chloroplast matrix as a plastoglobule (Lange, 1968; Johnson and An, 1991). Such extraplastidic carotenoids are also considered as secondary carotenoids (Grung et al., 1992). *Phaffia* and *Dunaliella* produce carotenoids during all growth phases, whereas in *H. pluvialis* the synthesis of ketocarotenoids depends on environmental conditions, mostly during the formation of resting cells (aplanospores) that occurs along with an active adaptation of the photosynthetic apparatus and a decline in photosynthetic activity (Hagen et al., 1993; Zlotnik et al., 1993). The accumulation of secondary carotenoids and those produced by light induction begins after growth has stopped and takes place in the perinuclear cytoplasm, conferring a characteristic red color to it (Yong and Lee, 1991). Carotenoid propagation outside chloroplasts occurs at the periphery of the cell during strong irradiation, causing an increase in the chloroplast vicinity (Hagen et al., 1994). Fan et al. (1998) suggest that astaxanthin formation results from a photoprotecting process, as it was observed that pigments return to the center of the cell when illumination stops (Yong and Lee, 1991).

Algae photosynthetic ability offers the possibility for autotrophic culture in simple mineral media for a number of biotechnological purposes. Attention has been paid to growing them in simple systems such as shallow open ponds (Pulz and Scheibenbogen, 1998).

From a biotechnological point of view, the induction issue associated with astaxanthin biosynthesis is of importance because the use of artificial induction may significantly shorten the fermentation cycle. For this reason, special attention was paid to studying the conditions regulating the transformation of normal flagellates into colored aplanospores (Czygan, 1970; Zlotnik et al., 1993).

Several researchers have investigated the carotenoid complex formed during *Haematococcus* growth and encysting. β-carotene, lutein, violaxanthin, and neoxanthin have been found in vegetative cells, while over 90% of carotenoids found in aplanospores correspond to astaxanthin mono- and di-esters (Grung et al., 1992). It is worth mentioning that astaxanthin esters are used less in aquaculture than free carotenoids (Storebakken et al., 1987). *Phaffia rhodozyma* apparently produces non-esterified astaxanthin

(Andrewes and Starr, 1976). Astaxanthin can be found as a number of configurational isomers. For example, 3S,3'S isomers are primarily found in *Haematococcus*, whereas 3R,3'R isomers are found in *Phaffia*, the latter having a lower bioavailability for cultured fish (Bjerkeng et al., 1997).

To date, efforts to stimulate astaxanthin production through the addition of specific compounds or carotenogenesis precursors have been virtually unsuccessful. Experiments to stimulate β-carotene synthesis in the fungus *Blakeslea trispora* have produced interesting results (Cerda-Olmedo and Hütterman, 1986). The effect of abscisic acid used as an agent for the adaptation of higher plants to water stress as well as in the culture of plant-cell tissues was examined in *Haematococcus*. The addition of 0.1-mM abscisic acid induces vegetative cells to grow in agar plates, but not in liquid media. These vegetative cells were transformed into red aplanospores more rapidly than in the control culture, but carotenogenesis remained unaffected when this hormone was added to cyst cells (Kobayashi et al., 1997).

Little has been done regarding strain improvement to optimize astaxanthin production (Margalith, 1999). An unlikely possibility involves using gametogenesis and sexual hybridization. Mutants resistant to carotenoid-biosynthesis inhibitors (e.g., norflurazon, fluoridone) have been obtained through hybridization by protoplast fusion, producing ploidy cells with a high carotenoid content (Tjahjono et al., 1994). Chumpolkulwong et al. (1997) demonstrated that *Haematococcus* mutants resistant to compactin (mevastatin), an inhibitor that competes with hydroxymethyl-glutaryl-CoA reductase (a key regulatory enzyme in isoprenoid biosynthesis), accumulated 1.4 to 2.0 more astaxanthin than the wild strain.

Molecular biology studies on genes and enzymes involved in astaxanthin biosynthesis reveal interesting findings. A number of carotenoids, including β-carotene and some ketocarotenoids, have been found in eubacteria (e.g., *Flavobacterium, Brevibacterium, Erwinia*), as well as in several archaebacteria (the halophilic bacterium *Halobacterium*).The carotenogenic genes of *Erwinia uredovora* have been introduced and expressed in non-carotenogenic bacteriae, including the colorless bacterium *Escherichia coli* (Misawa et al., 1990, 1991). Kajiwara and coworkers (1995) and Lotan and Hischberg (1995) used this transformed *E. coli* strain to clone the *bkt* gene of *H. pluvialis* which codifies for β-carotene ketolase, thereby obtaining an *E. coli* strain with the capability to express and synthesize canthaxanthin. This means that astaxanthin production may be demonstrated if the *bkt* gene is found in the presence of the *crtZ* gene of *E. uredovora* which codifies for β-carotene hydroxylase, a catalyzer for zeaxanthin formation (β-carotene, 3,3'-diol). In fact, it is feasible that the biosynthetic activity for axtaxanthin production might be improved through genetic manipulation (Kajiwara et al., 1995).

Industrial Applications of Astaxanthin

Haematococcus alga meal has been approved in Japan for use as a natural red food colorant and as a pigment for fish feeding. In addition, the U.S.

Food and Drug Administration (FDA) and Canada's Food Inspection Agency have both approved the use of the *Haematococcus* pigment as an additive in salmonid feed. Today, the FDA allows the use of *Haematococcus* astaxanthin as an additive in dietary supplements, while several European countries have approved its use for human consumption (Lorenz and Cysewski, 2000).

The steady growth of salmonid cultures has caused a growing demand for pigments. Natural astaxanthin derived from *Haematococcus* has the potential to be used as feed for cultured salmonids. The use of astaxanthin in the salmon diet derives not only from an interest in conferring a characteristic color to fish meat; Atlantic salmon requires this pigment for growth and survival. Astaxanthin at a concentration below 5.3 ppm in the salmon diet renders marginal growth rates, whereas levels above 5.3 ppm produce normal growth with a significantly higher amount of lipids. When salmon were fed with astaxanthin at a concentration lower than 1 ppm, more than 50% of them died and the survival rate dropped significantly. Survival rates for those groups fed high concentrations of astaxanthin reached values of at least 90%; hence, Atlantic salmon is one of the first species in which astaxanthin has been shown to be an essential vitamin. Fish fed minimum levels of 13.7 ppm of astaxanthin in the diet showed a 20% increase in lipid levels and had a better overall performance as compared to those fed 5.3 ppm of astaxanthin in the feed (Lorenz and Cysewski, 2000).

Sea bream (*Chrysophrys major, Pagnus major*, taid, and red snapper) are highly valued because of their skin pigmentation, which derives from the astaxanthin content. The lack of this pigment in the artificial diet causes a lighter color in cultured organisms, and cultures lacking astaxanthin supplementation contain a mere 5% of astaxanthin content compared with their wild counterparts. The stomachs of wild species have been found to contain *Squilla oratoria* and other crustaceans, which provide the required astaxanthin. However, these species have been shown not to develop pigmentation and convert these pigments into astaxanthin when other carotenoid types, such as β-carotene, lutein, canthaxanthin, and zeaxanthin, are included in the diet. Excretion, metabolic degradation, and an insufficient dietary intake, among other factors, produce a loss of carotenoids in the body and skin. In order to obtain a reddish pigmentation in sea bream, a dietary astaxanthin source must be provided. Nakazoe et al. (1984) fed cultures with diets containing β-carotene, zeaxanthin, lutein, canthaxanthin, free astaxanthin, or astaxanthin ester. Specimens fed β-carotene or canthaxanthin developed a reduction in integumentary carotenoid levels. Additionally, the results showed that the intake of astaxanthin ester may be reduced in up to 50% and yet achieve a carotenoid deposition 2.3 times as high as that of free astaxanthin.

Lorenz and Cysewski (2000) compared the diets of marine cultures to which 100 ppm of either free astaxanthin or its ester were added. Cultures fed free astaxanthin showed an improvement in the skin carotenoid content for a month, reaching a saturation point with no further variations. The other culture, fed astaxanthin ester, showed much greater significant increases of this pigment: After two months, the skin astaxanthin level was 1.7 times as high as that of fish fed free astaxanthin (13.23 mg/kg vs. 7.94 mg/kg). This demonstrated that in this type of culture, astaxanthin esters are used more efficiently than free astaxanthin. The same authors reported that astaxanthin from *Hamatococcus* has been successfully used in egg-yolk pigmenting.

Studies Focused on Carotenoid Production in Yeasts

Carotenoids are also synthesized by yeasts such as *Rhodotorula, Rhodosporidium*, and *Phaffia*. The latter produces astaxanthin as the major carotenoid type (An et al., 1989; Haard, 1988), whereas β-carotene, torulene, and torularhodin are the major carotenoids produced by *Rhodotorula* and *Rhodosporidium* (Frengova et al., 1997).

Frengova et al. (1997) introduced the production of carotene–protein and exopolysaccharides from *Rhodotorula glutinis* 22P grown in ultrafiltrated cheese serum, a natural substrate containing lactose. *R. glutinis* cannot assimilate lactose but easily assimilates glucose, galactose, and lactic aid, so that culturing it along with the homofermentative lactic acid bacterium *Lactobacillus helveticus* 12A creates suitable conditions for lactose assimilation by *R. glutinis*. In the *R. glutinis* 22P + *L. helveticus* 12A mixed culture, the homofermentative lactic acid bacterium transforms lactose into glucose, galactose, and lactic acid. It is in this way that the yeast *R. glutinis* 22P, which cannot assimilate lactose, vigorously grows and synthesizes carotenoids. Separately, the bacterium's growth is stimulated with the vitamins and amino acids produced by the yeast.

When grown on substrates containing synthetic mono- and disaccharides, yeasts are capable of synthesizing carotenoids (Tada and Shiroishi, 1982; Costa et al., 1987; Frengova et al., 1997) as they are when cultured on natural substrates such as molasses (Haard, 1988). Cultivating *R. glutinis* 22P + *L. helveticus* 12A in an MBR AG bioreactor (Zurich, Switzerland), the parameters for growth and carotenogenesis in cultured organisms, as described by Frengova et al. (1994), are incubation temperature, 30°C; baseline pH, 6.0 (not adjusted during the growth phase); air flow, 0.55 (Liters of air) · (Liters of medium)$^{-1}$ · (min)$^{-1}$; and stirring at 220 rpm for 7 days. The synthesis of exopolysaccharides and carotenoids was investigated for the microbial association between *R. glutinis* 22P and *L. helveticus* 12A, under the culture conditions described above. The days when the highest growth occurred were found not to coincide with those of the highest exopolysaccharide and carotenoid synthesis. The maximum concentration of carotenoids in cells occurred in the stationary phase of the yeast's growth cycle (263 mg/g cells in dry weight, day 6), whereas the highest cell production (30.9 g/L) and polysaccharide synthesis (8.2 g/L) reached a peak on day 4 of the yeast's growth cycle.

In a study of the biosynthesis of exopolysaccharides and carotenogenesis by a mixed culture of the yeast + *L. helveticus* 12, extrametabolic activities in the yeast were found, thus showing that milk serum is also used as a substrate for exopolysaccharide synthesis (Stauffer and Leeder, 1978). Comparatively, the mixed culture of the lactose-negative yeast *R. glutinis* 22P with the homofermentative lactic acid bacterium *L. helveticus* 12A produces a carotene–protein complex with much higher concentrations of carotenoids, protein, vitamins, minerals, and amino acids than the actual yeast biomass by *Rhodotorula* (Zalashko, 1990). Given the final composition of this complex, it can be used as a nutritional ingredient (Frengova et al., 1997).

Taking into account the results of the mixed culture of the lactose-negative yeast *R. glutinis* 22P with the homofermentative lactic acid bacterium *L. helveticus* 12A in ultrafiltered cheese whey, Frengova et al. (1997) proposed a system for the production of carotene–protein and exopolysaccharides. *Hansenulla holstii* and *Cryptococcus laurentii* were used as monocultures for biopolymer production. For this purpose, whey lactose is first hydrolyzed to produce glucose and galactose using a commercial β-galactosidase. Peak concentrations of biopolymers were obtained in 5 days using concentrations of 9.4 g/L for *H. holstii* and 5.7 g/L for *C. laurentii*.

Research has been conducted regarding the selection and production of carotenoid-hyperproducing mutants to find and develop new natural sources of carotenoids. Some of these works are mentioned in this chapter, but results obtained in this respect are still scarce. Relatively few methods of selection have been developed, including the use of antimycin, β-inone, flow cytometry, cell sorting, exposure to light, and the presence of rose bengal colorant. Furthermore, methods for selecting carotenoid-hyperproducing mutants based on their protecting activity have failed. These carotenoid-hyperproducing mutants have turned out to be more sensitive to antimycin, H_2O_2, thenoyl-tri-fluoro-acetone, cyanide, ultraviolet light, and duroquinone, all of which are related to oxygen radicals. This situation has resulted in reducing the selection methods available. An (1997) showed that hyperproducing mutants are more sensitive to photosensitization than the wild yeast strain. This is partly due to the low activity of the superoxide dismutase enzyme.

Low temperature and photosensitization can be used as agents for the selection of carotenoid-hyperproducing mutants. At 20°C, the *Phaffia rhodozyma* Ant-1 strain produces 3.2 times more carotenoids (960 mg carotenoid per g yeast) compared to the 67-385 wild strain, which produces 300 mg carotenoid per g yeast. The carotenoid-hyperproducing mutants show hypersensitivity to photosensitization due to lower activity of the superoxide dismutase enzyme. The Ant-1 and 2A2N carotenoid-hyperproducing mutants also showed a reduced enzyme activity compared to the 67-385 wild strain. Limited cell metabolism caused by low temperatures (2°C), together with photosensitization, represents an example of conditions for the selection of carotenoid-hyperproducing strains of *P. rhodozyma*.

Parajó et al. (1998) obtained carotenoids by growing *Phaffia rhodozyma* in media prepared from hemicellulosic hydrolyzed substrate of *Eucalyptus globulus* wood. These researchers produced xylose-rich solutions as fermentation medium of *Phaffia*. Acid hydrolysis was combined with active-carbon absorption. Based on findings from previous experiments and in order to study cell growth dynamics and carotenoid production by *Phaffia*, fermentation trials were conducted optimizing the medium and vacuum concentrating the hydrolyzed products. After neutralization with caustic soda, detoxification with active carbon, and supplementation with 3 g peptone per L, the medium suitability for yeast growth and carotenoid production was demonstrated. Using hydrolyzed products concentrated through evaporation, the biomass concentration was improved to 23.2 g/L. Total carotenoids reached 12.9 mg/L and astaxanthin increased to 10.4 mg/L. The highest volumetric rates of carotenoid production were obtained by using 29 to 40.8 g xylose per L in the culture medium.

Del Campo et al. (2001) studied biomass and lutein production by the unicellular green algae *Muriellopsis* sp. The experiment was conducted for one year using a tubular photo-bioreactor placed outdoors, where the effects of the dilution rate, mixing, and daily solar cycles on lutein and biomass productivity were investigated throughout the year. The highest productivity was obtained in May for lutein (approximately 180 mg/m^2/day) and in July for biomass (approximately 40 g dry weight/m^2/day). The values corresponding to the optimum dilution rate fluctuated, being lower in May (0.06/hr) compared to those of November (0.09/hr). Photosynthetic efficiency figures (approximately 4%, with 40 g dry weight/m^2/per day) were recorded throughout the year. A rapid increase in lutein content was observed in *Muriellopsis* sp. in response to radiation during the first hours of the day, with a peak in lutein content at noon (6 mg/g dry weight), followed by a steady decline in the production rate. An increase in cell growth was observed after the peak lutein/chlorophyll II ratio was established, showing that lutein may play a photoprotecting agent role.

Genes from epiphytic *Erwinia* species, the marine bacterium *Agrobacterium aurantiacum* and the *Alcaligenes* sp. PC-1 strain, responsible for carotenoid synthesis including lycopene, β-carotene, and astaxanthin, have been isolated and their roles elucidated (Misawa et al., 1990, 1993). The compound farnesyl pyrophosphate (diphosphate) (FPP) is the first substrate of the encoded enzymes for carotenoid biosynthesis and is a common precursor of a wide range of isoprenoid compounds, including sterols, hopanols, dolicols, and quinones, as mentioned earlier.

The yeast *Saccharomyces cereviseae*, carrier of the *Erwinia uredovora* carotenogenic genes, along with FPP, a compound that is innocuous for yeasts, have been used for the production of lycopene and β-carotene (Yamano et al., 1994), but the amounts produced by these microorganisms are only 0.1 mg of lycopene and 0.1 mg of β-carotene per g (dry weight) of cells.

Candida utilis is a substance recognized as GRAS by the U.S. Food and Drug Administration. This has been one of the reasons for its large-scale production using relatively low-cost sugars as carbon sources for the production of unicellular protein. In addition to various chemical substances such as glutathione and RNA (Boze et al., 1992; Ichii et al., 1993), *Candida utilis* concentrates large amounts of ergosterol within the cell during the stationary phase (6 to 13 mg/g dry weight of cells) (Miura et al., 1998). Given the fact that it accumulates large amounts of ergosterol by redirecting the carbon flow for ergosterol biosynthesis toward a non-endogenous pathway for carotenoid synthesis via FPP, this yeast might synthesize large amounts of carotenoids. The *Candida utilis* strain carrying the *crtE*, *crtB*, and *crtI* genes was previously tested for lycopene production (Miura et al., 1998). Using six carotenogenic genes synthesized according to the normally used codon of the glyceraldehyde-3-phosphate dehydrogenase (GAP) gene of *C. utilis*, which is expressed at three levels, it was possible for the yeast to synthesize lycopene, β-carotene, and astaxanthin. Production of carotenoids was increased in this way.

The pCLEBI13-2 plasmid containing the original *crt* genes from *E. uredovora*, flanked by the GAP, PGK, and PMA promoters and terminators of *Candida utilis* (Miura et al., 1998), was expressed in the yeast *C. utilis*, which was capable of producing 0.8 mg lycopene and 0.4 mg phytoene per g of dry weight. The modified *crt* gene increased lycopene and phytoene production by approximately 1.5 and 4.3 times. The *C. utilis* strain carrying the pCLR1EBI-3 plasmid produced 1.1 mg lycopene and 1.7 mg phytoene per g dry weight during the stationary phase of growth. The transformed yeast increased its lycopene content according to the incubation time, with a corresponding decline in phytoene content. This change in the lycopene/phytoene ratio illustrates the conversion of lycopene into phytoene during culture. However, because relatively high amounts of lycopene are stored during the stationary phase, this probably indicates that the conversion of phytoene into lycopene represents a rate-limiting step in the yeast, perhaps due to the fact that the phytoene-to-lycopene conversion stages are insufficient around the membrane of the yeast due to a lack of electron carriers for the dehydrogenation reactions. This behavior was not observed during the production of carotenoids by the bacteria *Escherichia coli* carrying the *crt* genes from *Erwinia* (Misawa et al., 1990).

Misawa et al. (1990) also developed two *Candida utilis* strains, one of them a producer of β-carotene and the other of astaxanthin. The genes *crtE*, *crtB*, *crtI*, and *crtY* were introduced in the case of the β-carotene-producing strain and *crtE*, *crtB*, *crtI*, *crtY*, *crtZ*, and *crtW* for the astaxanthin-producing strain. Under these conditions, the same research team made the astaxanthin-producing strain to reach production levels very similar (0.4 mg/g dry weight) to those of *Phaffia rhodozyma*, a natural astaxanthin producer (Johnson and An, 1991). Miura et al. (1998) have proposed the use of bacterial *crt* genes in *P. rhodozyma* to increase astaxanthin production. Also, Wery et al. (1997) have developed a transformation system for *P. rhodozyma*.

Verdoes et al. (1999) described the isolation of a gene involved in astaxanthin biosynthesis, *crtI*, that codes for phytoene denaturase from the yeast *Xanthophyllomyces dendrorhous* in *Escherichia coli* by heterologous supplementation. The region for which the *crtI* gene codes is interrupted by 11 introns. These authors determined the function of cDNA through chromatographic analysis of the pigment in the different transforming *E. coli* strains. The insertion in *E. coli* of genes from *Erwinia uredovora* coding for geranyl-geranyl diphosphate synthetase and phytoene synthetase contained in a plasmid, plus the cDNA coding for phytoene denaturase from *Xanthophyllomyces dendrorhous*, resulted in the accumulation of lycopene in the transforming strain. The sequence of amino acids showed a high homology with bacteria and particularly with fungi. The stages through which phytoene becomes lycopene in *X. dendrorhous* are carried out using a unique genetic product.

Metabolic Engineering of Carotenoids in *Escherichia coli*

By using the isoprenoid-pathway engineering, several researchers have improved the production of carotenoids in *Escherichia coli*. This pathway makes use of glycolytic intermediates and, similar to which happens in other biosynthetic pathways, this one is controlled by the precursor supplement among other factors. Matthews and Wurtzel (2000) increased the precursor content in non-carotenogenic *Escherichia coli* by regulating the isoprenoid pathway through the overexpression of D-1-deoxyxyllulose 5-phosphate synthetase (*dxs*) (the enzyme responsible for the transformation of pyruvate and glyceraldehyde 3-phosphate into D-1-deoxyxyllulose 5-phosphate) of *E. coli* to increase deoxyxyllulose 5-phosphate. The overexpression of *dxs* in *E. coli* and the insertion of the set of genes coding for the carotenoid biosynthetic enzymes from *Erwinia uredovora* resulted in an increase in the accumulation of lycopene or zeaxanthin, compared to controls that did not express *dxs*. The overexpression of D-1-deoxyxyllulose 5-phosphate synthetase (*dxs*) increased the levels and accumulated carotenoids up to 10.8 times compared to controls lacking overexpression, so that colonies accumulated large amounts of lycopene (1333 mg/g dry weight) and zeaxanthin (592 mg/g dry weight). During an 11-day period, these researchers observed that zeaxanthin-producing colonies grew twice as fast as lycopene-producing colonies (Matthews and Wurtzel, 2000).

The *dxs* gene was cloned into *Escherichia coli* by Sprenger et al. (1997), and Wurtzel et al. (1997) studied the variation in the expression of the carotenoid genes in *E. coli* strains. The TOP10F' *E. coli* strain was observed to be a strong and stable-after-cloning strain. Later work by Lois et al. (1998) showed that the *E. coli* strain transformed with the pTAC-ORF2 plasmid containing the *dxs* gene controlled by a strong promoter accumulated high concentrations of D-1-deoxyxyllulose 5-phosphate synthetase (*dxs*). More recently, Matthews and Wurtzel (2000) inserted into the TOP10F' *E. coli* strain the pTAC-ORF2 plasmid containing the *Erwinia uredovora* genes required to carry out carotenoid biosynthesis.

Matthews and Wurtzel (2000) performed at least five different assays with several plasmid-carrying strains with or without the *dxs* gene. Plasmids tested were pTAC-ORF2, which codes for D-1-deoxyxyllulose 5-phosphate synthetase; pACCRT-*E1B*, which codes for geranyl-geranyl pyrophosphate synthetase (GGPPS), phytoene synthetase (PSY), and phytoene denaturase (CRT1); pACCAR25D*crtX*, which codes for GGPPS, phytoene synthetase (PSY), phytoene denaturase (CRT1), lycopene β-cyclase (LCY), and β-carotene hydroxylase (HYD); and pACCAR25D*crtB*, which codes for GGPPS, phytoene denaturase (CRT1), lycopene β-cyclase (LCY), and β-carotene hydroxylase (HYD).

The coexpression of pACCRT-*E1B* and pTAC-ORF2 resulted in increased lycopene production from 160 (absence of pTAC-ORF2) to 1106 mg/g of dry weight, and the coexpression of pACCAR25D*crtX* and pTAC-ORF2 resulted in increased zeaxanthin production from 186 (absence of pTAC-ORF2) to 526 mg/g of dry weight. No synthesis of pigments took place under the presence solely of pACCAR25D*crtB*. Finally, it is worth mentioning that these researchers considered that success in managing these coupled pathways in bacteria can also be applied to plants.

Farmer and Liao (2001) investigated the interaction between central metabolism and the isoprenoid pathway focusing on the flow-transfer line between the two pathways. They studied isoprenoid production in *Escherichia coli* through a rational management of glycolysis, finding that the distribution of the flow between pyruvate and G3P played the major role in the production of these compounds. They demonstrated that alterations such as the overexpression of Pps (phosphoenol pyruvate synthetase) and Pck (PEP carboxykinase) or the inactivation of the pykFA genes, which promote the flow of G3P, may increase production of lycopene while those deviating the flow towards G3P, like the overexpression of Ppc (PEP carboxylase), decreased lycopene production. Results indicated that G3P is metabolized through the glycolytic pathway to produce pyruvate so that it does not constitute a selective channel of the isoprenoid pathway.

Novel Carotenoids by Gene Combination

The synthesis of a specific carotenoid requires the presence of genes that may be combined within a single plasmid or may be included in several plasmids. A wide variety of genes are available which can be chosen to synthesize C_{40}-carotenoids. The *crtE* and *crtB* genes from *Erwinia* are always expressed in *Escherichia coli* for the synthesis of the first compound in the C_{40}-pathway. In this respect, several genes for phytoene-denaturase are also available, allowing us to achieve several degrees of insaturation of acyclic molecules. A number of genes may be selected to obtain cyclic lycopene molecules with β and ε rings. Likewise, genes for the insertion of functional groups (e.g., keto- and hydroxy- groups) in carotenoid molecules, in rings, and in linear chains can also be found. Regarding the C_{30}-pathway, the only genes that may be cloned are those coding for diapophytoene and diaponeurosporene. Most carotenogenic genes and plasmids have been expressed in *E. coli*, except for the gene coding for zeaxanthin epoxidase in *Nicotiana*, because reduced ferredoxin is required to form the epoxide and this cannot be provided by *E. coli*.

It has been noted that the major regulatory phase for the formation of a number of terpenoids in *E. coli* is the isomerization of isopentenyl pyrophosphate (IPP) to dimethyl-allyl pyrophosphate (a substrate that serves for chain elongation), which takes place only in the terpenoid pathway. The gene coding for IPP isomerase has been cloned in *E. coli*, and the amount of synthesized carotenoids depends on the activity of the expressed IPP isomerase.

In *E. coli*, high expression levels of the *crtB* gene for the active phytoene synthetase reduce the growth of the transformed strain, as a result of the deviation of geranyl-geranyl pyrophosphate (GGPP), a substrate of phytoene synthetase, from the endogenous metabolic pathway and of the low synthetic capacity of the strain's GGPP. In addition, high expression levels of the GGPP-synthetase gene produced an increment of up to 20% in carotenoid synthesis as compared with a low expression level. This demonstrates that the second limiting phase for carotenoid synthesis is the amount of available GGPP. Additionally, the levels of GGPP-synthetase and IPP-isomerase were shown to increase when the genes coding for both enzymes are introduced; consequently, a sufficient amount of GGPP and other precursors is available for both the endogenous metabolism and synthesis of carotenoids (Sandmann et al., 1999).

Sandmann et al. (1999) expressed in *Escherichia coli* several carotenogenic genes from *Erwinia uredovora* and two *Rhodobacter* species contained in compatible plasmids, which made possible the synthesis of eight different carotenoid hydroxy-derivatives, in addition to the following acyclic carotenoids: 1-hydroxyneurosporene, 1-hydroxylycopene, 1,1′-dihydroxy-lycopene, and dimethyl-spheroidene, as well as the cyclic carotenoids: 3-hydroxy-β-zea-carotene, 7-8 dihydro-zeaxanthin, 3- or 3′-hydroxy-7,8-dihydro-β-carotene, and 1′-hydroxy-γ-carotene. Most of these carotenoids are found in trace levels in the original sources or are new and uncommon and must be identified and characterized through infrared (IR) spectroscopy, nuclear magnetic resonance (NMR), mass spectrometry, etc. (Sandmann et al., 1999).

Albrecht et al. (2000) synthesized four novel strongly antioxidant lipophilic carotenoids. They expressed in *Escherichia coli* three different carotenoid-denaturases combined with carotenoid-hydratase, carotenoid-cyclase, and carotenoid-hydrolase using compatible plasmids and combinatorial biosynthesis to synthesize novel lipophilic carotenoids which are powerful cellular antioxidants. Carotenoids were designated as 1-OH-3′,4′-didehydro-lycopene, 3,1′-(OH)$_2$-γ-carotene, 1,1′-(OH)$_2$-3,4,3′,4′-tetradehydro-lycopene, and 1,1′-(OH)$_2$-3,4-didehydro-lycopene. The differences among these structurally related compounds derive from the length of the hydrocarbon chain and the position of hydroxy groups (C1′). The strong antioxidant activity steadily increases with the length of conjugated double bonds (from 11 to 13) in the monohydroxylated acyclic polyenic chain. However, the monocyclic 3,1′-(OH)$_2$-γ-carotene having 10 conjugated bonds in the hydrocarbon chain plus a terminal β group achieved a better protection than 1-OH-3,4-didehydro-lycopene having 12 conjugated double bonds. Further, Albrecht et al. (2000) determined the antioxidant activity of each compound using a membrane-liposome model system and demonstrated the capacity of a compound to protect against photooxidation and radical-mediated peroxidation reactions. They found that the monohydroxylated acyclic derivatives tend to achieve a better protection against peroxidation, whereas the monocyclic derivative 3,1'-$(OH)_2$-γ-carotene inhibits both reactions to the same extent. These compounds achieved a higher protection against membrane degradation, compared to OH- and 1,1'-$(OH)_2$-lycopene derivatives.

Conclusions

Rising demand for natural colorants as well as a deeper understanding of the health benefits of carotenoids have driven research about carotenogenesis in microorganisms. Manipulation of metabolic pathways has allowed us to obtain novel compounds with increased activity. Hyperproductive strains have been engineered that can produce a wide variety of carotenoids with the desired biological activity. Yields of these compounds are also subjected to study. Coupling of pathways is a strategy that has successfully been followed to elucidate and improve biosynthesis of these compounds. The challenge today is to produce strains capable of synthesizing stable compounds in terms of external effects such as light, pH, etc., some of which would be very difficult to produce by means of chemical synthesis.

References

Albrecht, M., Takaichi, S., Steiger, S., Wang, Z., and Sandmann, G. (2000) Novel hydroxycarotenoids with improved antioxidative properties produced by gene combination in *Escherichia coli*, *Nat. Biotechnol.*, 28, 843–846.

An, G.-H., Schumann, D., and Johnson, E. (1989) Isolation of *Phaffia rhodozyma* with increased astaxanthin content, *Appl. Environ. Microbiol.*, 55, 116–124.

An, G.-H. (1997) Photosensitization of the yeast *Phaffia rhodozyma* at a low temperature for screening carotenoid hyperproducing mutants, *Appl. Biochem. Biotechnol.*, 66, 262–268.

Andrewes, A.G. and Starr, M.P. (1976) (3r, 3R')Astaxanthin from the yeast *Phaffia rhodozyma*, *Phytochemistry*, 15, 1009.

Bjerkeng, B., Follig, M., Lagocki, S., Storebakken, T., Olli, J.J., and Alsted, N. (1997) Bioavailability of all-E-astaxanthin in rainbow trout *Oncorhynchus mykiss*, *Aquaculture*, 157, 63–82.

Boze, H., Moulin, G., and Galzy, P. (1992) Production of food and fooder yeasts, *Crit. Rev. Biotechnol.*, 12, 65–86.

Cerda-Olmedo, E. and Hütterman, A. (1986) Stimulation and inhibition of carotene biosynthesis in Phycomyces by aromatic compounds, *Angew Bot.* , 60, 59–70.

Chen, F. (1996) High cell density culture of microalgae in heterotrophic growth, *Trends Biotechnol.*, 14, 421–426.

Chumpolkulwong, N., Kakizono, T., Handa, T., and Nishio, N. (1997) Isolation and characterization of compactin resistant mutants of an astaxanthin synthesizing green alga *Haematococcus pluvialis*, *Biotechnol. Lett.*, 19, 299–302.

Cormier, F. (1997) *Food Colorants from Plant Cell Cultures: Functionality of Food Phytochemicals*, Plenum Press, New York, pp. 201–222.

Costa, I., Martelli, H., Da Silva, I., and Pomeroy, D. (1987) Production of β-carotene by *Rhodotorula* strain, *Biotechnol. Lett.*, 9, 373–375.

Czygan, F.C. (1970) Blood-rain and blood-snow: nitrogen deficient cells of *Haematococcus pluvialis* and *Chlamydomonas nivalis*, *Arch. Microbiol.*, 74, 69–76.

Del Campo, J.A., Rodríguez, H., Moreno, J., Vargas, M.A., Rivas, J., and Guerrero, M.G. (2001) Lutein production by *Muriellopsis* sp. in an outdoor tubular photobioreactor, *J. Biotechnol.*, 85, 289–295.

Downham, A. and Collins, P. (2000) Colouring our foods in the last and next millenium., *Int. J. Food Sci. Technol.*, 35, 5–22.

Fan, L., Vonshak, A., Zarka, A., and Boussiba, S. (1998) Does astaxanthin protect *Haematococcus* against light damage?, *Z. Naturforsh.*, 53c, 93–100.

Farmer, W.R. and Liao, J.C. (2001) Precursor balancing for metabolic engineering of lycopene production in *Escherichia coli*, *Biothecnol. Prog.*, 17, 57–61.

Frengova, G., Simova, E., Pavlova, K., Beshkova, D., and Grigorova, D. (1994) Formation of carotenoids by *Rhodotorula glutinis* in whey ultrafiltrate, *Biotechnol. Bioeng.*, 44, 888–894.

Frengova, G., Simova, E., and Beshkova, D. (1997) Carotene-protein and exopolysaccharide production by cocultures of *Rhodotorula glutinis* and *Lactobacillus helveticus*, *J. Ind. Microbiol. Biotechnol.*, 18, 272–277.

Grung, M., D'Souza, F.M.L., Borowitzka, M., and Liaaen-Jensen, S. (1992) Algal carotenoids 51. Secondary carotenoids 2. *Haematococcus pluvialis* aplanospores as a source of (3S, 3'S)-astaxanthin esters, *J. Appl. Phycol.*, 4, 164–171.

Haard, N. (1988) Astaxanthin formation by the yeast *Phaffia rhodozyma* on molasses, *Biotechnol. Lett.*, 10, 609–614.

Hagen, C., Braune, W., Birckner, E., and Nuske, J. (1993) Functional aspects of secondary carotenoids in *Haematococcus lactustris* (Girod) Rostafinski (Volvocales). I. The accumulation period as an active metabolic process, *New Phytol.*, 125, 624–633.

Hagen, C., Braune, W., and Bjorn, L.S. (1994) Functional aspects of secondary carotenoids in *Haematococcus lactustris* (Girod) Rostafinski (Volvocales). III. Action as a sunshade, *J. Phycol.*, 30, 241–248.

Havaux, M. (1998) Carotenoids as membrane stabilizers in chloroplasts, *Trends Plant Sci.*, 4, 147–151.

Hirschberg, J. (1999) Production of high-value compounds: carotenoids and vitamin E, *Curr. Opin. Biotechnol.*, 10, 186–191.

Ichii, T., Takenaka, H., Konno, T., Ishida, H., Sato, A., Suzuki, and Yamazumi, K. (1993) Development of a new commercial-scale airlift fermentor for rapid growth of yeast, *J. Ferment. Bioeng.*, 75, 375–379.

Johnson, E.A. and An, G.-H. (1991) Astaxanthin from microbial sources, *Crit. Rev. Biotechnol.*, 11, 297–326.

Kajiwara, S., Kakizono, T., Saito, T., Kondo, K., Ohtani, T., Nishio, N., Nagai, S., and Misawa, N. (1995) Isolation and functional identification of a novel cDNA for astaxanthin biosynthesis from *Haematococcus pluvialis* and astaxanthin synthesis in *Escherichia coli*, *Plant Mol. Biol.*, 29, 343–352.

Kobayashi, M., Hirai, N., Kurimura, Y., Ohigashi, H., and Tsuji, Y. (1997) Abscisic acid-dependent algal morphogenesis in the unicellular green alga *Haematococcus pluvialis*, *Plant Growth Regul.*, 22, 79–85.

Lang, N.J. (1968) Electron microscopic studies of extraplastidic astaxanthin in *Haematococcus*, *J. Phycol.*, 4, 12–19.

Lois, L.M., Campos, N., Putra, S.R., Danielsen, K., Rohmer, M., and Boronat, A. (1998) Cloning and characterization of a gene from *Escherichia coli* encoding a tranketolase-like enzyme that catalyzes the synthesis of D-1-deoxyxylulose 5-phosphate, a common precursor for isoprenoid, thiamin, and pyridoxol biosynthesis, *Proc. Natl. Acad. Sci. USA*, 95, 2105–2110.

Lorenz, R.T. and Cysewski, G.R. (2000) Commercial potential for *Haematococcus* microalgae as a natural source of astaxanthin, 18, *TIBTECH*, 160–166.

Lotan, T. and Hirschberg, J. (1995) Cloning and expression in *E. coli* of the gene encoding β-ε-4-oxygenase that converts β-carotene to the ketocarotenoid canthaxanthin in *H. pluvialis*, *FEBS Lett.*, 363, 125–128.

Margalith, P.Z. (1999) Production of ketocarotenoids by microalgae, *Appl. Microbiol. Biotechnol.*, 51, 431–438.

Mathews-Roth, M.M. (1993) Carotenoids in erythropoietic protoporphyria and other photosensitivity diseases, *Ann. N.Y. Acad. Sci.*, 691, 127–138.

Matthews, P.D. and Wurtzel, E.T. (2000) Metabolic engineering of carotenoid accumulation in *Escherichia coli* by modulation of the isoprenoid precursor pool with expression of deoxyxylulose phosphate synthase, *Appl. Microbiol. Biotechnol.*, 53, 396–400.

Misawa, N., Nagakawa, M., Kobayashi, K., Yamano, S., Izawa, Y., Nakamura, K., and Harashima, K. (1990) Elucidation of the *Erwinia uredovora* carotenoid biosynthetic pathway by functional analysis of gene products expressed in *Escherichia coli*, *J. Bacteriol.*, 172, 6704–6712.

Misawa, N., Yamano, S., and Ikenaga, H. (1991) Production of β-carotene in *Zymomonas mobilis* and *Agrobacterium tumefaciens* by introduction of the biosynthesis genes from *Erwinia uredovora*, *Appl. Environ. Microbiol.*, 57, 18747–1849.

Misawa, N., Yamano, S., Linden, H., De Felipe, M.R., Lucas, M., Ikenaga, H., and Sandmann, G. (1993) Functional expression of the *Erwinia uredovora* carotenoid biosynthesis gene *crtI* in transgenic plants showing an increase of β-carotene biosynthesis activity and resistance to the bleaching herbicide norflurazon, *Plant J.*, 4, 833–840 [erratum published in *Plant J.*, 5(2) 309, 1994].

Miura, Y., Kondo, K., Saito, T., Shimada, H., Fraser, P.D., and Misawa, N. (1998) Production of the carotenoid lycopene, β-carotene, and astaxanthin in food yeast *Candida utilis*, *Appl. Environ. Microbiol.*, 64(4), 1226–1229.

Nakazoe, J., Ishii, S., Kamimoto, M., and Takeuchi, M. (1984) Effects of supplemental carotenoid pigments on the carotenoid accumulation in young sea bream (*Chysophry major*), *Bull. Tokai Reg. Fish. Res. Lab.* 113, 29–41.

Olson, J.A. and Krinsky, N.I. (1995) Introduction: the colorful, fascinating world of the carotenoids, important physiologic modulators, *FASEB J.*, 9, 1547–1550.

Parajó, J., Santos, C.V., and Vázquez, M. (1998) Production of carotenoids by *Phaffia rodozyma* growing on media made from hemicellulosic hydrolysates of *Eucaliyptus globulus* wood, *Biotechnol. Bioeng.*, 59, 501–506.

Pulz, O. and Scheibenbogen, K. (1998) Photobioreactors: design and performance with respect to the light energy input, *Adv. Biochem. Eng. Biotechnol.*, 59, 124–152.

Sandmann, G., Albrecht, M., Schnurr, G., Knörzer, O., and Böger, P. (1999) The biotechnological potential and design of novel carotenoids by gene combination in *Escherichia coli*, *TIBTECH*, 17, 233–237.

Schmidt-Dannert, C. (2000) Engineering novel carotenoids in microorganisms, *Curr. Opin. Biotechnol.*, 11, 255–261.

Sprenger, G.A., Schorken, U., Wiegert, T., Grolle, S., de Graaf, A.A., Taylor, S.V., Begley, T.P., Bringer-Meyer, S., and Sahm, H. (1997) Identification of a thiamin-dependent synthase in *Escherichia coli* required for the formation of the 1-deoxyxylulose 5-phosphate precursor to isoprenoids, thiamin, and pyridoxol, *Proc. Natl. Acad. Sci. USA*, 94, 12857–12862.

Stahl, W. and Sies, H. (1998) The role of carotenoids and retinoids in gap junctional communication, *Int. J. Vit. Nutr. Res.*, 68, 354–359.

Stauffer, K. and Leeder, Y. (1978) Extracellular microbial polysaccharide production by fermentation on whey, *J. Food Sci.*, 43, 756–758.

Storebakken, T., Foss, P., Schiedt, K., Austreng, E., Liaeen-Jensen, S., and Manz, V. (1987) Carotenoids in diets for salmonids. IV. Pigmentation of Atlantic salmon with astaxanthin, astaxanthin dipalmitate and cantaxanthin, *Aquaculture*, 65, 279–292.

Tada, M. and Shiroishi, M. (1982) Mechanism of photoregulated carotenogenesis in *Rhodotorula minuta*. I. Photocontrol of carotenoid production, *Plant Cell Physiol.*, 23, 541–547.

Tjahjono, A.E., Hayama, Y., Kakizono, Y., Nishio, N., and Nagai, S. (1994) Isolation of resistant mutants against carotenoid biosynthesis inhibitors for a green alga *Haematococcus pluvialis*, and their hybrid formation by protoplast fusion for breeding of higher astaxanthin producers, *J. Ferment. Bioeng.*, 77, 352–357.

Van den Berg, H., Faulks, R., Granado, H.F., Hirschberg, J., Olmedilla, B., Sandmann, G., Southon, S., and Stahl, W. (2000) The potential for the improvement of carotenoid levels in foods and the likely systemic effects, *J. Sci. Food Agric.*, 80, 880–912.

Verdoes, J.C., Misawa, N., and Van Ooyen, A.J.J. (1999) Cloning and characterization of the astaxanthin biosynthetic gene encoding phytoene desaturase of *Xanthophyllomyces dendrorhous*, *Biotechnol. Bioeng.*, 63, 750–755.

Wery, J., Gutker, D., Renniers, A.C.H.M., Verdoes, J.C., and Van Ooyen, A.J.J. (1997) High copy number integration into the ribosomal DNA of the yeast *Phaffia rodozyma*, *Gene*, 184, 89–97.

Wurtzel, E.T., Valdez, G., and Matthews, P.D. (1997) Variation in expression of carotenoid genes in transformed *E. coli* strains, *Biores. J.*, 1, 1–11.

Yamano, S., Ishii, T., Nakagawa, M., Ikenaga, H., and Misawa, N. (1994) Metabolic engineering for production of β-carotene and lycopene in *Saccharomyces cereviseae*, *Biosci. Biotechnol. Biochem.*, 58, 112–1114.

Yong, Y.Y.R. and Lee, Y.K. (1991) Do carotenoids play a photoprotective role in the cytoplasm of *Haematococcus lacustris* (Chlorophyta)?, *Phycologia*, 30, 257–261.

Zalashko, M. (1990) Processing of milk whey based on lactose oxidation, in *Biotechnology of Milk Whey Processing*, E. Sokolova, Ed., Science Press, Moscow, Russia, pp. 161–163.

Zlotnik, I., Sukenik, A., and Bubinsky, Z. (1993) Physiological and photosynthetic changes during the formation of red aplanospores in the chlorophyte *Haematococcus pluvialis*, *J. Phycol.*, 29, 463–469.

9

Studies on the Reverse Micellar Extraction of Peroxidase from Cruciferae Vegetables of the Bajio Region of México

Carlos Regalado-González, Miguel A. Duarte-Vázquez, Sergio Huerta-Ochoa, and Blanca E. García-Almendárez

CONTENTS

1-56676-892-6/03/$0.00+$1.50
© 2003 by CRC Press LLC

Introduction

The impressive advances in genetic engineering have made it possible to produce proteins important for pharmaceutical and industrial purposes; however, the technology for separating and purifying proteins has not advanced at the same pace (Sadana and Raju, 1990). It has been estimated that up to 90% of new product costs are attributable to downstream processing (Kadam, 1986). Thus, biotechnology requires separation techniques that can be scaled up into feasible, more economical downstream technology. An alternative to bioseparation is the selective solubilization of proteins using reverse micelles. They offer an essentially nondenaturing environment, thus opening up the possibility of using liquid–liquid extraction technology, which may be used continuously to control the selectivity of protein extraction from solution mixtures or fermentation broths without any significant loss of biological activity. This work reviews the fundamental and technological aspects of reverse micelles in the context of bioseparations. The reverse micellar system is discussed, followed by reports of protein solubilization in reverse micelles and the kinetics of protein partitioning. Finally, we report our results on peroxidase purification using reverse micelles from some vegetables of the *Cruciferae* family available in the Bajio region of México.

Reverse Micellar Systems

Reverse micelles are thermodynamically stable, optically transparent, isotropic dispersions of water in oil (w/o) stabilized by a surfactant. Examples of ternary systems (surfactant, cosurfactant, solvent) forming reverse micelles and used for protein solubilization purposes are given in Table 9.1; more information on these systems is given by Luisi and Steinmann-Hofmann (1987). The anionic surfactant AOT (aerosol-OT [sodium *bis*{2-ethylhexyl} sulfosuccinate]) is very efficient in reducing the interfacial free energy (Eicke and Kvita, 1984), and its molecular flexibility allows the formation of very stable reverse micelles in isooctane media (Maitra, 1984). Because the anionic surfactant AOT does not require a cosurfactant, it has been widely used. Trioctylmethylammonium chloride (TOMAC) at concentrations higher than 12 mM and one-alkyl-chain cationic surfactants, such as cetyltrimethyl ammonium bromide (CTAB), require a short-chain alcohol ($\geq C_4$) to form

TABLE 9.1

Examples of Surfactant Systems Reported to Form Reverse Micelles

Surfactant	Solvent
Cationic	
Trimethyloctylammonium chloride (TOMAC)	Cyclohexane, isooctane/octanol, isooctane/octanol/rewopal HV5 (a nonionic cosurfactant)
Aliquat 336 (mixture of trialkyl [$C_8\text{-}C_{10}$] methyl ammonium chloride salts)	Isooctane/isotridecanol, isooctane/n-butanol
Hexadeciltrimethylammonium bromide (CTAB)	Isooctane/hexanol (plus butanol to obtain an aqueous phase in equilibrium with a reverse micelle phase), n-octane/hexanol, n-octane/chloroform
Anionic	
Sodium *bis*(2-ethylhexyl) sulfosuccinate (AOT)	Isooctane, n-hydrocarbons ($C_6\text{-}C_{10}$), cyclohexane, carbon tetrachloride, benzene
Sodium dodecylbenzensulfonate	Decane/butanol
Nonionic	
Ethoxylated nonylphenols	n-Octane
Triton X-100	Cyclohexane/hexanol
Zwitterionic	
Phosphatidylcholine, phosphatidyletanolamine	Benzene, n-heptane, n-hexane

reverse micelles in apolar solvents. The size of the aqueous core of reverse micelles is approximately proportional to the Wo (molar ratio of water to surfactant) values. This has been checked by small-angle neutron scattering (SANS) (Robinson et al., 1984). Based on geometrical considerations, and assuming that all of the surfactant is located in the interface of the spherical reverse micelle, the radius of the aqueous core of the reverse micelle (r_c) is given by:

$$r_c = \frac{3V_w Wo}{A_s} \qquad (9.1)$$

where V_w is the molar volume of water (18×10^{-6} m^3 mol^{-1}); A_s is the surfactant headgroup area (3.0×10^5 m^{2}mol^{-1} for AOT and TOMAC; 3.9×10^5 m^{2}mol^{-1} for CTAB). The latter physical parameters can be assumed to be constant at room temperature (Chen et al., 1986).

The size of AOT reverse micelles in equilibrium with a bulk aqueous phase is a strong function of both cation concentration and type (Na$^+$ > K$^+$ ≈ Ca^{2+}, for 0.1 M concentration and 0.1 M AOT) (Leodidis and Hatton, 1989). In contrast, Krei and Hustedt (1992) found that the reverse micelle size of cationic surfactants (e.g., CTAB) is a function of both anion concentration and type (F$^-$ > CH$_3$COO$^-$ > Cl$^-$ > Br$^-$, for 0.1 M concentration and 0.2 M CTAB). In the AOT/isooctane/water system, an increase in the Wo value from 4 to 50 results in an increase of r_c from 1.0 to 7.9 nm; the average aggregation number (N_{agg}) increases from 35 to 1380, while the area of the surfactant head changes from 0.36 to 0.57 nm^2 (Maitra, 1984).

Dynamic Aspects of Reverse Micelles

Fletcher et al. (1987) showed that the exchange rate of material between reverse micelles depended on droplet size (Wo) and temperature. They showed that reverse micelles exist in a dynamic equilibrium involving fusion of the droplets to form short-lived droplet dimers followed by their resepa-ration, resulting in a random redistribution of the micellar contents and surfactant shell. Only about one in every thousand collisions leads to an exchange, and this process takes place in the millisecond to microsecond time range (Clark et al., 1990). The microemulsion system is then kinetically active, but the fusion process maintains the equilibrium properties (droplet size, polydispersity) of the microemulsion (Eastoe et al., 1991).

Solubilization of Proteins in Reverse Micelles

Electrostatic interactions have been identified as the main driving force of protein solubilization. Protein solubilization occurs via a cooperative mech-anism in which the interface deforms around the protein to form the micelle (Dungan et al., 1991). There is a strong dependence of solubility and aqueous phase pH and ionic strength. Luisi et al. (1979) were the first to recognize the potential for separating and purifying proteins from an aqueous solution. Table 9.2 shows that most of the studies have been done with AOT and, to a lesser extent, other cationic and nonionic surfactants. However, the use of anionic (AOT) or cationic (CTAB, TOMAC, etc.) surfactants limits partition-ing studies to proteins small enough to fit inside reverse micelles, which, in practice, means relatively low-molecular-weight molecules (<100 kDa) (Dek-ker, 1990). Table 9.2 shows a trend toward studying mixtures of anionic/ nonionic or nonionic surfactants which give droplet sizes from 15 to 30 nm in diameter (Aveyard et al., 1989). However, to date we have little under-standing of the size limits or the main driving forces necessary for successful protein solubilization using nonionic surfactants. Despite all the efforts to model protein partitioning into reverse micelles, it is still a fact that the use of reverse micelles as a tool for the separation of proteins is mainly based on trial-and-error experiments.

Factors Affecting Protein Solubilization

Unlike many other liquid–liquid extraction systems, reverse micellar extrac-tions may exhibit very high partition coefficients with water (i.e., almost quantitative protein transfer). Protein partitioning into a reverse micellar phase is influenced by the size, isoelectric point, and charge distribution of the protein; the size of the reverse micelle; the type of solvent used; the nature of the surfactant used; the free energy changes associated with the change in size of the reverse micelles due to protein solubilization; the presence of affinity ligands; and the pH, ionic strength, and type of salt of

TABLE 9.2

Summary of Recent Research on Reverse Micelles and the Factors Influencing Protein Partitioning[a]

Authors	Surfactant Used	Protein Studied	Main Areas Studied and Results
Adachi et al. (1996, 1997, 1998)	AOT, TOEDE	Concanavalin A, trypsin	Effect of pH, IS, and AOT on FT bioaffinity separation of concanavalin A; trypsin extraction from pancreatin and recovery (50%) using bioaffinity.
Almeida et al. (1998)	AOT	α-chym, δ-chym	Stability and activity of two chymotrypsins inside RM.
Bartsotas et al. (2000)	AOT	Albumin, β-gal, *B. st.* FB	Effect of buffers, cations, proteins, and *B. st.* FB fractionated proteins on emulsion formation and stability.
Bernardi and Palmieri (1996)	AOT	Trypsin-like endoproteinase	Solubilization of *O. nubilatis* endo-proteinase from the larvae powder at different Wo and inhibition studies.
Bosetti et al. (1997)	DTAB, AOT	Glutaryl-7-ACA acylase	FT and BT under fixed Wo, but taking few hours, with low activity yields.
Brandani et al. (1996a,b)	TOMAC	α-amylase	Effect of pH, IS, and surfactant on FT and BT; effect of partition coefficient on protein transfer modeling.
Cardoso et al. (1999)	TOMAC	Aspartic acid, phenilalanine, tryptophan	Effect of IS, pH, and TOMAC on FT driving force, solute RM interfacial interaction, and location in the RM.
Chang and Chen (1996); Chang et al. (1997)	AOT/ALQ-336	α-amylase	Effect of pH, IS, and co-solvent (alkyl alcohols) on α-amylase purification.
Goto et al. (1997)	DOLPA, various	Various	Surfactants forming a close-packed complex with protein are good protein-solubilizing agents.
Hashimoto et al. (1998)	AOT	Ribo-a	Refolding of denatured ribo-a by solid–liquid extraction into RM.
Hayes (1997); Hayes and Marchio (1998)	AOT	Ribo-a; lysozyme, α-chym, pepsin, BSA, catalase	Mechanistic model for the solid-phase protein extraction. Protein BT increased with cosurfactant (1-alkanols) chain length and Wo. Protein release mechanisms are discussed.

-- continued

TABLE 9.2 (continued)

Summary of Recent Research on Reverse Micelles and the Factors Influencing Protein
Partitioning[a]

Authors	Surfactant Used	Protein Studied	Main Areas Studied and Results
Hossain et al. (1999)	AOT, AOT/ Tween 85	Lipase	Kinetic modeling of catalysis in RM; activity increased at higher Wo.
Hu and Gulari (1996a,b)	NaDEHP	Cyt-c, α-chym, antibiotics	Effect of pH and IS on protein and antibiotic solubilization and of Ca^{2+} on their recovery.
Huang et al. (1996).	DAB	Ribo-a	Conformation and activity of solubilized ribo-a.
Jarudilokkul et al. (1999, 2000a,b)	TOMAC, DTAB, AOT	Ribo-a, cyt-c, lysozyme	Protein BT is faster with higher yield by using a counterionic surfactant. Type and concentration of demulsifiers to suppress emulsion formation in RM; kinetics of protein FT and BT in a Graesser contactor.
Kawakami and Dungan (1996)	AOT	α-lacta, β-lacto	Effect of hydrophobic and electrostatic interactions on protein solubilization over a wide range of pH and IS.
Krieger et al. (1997)	AOT	Lipase	Effect of IS, pH, and AOT on protein FT and BT, from a crude extract.
Lazarova and Tonova (1999)	CTAB	α-amylase	Kinetics of protein FT and BT in a stirred cell with two separate compartments.
Lye et al. (1996)	AOT	Lysozyme	Modeling kinetics of lysozyme extraction using a spray column.
Melo et al. (1996, 1998)	AOT, CTAB	Cutinase	Cutinase reversible inactivation faster in AOT than CTAB RM. Modeling cutinase denaturation in RM of AOT and CTAB.
Nagayama et al. (1999)	AOT	Lipase	High efficacy of protein and activity recoveries by adding GuHCl (80 mM).
Naoe et al. (1999)	DK-F-110	Cyt-c	Effect of pH, buffer, and IPA on cyt-c FT to nonionic surfactant sugar ester RM.
Ono et al. (1996)	DOLPA	Hemoglobin	New anionic surfactant; no need for co-surfactant.
Orlich and Schomaecker (1999) Orlich et al. (2000)	$C_{13}EO_6$ (hexa-ethoxylene-tridecanol)	LADH, YADH, CR	Effect of pH and T on kinetics of a ketone reduction with cofactor regeneration in RM; semibatch LADH catalysis with cofactor regeneration from CR in RM; effect of RM composition on activity.

-- continued

TABLE 9.2 (continued)

Summary of Recent Research on Reverse Micelles and the Factors Influencing Protein
Partitioning[a]

Authors	Surfactant Used	Protein Studied	Main Areas Studied and Results
Rariy et al. (1998)	AOT	α-chym	Increase of α-chym stability because glycerol increased RM structural order.
Regalado et al. (1996); Pérez et al. (1999)	AOT, CTAB	Horseradish peroxidase	Two-stage RM purification of HRP from horseradish roots; purification of peroxidase from broccoli.
Shiomori et al. (1996, 1998)	AOT	Lipase, BSA	Kinetics of lipase catalysis in RM; successful solubilization of BSA.
Spirovska and Chaudhuri (1998)	AOT	Lysozyme, ribo-a	BT recovery of 95% achieved with KCl 2.5 M, and 60–80% with an ion-exchange matrix. Glucose, sucrose, and AOT enhance ribo-a FT and BT.
Stobbe et al. (1997)	AOT, Span 80	α-chym	New liquid emulsion membrane where protein extraction is conducted by reverse micelles.
Sun et al. (1998)	Lecithin	Lysozyme, BSA, cyt-c	High selectivity of immobilized CB at low IS; effect of pH, T, and IS on lysozyme affinity extraction.
Tong and Furusaki (1997)	AOT	Lysozyme	Modeling of lysozyme extraction into RM, using a rotating disc contactor at 30–48 rpm.
Vasudevan and Wiencek (1997)	Alkyl sorbitan esters, Neodol 91-2.5	Cyt-c, concanavalin A, BSA	Role of the interface in protein extraction using nonionic surfactants.

[a] From publications appearing in indexed journals from 1996 to 2000.

Abbreviations: ALQ-336: aliquat-336; BDBAC: benzyl-dodecyl-bis(hydroxyethyl)-ammonium chloride; BSA: bovine serum albumin; *B. st.* FB: *Bacillus stearothermophilus* fermentation broth; BT: backward transfer; α-, δ-chym: α- or δ- chymotrypsin; CB: cibacron blue; CR: carbonyl reductase; CTAB: cetyltrimethylammonium bromide; cyt-c: cytochrome-c; DAB: dodecylammonium butyrate; DEHPA: di(2-ethylhexyl) phosphoric acid; DOLPA: dioleyl phosphoric acid; DTAB: dodecyltrimethylammonium bromide; FT: forward transfer; β-gal: β-galactosidase; glutaryl-7-ACA: glutaryl-7-aminocephalosporanic acid; GuHCl: guanidine hydrochloride; HRP: horseradish peroxidase; IPA: isopropyl alcohol; IS: aqueous-phase ionic strength; α-lacta: α-lactalbumin; β-lacto: β-lactoglobulin; LADH: horse liver alcohol deshydrogenase; NaDEHP: sodium *bis*(2-ethylhexyl) phosphate; RM: reverse micelles; ribo-a: ribonuclease-a; SLM: supported liquid membrane; Span 60: sorbitan monoestearate; T: temperature; TOEDE: tetra-oxyethylene decylether; TOMAC: trioctylmethylammonium chloride; Tween 85: polyoxyethylene sorbitan trioleate; YADH: yeast alcohol deshydrogenase.

the aqueous phase. Relevant reviews are found in Luisi and Magid (1986) and Hatton (1989). Hydrophobic interactions between apolar regions at the surface of the protein and the surfactant may also be important (Pires and Cabral, 1993).

Back-Extraction Process

It is usually sufficient to adjust the pH and/or ionic strength of the stripping solution to conditions that are unfavorable for protein solubilization to extract the protein from the reverse micellar phase. However, there are some exceptions to this, because of strong protein–surfactant interactions, which can be overcome by adding a small amount of short-chain alcohol (e.g., isopropanol) to the stripping phase without protein denaturation and with no destruction of the functional integrity of the system (Carlson and Naga-rajan, 1992). Dekker et al. (1991) reported the advantage of increasing the temperature (35°C) of the organic phase to desolubilize α-amylase in the excess aqueous phase. An ion-exchange matrix (Chaudhuri and Spirovska, 1994), an adsorption process with silica (Leser et al., 1993), and micelle dehy-dration with aluminosilicate molecular sieves (Gupta et al., 1994) have been used to recover proteins successfully from reverse micelles. More recent reports are shown in Table 9.2.

Kinetics of Protein Partitioning

The kinetics of protein extraction is important, because it enables us to elucidate the mechanism of mass transfer and to obtain basic information for the design and scale-up of suitable liquid–liquid contactors (Bausch et al., 1992). Contacting equipment used to study the mass transfer rate of protein extraction include Lewis-type stirred cells (Nitsch and Plucin-sky, 1990), vibrating plates (Hentsch et al., 1992), mixer-settlers (Dekker, 1990), hollow-fiber modules (Dahuron and Cussler, 1988), spray columns (Lye et al., 1996), packed columns (Nishii et al., 1999), rotating disk con-tactors (Tong and Furusaki, 1997), centrifugal contactors (Lu et al., 1998), and a Graesser ("raining bucket") contactor (Jarudilokkul et al., 2000b). The transfer of protein and other solutes into and out of reverse micelles has been modeled, in all cited reports, on the basis of a modified form of the two-film theory (with significant or negligible interfacial resistance), the penetration-surface renewal theory, or a diffusion model. These reports suggest that the rates of forward and backward protein transfers are apparently controlled by a different mechanism: a diffusion-controlled, rate-determining step exists on the aqueous side for protein solubilization, while the re-extraction process is governed by the interfacial resistance to protein transfer (Plucinsky and Nitsch 1989). The forward-transfer rate is usually one or two orders of magnitude faster than the back-extraction process.

Peroxidase

The peroxidases (E.C. 1.11.1.7) are oxido-reductases ubiquitous in the plant and animal kingdoms. Most are glycoproteins with a polypeptidic chain of about 300 amino-acid residues having a heme group in the active site (Campa, 1991). As many as 40 isoenzymes have been detected in impure preparations (Dunford, 1991). Individual isoenzymes differ in physicochemical properties, amino-acid composition, and catalytic properties (Paul and Stigbrand, 1970). Peroxidase is used primarily to produce test kits for determination of glucose, cholesterol, etc., and in a more purified form for enzyme immunoassays and immunohistochemistry (Tijssen, 1985).

Peroxidase Purification Using Reverse Micelles: Natural Sources

Paradkar and Dordick (1993) proposed an affinity-based reverse micellar extraction for the purification of peroxidase from soybean hulls. A high purification factor with a nearly pure peroxidase was obtained after regeneration of the concanavalin A ligand. Unfortunately, the ligand could not be completely recovered, thus adding to the cost of the process. Huang and Lee (1994) purified peroxidase from horseradish roots. A single-step extraction at pH 11 was used to remove contaminant proteins. A purification factor of 24 indicated a partly purified peroxidase, but no evidence was provided regarding the purity achieved. Regalado et al. (1996) purified the same enzyme (80% pure) using a two-stage reverse micellar extraction with organic phase recycle from the dialyzed aqueous extract, obtaining a purification factor of 80.

Vegetables of the Bajio Region of México

About 90% of peroxidase world production is obtained from horseradish roots (*Armoracia rusticana*) (Krell, 1991), a plant that belongs to the *Cruciferae* family and does not grow well in México. Other *Cruciferae* plants extensively produced in the Bajio region that may be used as alternative sources of peroxidase include broccoli, Brussels sprouts, turnip roots, and radishes. Pérez-Arvisu et al. (1999) purified peroxidase from broccoli using 0.2-M CTAB reverse micelles in isooctane/n-pentanol (95:5) with 60% yield and a purification factor of 6.7. These values were similar to those obtained by ammonium sulfate and acetone precipitation. Thus, reverse micellar extraction can be of potential advantage when used during prepurification stages of proteins. Here we present the use of reverse micelles to purifiy peroxidase from turnip roots (*Brassica napus* L. var. esculenta, D.C.), Brussels sprouts (*Brassica oleracea* var. gemnifera), and radishes (*Raphanus sativus paruus*).

Materials and Methods

Materials

Turnip roots, radishes, and Brussels sprouts were purchased from a local market. These vegetables were washed, rinsed with distilled water, drained, frozen, and stored at –20°C until use. AOT, analytical-grade *n*-pentanol, 12-kDa molecular weight cut off (MWCO), dialysis membranes, ABTS (2,2'-*azinobis*[3-ethylbenzthiazoline-6-sulfonic acid]), hydrogen peroxide (30% v/v), and spectrophotometric-grade isooctane (99%) were purchased from Sigma (St. Louis, MO). CTAB (>99%) was supplied by Fluka (Buchs, Switzerland). Deionized water (Milli-Q) was used to prepare all aqueous solutions.

Methods

- **Firmness.** Firmness of the vegetables was determined using a texture analyzer TA-XD (Texture Technologies Corp., Scarsdale, NY) with a 5-mm-diameter cylindrical probe with cross-sectional area of 19.63 cm^2 and traveling at a downward speed of 1 mm/sec. The maximum force exerted for penetration was recorded.

- **Color.** The color was measured using a Minolta spectrophotometer (model CM-2002, Japan), with standard illuminant C at 2°. The CIELAB system was used involving values of L (lightness), a (green-red), and b (blue-yellow). For objective color measurement, the method of McGuire (1992) expressing chromaticity (strength of the color) as $(a^2 + b^2)^{1/2}$ and hue as tan^{-1} (b/a) was used.

- **Moisture content.** Moisture was measured gravimetrically according to AOAC (1990).

- **Crude extract.** The vegetables (700 g) were removed from frozen storage and homogenized at 4°C using 1 L of 10-mM potassium acetate buffer (pH 6.0). The extract was centrifuged at 12,000 g (Beckman J21R), and the supernatant was used for further purification.

- **Protein determination.** Protein assays in the aqueous extracts were conducted using the Bicinchoninic acid method (Smith et al., 1985), with bovine serum albumin (BSA) as standard.

- **Peroxidase activity.** Activity was evaluated using ABTS as the hydrogen donor, measuring the change in absorbance at 414 nm (Childs and Bardsley, 1975). One activity unit (U) was defined as the micrograms of ABTS consumed per minute.

- **Reverse micellar extraction.** The crude extract from all vegetables was dialyzed against deionized water for 48 h prior to reverse micellar extraction. When AOT (0.10 M) in isooctane was used, the aqueous-phase pH was adjusted with malonic acid buffer (30 mM) between 3.0 and 4.0. Ionic strength (IS) was adjusted between 0.05 and 0.20 M using NaCl. Phase transfer was conducted in well-stirred closed vessels, at room temperature: 25±2°C. Forward and backward transfers (5 mL of each phase) were performed until equilibrium was reached (8 and 15 min, respectively). An Eppendorf benchtop centrifuge (model 5804R; Hamburg, Germany) was used to achieve phase separation (5000 g for 2 min at room temperature). Back extraction was carried out with an equal volume of fresh aqueous buffer (80 mM KH_2PO_4, pH 8.0) containing 1.0 M KCl. An Orion Karl Fischer titrator (model AF7; Beverly, MA) was used to measure water content in the organic phase. CTAB (0.20 M) was dissolved in a mixture of isooctane-n-pentanol or hexanol (90/10). pH varied from 7 to 12, using 30-mM phosphate buffer. IS was adjusted between 0.05 and 0.60 M with KCl or NaCl. Back extraction was conducted as above, but the aqueous phase contained 2.0 M KCl (pH 4.0, adjusted with HCl).

Results and Discussion

Raw Material Characterization

Peroxidase concentration and proportion of isoenzymes can vary according to the degree of ripeness of the vegetable (Paul and Stigbrand, 1970; Contreras-Padilla and Yahia, 1998). The vegetables were purchased at commercial maturity. Color, penetration resistance, and moisture were measured as indicators of maturity index (Table 9.3). To evaluate the potential of each vegetable as an alternative source of peroxidase, we measured initial protein content and activity of each crude extract (Table 9.4). This table shows that turnip roots had the highest peroxidase activity per unit weight of fresh material.

Extraction from Dialyzed Extracts

Earlier studies on HRP purification showed reduced protein solubilization into reverse micelles when the crude extract was used without previous dialysis (Regalado et al., 1996). This was attributed to the mineral content, especially Ca^{2+}, Mg^{2+}, and K^+, which caused electrostatic screening of the surfactant headgroups, hindering interaction with the protein. Because these minerals are present in good amounts in the *Cruciferae* vegetables used in this study (Table 9.4), we decided to start the extraction from the dialyzed extract.

TABLE 9.3

Maturity Indices Tested on Raw Vegetables Used for Peroxidase Purification Using Reverse Micelles[a]

	Penetration Test		Color			
Vegetable	ΔP[b] (kPa)	d[c] (mm)	L[d]	Hue	C[e]	Moisture (%)
Brussels sprouts	47.5	4.8	47.2	−1.1	29.7	94.3
Radish	17.0	3.8	38.0	0.3	40.0	96.5
Turnip	26.5	4.6	81.5	−1.5	17.6	93

[a] Average values of three replicates, with SE within 5% of the mean.
[b] Required pressure for tissue rupture.
[c] Maximum penetration length before tissue rupture (maximum peak in a plot of penetration strength against penetration length).
[d] Lightness (overall relationship of light reflection to absorption).
[e] Chromaticity (color intensity strength or purity).

TABLE 9.4

Peroxidase Activity, Protein, and Mineral Content of Vegetables Studied

Vegetable	Activity[a] (U/g fresh material)	Protein[a] (mg/mL)	Ca[b] (mg/100 g)	Mg[b] (mg/100 g)	K[b] (mg/100 g)
Brussels sprouts	7.9	3.5	53	23	390
Radish	6.6	0.6	30	9.0	268
Turnip	10.8	1.2	20	20	232

[a] Values are the average of three replicates, with SE within 5% of the mean.
[b] From Muñoz de Chavez et al. (1996).

Peroxidase Purification from Radish

Peroxidase from this source maintained high activity at pH values between 4.0 and 11.0, where the relative activity was 80% of the highest value (pH 6.0). However, at pH 3.25, the relative activity decreased to 25% while at pH 12.0, it was 60%. When AOT reverse micelles were used, significant peroxidase solubilization could be achieved at lower than 4.0, with the consequent inactivation of most of the recovered peroxidase after back extraction (Figure 9.1). This figure also shows the effect of changing ionic strength.

The best extraction occurred at pH 3.8, where 21% yield was obtained after back extraction, with a purification factor of 17. This suggests that a small fraction of radish peroxidase isoenzymes acquired enough positive charge at this pH to produce electrostatic interactions with the AOT and were solubilized in the reverse micelles. A change from Na$^+$ to K$^+$ to adjust IS resulted in little peroxidase extraction because K$^+$ has a greater ionic radius than Na$^+$, promoting smaller reverse micelles (Kawakami and Dungan, 1996) unable to fit peroxidase.

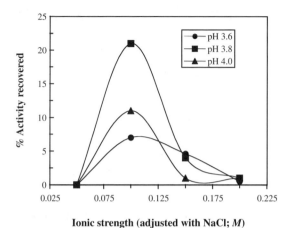

FIGURE 9.1
Effect of forward-transfer pH and ionic strength on peroxidase purification from radish using reverse micelles of 0.10 M AOT in isooctane; mean values of triplicate runs with SEM within 5%.

Two-Step Forward Transfer

We decided to carry out forward transfer in two extraction steps. A similar process was successfully used for peroxidase extraction from horseradish (Regalado et al., 1996). The first extraction was conducted at a pH of 4.2, IS = 0.10 M, where essentially no peroxidase was solubilized, but other proteins migrated into the AOT reverse micelles. In the second step, the aqueous-phase pH was adjusted to 3.8 with concentrated HCl, and peroxidase was solubilized into the organic phase. Back extraction was conducted as described above. The activity recovered doubled (40%), but the purification factor decreased sharply (3.0).

Extraction Using CTAB

To solubilize more peroxidase from the dialyzed radish extract, we used the cationic surfactant CTAB to solubilize peroxidase at higher pH values, an approach considered due to the high pH stability (11.0) of peroxidase. We expected stronger electrostatic interactions from a larger fraction of peroxidase isoenzymes negatively charged at high pH values with 0.20 M CTAB and therefore more peroxidase solubilization. The effect of ionic strength (adjusted with NaCl or KCl) and pH of forward transfer on peroxidase recovery after back extraction is shown in Figure 9.2. Low ionic strength at relatively high pH values (9.0 to 11.0) of forward extractions favored a better peroxidase recovery. This was attributed to the size of the reverse micelles, which were big enough to accommodate peroxidase and some solubilizing water.

When IS increased, the repulsion among the headgroups of CTAB decreased, leading to smaller reverse micelles and reduced activity recovered. Increased ionic strength reduces electrostatic interactions between

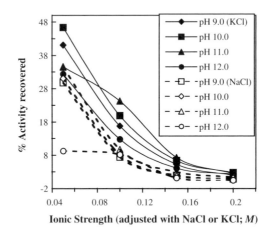

FIGURE 9.2
Effect of forward-transfer pH and ionic strength on peroxidase purification from radish using reverse micelles of 0.20 M CTAB in isooctane/pentanol (90:10). Back extraction with a fresh aqueous phase containing 2.0 M KCl, pH 4.0; mean values of triplicate runs with SEM within 5%.

protein and surfactant headgroups (Goklen and Hatton, 1985); this further supports the reduction in activity recovery. Adjustment of IS with KCl led to higher peroxidase activity recovery (Figure 9.2), probably due to the better emulsion breakage effect of K+ than Na+, as reported by Bartsotas et al. (2000) for AOT–isooctane–water emulsion stability. The low activity recovery shown after peroxidase extraction at pH 12 (Figure 9.2) was attributed to peroxidase inactivation, despite stronger electrostatic interactions between peroxidase and CTAB. Using IS = 0.05 M (adjusted with KCl), pH 10.0, the Wo value of the organic phase was 30.

According to Eq. (9.1), the aqueous core radius (r_c = 4.1 nm) of the CTAB reverse micelles should be enough to accommodate peroxidase, which is a globular protein (2.5 nm in radius) (Clementi and Palade, 1969). Thus, solubilization is apparently not size limited but probably is more related to the extent of surface charge of the most abundant isoenzymes. We have obtained in preliminary results several different radish isoenzymes covering a wide pH range, especially acidic. The relative abundance of these isoenzymes probably modulates the overall net charge of peroxidase at different pH values, leading to the results already discussed. A similar behavior was found for HRP by Regalado et al. (1994).

Under the best conditions (KCl-adjusted IS = 0.05 M, pH 10), the activity yield after back extraction was 48.2% (Figure 9.2), with specific activity of 9.3 U/mg and a purification factor of 5.7. Acetone precipitation of radish extract produced a higher activity yield (94.7%) but lower purification factor (3.4). Ammonium sulfate precipitation gave lower values. This suggests that CTAB reverse micelles may be used instead of precipitation for peroxidase prepurification from radish extract before chromatographic steps are conducted to obtain higher purity.

Peroxidase Purification from Turnip Roots

Peroxidase from turnip roots had pH activity behavior similar to that of radish, with high peroxidase activity in the range of 3.2 to 11.

Extraction Using AOT

AOT reverse micelles significantly solubilized peroxidase at pH lower than 3.2, but after back extraction most of the recovered peroxidase was denatured. An increase in IS produced increasingly smaller activity recovery after back extraction. The best activity recovery was found at IS = 0.05 M, pH 3.1 (Table 9.5). The IS was adjusted with NaCl, as KCl adjustments did not permit peroxidase solubilization because of the reduced size of the reverse micelles (see above). This extract had more than three isoenzymes; one of them was cationic (isoelectric point pI = 8.5), while two others were acidic (pI ~ 3) (Duarte-Vázquez et al., 2000), and most likely others were slightly acidic. Therefore, because most isoenzymes had a low or relatively low pI, it was expected that extraction at a pH higher than the pI of most of the isoenzymes could produce better results.

Extraction Using CTAB

The cationic surfactant CTAB was used together with n-pentanol and n-hexanol as cosurfactants in isooctane; however, peroxidase solubilization only occurred when the former was employed. The effect of ionic strength (adjusted with NaCl) and pH on the recovery of peroxidase after back extraction is shown in Figure 9.3. This figure shows that electrostatic interactions apparently increased with the pH until a value of 12, resulting in greater peroxidase solubilization and higher activity recovery after back extraction. However, at this latter pH, peroxidase was partly inactivated and a lower activity recovery was observed. At increasingly high IS the observed differences of activity recovered had a tendency to decrease until they completely disappeared at 0.20 M, where peroxidase was not solubilized (Figure 9.3). When KCl was used to adjust IS during forward transfer, at pH values from 10 to 12, similar

TABLE 9.5

Results of Turnip Roots Peroxidase Purification Using Reverse Micelles of 0.10-M AOT in Isooctane[a]

Purification Step	Protein (mg)	Activity (U)	Specific Activity (U/mg)	Purification Factor	Yield (%)
DE[b]	1200	7800	6.5	1.0	100
PRP[c]	0.086	1.7	20	3.1	22

[a] Average of three replicates, with SE within 5% of the mean.
[b] Dialyzed turnip roots extract.
[c] Purified recovered peroxidase after back extraction from reverse micelles. Forward transfer: 0.05 M NaCl, pH 3.1; back extraction: 2.0 M KCl, pH 8.0 (80 m M phosphate buffer).

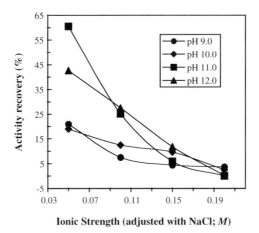

FIGURE 9.3
Effect of forward-transfer pH and ionic strength on peroxidase purification from turnip roots using reverse micelles of 0.20 M CTAB in isooctane/n-pentanol (90:10) and 10% butanol. Back extraction with a fresh aqueous phase containing 2.0 M KCl, pH 4.0; mean values of triplicate runs with SEM within 5%.

(but lower than using NaCl) peroxidase activity was recovered (48%), at low IS (0.05 M) (results not shown). The effect of increasing IS was a gradual reduction in peroxidase activity recovery: 40% recovery at IS = 0.10 M, pH 11, which decreased to 0 at IS = 2.0 M. The size of CTAB reverse micelles is controlled by the type and concentration of the anion used to adjust the IS (Krei and Hustedt, 1992). Similar activity recovery patterns were expected when NaCl and KCl were used to adjust IS in CTAB reverse micellar extraction. However, Cl⁻ showed a different peroxidase activity recovery behavior, which could be attributed to the better activity protection of Na⁺ than that of K⁺. This factor probably overcame the better emulsion destabilizing effect of K⁺. On the other hand, it is known that the ionic species tend to migrate to the aqueous core of the reverse micelles, and that K⁺ has a greater ionic radius than Na⁺. This probably resulted in a partial displacement of the protein. Based on extraction conditions at IS = 0.05 M, pH 11, the purification factor was significantly higher, but the activity yield lower (Table 9.6) than those obtained by acetone and ammonium sulfate precipitation (purification factor 3.4, yield 76%). Therefore, reverse micelles can be used to obtain a partly purified turnip root peroxidase which could be used in applications where high purity is not required, such as analysis of glucose, cholesterol, etc. Further purification to homogeneity was achieved using anion-exchange chromatography which, due to its high specific activity, is suitable for enzyme immunoassays (Duarte-Vázquez et al., 2000).

TABLE 9.6

Peroxidase Purification from Turnip Roots Using Reverse Micelles of 0.20 M CTAB in Isooctane/Pentanol (90:10)[a]

Purification Stage	Protein (mg)	Activity (U)	Specific Activity (U/mg)	Purification Factor	Yield (%)
DCE[b]	1200	7800	6.50	1.00	100
RM (NaCl)[b]	40.4	4720	116	17.8	60.5
RM (KCl)[c]	22.5	3900	173	26.6	50.0

[a] Average of three replicates, with SE within 5% of the mean.
[b] Reverse micellar peroxidase extraction adjusting IS = 0.05 M with NaCl, pH 11.0.
[c] Reverse micellar peroxidase extraction adjusting IS = 0.05 M with KCl, pH 11.0.
Note: Back extraction in aqueous solution containing 2.0 M KCl, pH 4.0.

Peroxidase Purification from Brussels Sprouts

The pH range where peroxidase activity was close to the highest (pH = 4.5) was 3.6 to 8.5. Outside this range, peroxidase activity decreased sharply.

Extraction Using AOT

Using AOT reverse micelles at pH 3.8, IS = 0.10 M (NaCl) resulted in low activity yield (8%). We tried the two-step forward transfer strategy, with the first extraction at IS = 0.10 M, pH 4.2, and peroxidase solubilization in the second step at the same IS, pH 3.5. After back extraction, the activity recovered was only 12.5%, with a purification factor of 3.0. Using biochemical methods, we purified two acidic Brussels sprouts peroxidase isoenzymes (Regalado et al., 1999), and according to our purification procedure more anionic isoenymes were expected; therefore, we decided to use peroxidase extraction with a cationic surfactant to solubilize these acidic isoenzymes.

Extraction Using CTAB

Low yields were obtained when CTAB reverse micelles were used to solubilize peroxidase at IS = 0.1 to 0.3 M (adjusted with KCl or NaCl), and pH from 6.0 to 10.0. The best activity recovery was 30% with a purification factor of 4.5. Different short-chain alcohols were tested as cosolvents, such as *n*-butanol, *n*-pentanol, *n*-hexanol, and isooctanol. The combinations of isooctane with butanol or isooctanol resulted in enzyme inactivation. The cosolvent that resulted in better extraction yields was *n*-hexanol. Different types of anions were also tested, such as KBr, KF, and KCl. Using the first two anions to adjust IS, most of the organic phase dissolved in the aqueous phase, with about a quarter of the initial volume of the organic phase remaining after centrifugation, which upon back extraction gave a 6% activity yield. Using only Cl⁻, the organic and aqueous phases preserved their volumes after phase separation. This is in contrast with the report of Krei and Hustedt (1992), who obtained a maximum α-amylase protein and activity recovery using KF to adjust IS; however, that research was conducted using a model system.

Two-Step Extraction with CTAB

In an attempt to improve the activity recovery, the dialyzed Brussels sprouts crude extract was subjected to a forward transfer divided into two steps. First, the organic phase was contacted with the dialyzed crude extract at pH 6.0 (30-mM phosphate buffer) at different IS that varied from 0.10 to 0.60 M, adjusting with KCl. Under these conditions, peroxidase was barely solubilized, as assessed by activity measurements in the remaining aqueous phase; however, other proteins were removed into the organic phase. After centrifugation, the organic phase was regenerated by contacting with a fresh aqueous phase containing 0.4-M KCl, pH 4.0. Under these back-transfer conditions, the contaminating proteins were reduced to an insignificant level. The organic phase could be recycled four times without significant loss of efficiency. The remaining aqueous phase was used for the second forward transfer. Here, peroxidase extraction was conducted by keeping the IS constant but adjusting the pH between 8.0 and 10.0 with concentrated NaOH. Back extraction was conducted as previously described. The results are illustrated in Figure 9.4, which shows that peroxidase was inactivated at pH > 9.0. The effect of increasing IS was to increase protein solubilization (from 0.10 to 0.20 M), but further increases in IS resulted in more reduction of protein activity recovery after back extraction. The purification achieved using this process is shown in Table 9.7. Precipitation of the crude extract with acetone or ammonium sulfate or both gave lower purification results; for instance, after treatment with both precipitants, the purification factor was 11 with a yield of 44%. Therefore, reverse micelles can be used for the initial stage of peroxidase purification from Brussels sprouts, a process that gave 5.5 times better purification factor and double activity yield than the precipitation methods.

Conclusions

Purification using CTAB reverse micelles produced a turnip peroxidase extract having a high specific activity (173 U/mg) suitable for the analysis of glucose, cholesterol, etc., where high purity is not important. A two-step forward extraction with reverse micelles of 0.20-M CTAB in isooctane/hexanol (90:10), followed by back extraction, produced a Brussels sprouts peroxidase with a 5.5-fold purification factor and twice the activity yield when compared to acetone and ammonium-sulfate precipitations. Similarly, turnip root peroxidase purification using 0.20-M CTAB reverse micelles in isooctane/pentanol (90:10) produced about 8 times the purification factor

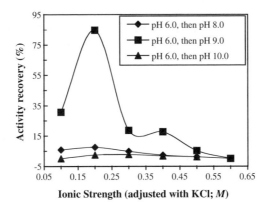

FIGURE 9.4
Effect of pH and ionic strength on reverse micellar peroxidase purification from Brussels sprouts using a two-stage forward extraction with 0.20 M CTAB in isooctane/hexanol (90:10). The first forward-transfer step was conducted at the pH shown on the left side of the legend, while the second step pH used is shown on the right side of the legend; mean values of triplicate runs with SEM within 5%.

TABLE 9.7

Summary of the Purification Stages of Brussels Sprouts Peroxidase Using CTAB Reverse Micelles in Isooctane/Hexanol (90:10)[a]

Purification Stage	Protein (mg)	Activity (U)	Specific Activity (U/mg)	Purification Factor	Yield (%)
DCE[b]	3.20	1.40	0.45	1.00	100
PP[c]	0.045	1.20	26	61	85

[a] Average of three replicates, with SE within 5% of the mean.
[b] Dialyzed crude extract from Brussels sprouts.
[c] Purified peroxidase after back extraction. Forward transfer: 0.2 M KCl, pH 6.0 (first stage), pH 9.0 (second stage). Back extraction: aqueous solution containing 2.0 M KCl, pH 4.0.

achieved with ammonium sulfate with similar activity yield. Therefore, reverse micelles can be advantegously used for the initial stage of peroxidase purification from these sources.

Peroxidase isoenzymes can have different amino acid and carbohydrate composition, leading to pI in a wide pH range. This complicates reverse micellar extraction because of weak electrostatic interactions at low or high pH suitable for extraction with anionic or cationic surfactants without denaturation.

The effect of K[+] on forward transfer was important in achieving better activity recovery after back extraction for radish and Brussels sprouts extracts. For turnip roots, the use of Na[+] instead of K[+] to adjust IS during forward transfer allowed 50% better purification factor, but about 20% less activity yield.

Because liquid–liquid extraction techniques can be applied to reverse micellar extraction, the advantages and limitations of this technology apply to the large-scale recovery of bioproducts. Adequate extraction equipment must be selected, bearing in mind the objectives of high capacity, high mass transfer rate, and low axial mixing. Typical extraction equipment meeting these requirements are mixer-settler units, centrifugal extractors, and column contactors. Fundamental studies over a wide range of process conditions must be conducted to support the scaling-up of a reverse micellar extraction. This technology has yet to prove its value on a pilot-scale basis over a complete recovery cycle before it can be ranked as a genuine alternative method of protein purification.

References

Adachi, M., Harada, M., Shioi, A., Takahashi, H., and Katoh, S. (1996) Selective separation of concanavalin A in a reverse micellar system, in *Proc. of ISEC '96*, D.C. Shallcross, R. Paimin, and L.M. Prvcic, Eds., The University of Melbourne, pp. 1393–1398.

Adachi, M., Yamazaki, M., Harada, M., Shioi, A., and Katch, S. (1997) Bioaffinity separation of trypsin using trypsin inhibitor immobilized in reverse micelles composed of a nonionic surfactant, *Biotechnol. Bioeng.*, 53, 406–408.

Adachi, M., Shibata, K., Shioi, A., Harada, M., and Katoh, S. (1998) Selective separation of trypsin from pancreatin using bioaffinity in reverse micellar system composed of a nonionic surfactant, *Biotechnol. Bioeng.*, 58, 649–653.

Almeida, F.C.L., Valente, A.P., and Chaimovich, H. (1998) Stability and activity modulation of chymotrypsins in AOT reversed micelles by protein-interface interaction: interaction of α-chymotrypsin with a negative interface leads to a cooperative breakage of a salt bridge that keeps the catalytic active conformation (Ile[16]-Asp[194]), *Biotechnol. Bioeng.*, 59, 360–363.

AOAC (1990) *Official Methods of Analysis*, Vol. 1, 15th ed., K. Helrich, Ed., AOAC, Arlington, VA, pp. 47–48.

Aveyard, R., Binks, B.P., and Fletcher, P.D.I. (1989) Interfacial tensions and aggregate structure in $C_{12}E_5$/oil/water microemulsion systems, *Langmuir*, 5, 1210–1217.

Bartsotas, P., Poppenborg, L.H., and Stuckey, D.C. (2000) Emulsion formation and stability during reversed micelle extraction, *J. Chem. Technol. Biotechnol.*, 75, 738–744.

Bausch, T.E., Plucinsky, P.K., and Nitsch, W. (1992) Kinetics of the re-extraction of hydrophilic solutes out of AOT-reversed micelles, *J. Colloid Interface Sci.*, 150, 226–234.

Bernardi, R. and Palmiere, S. (1996) Isolation of a trypsin-like activity from larvae of corn borer (*Ostrinia nubilatis*) using reverse micelles, *Biotechnol. Lett.*, 18, 663–666.

Bosetti, A., Tassinari, R., and Battistel, E. (1997) Partial purification of glutaryl-7-ACA acylase from crude cellular lysate by reverse micelles, *Appl. Biochem. Biotech.*, 66, 173–183.

Brandani, S., Brandani, V., DiGiacomo, G., and Spera, L. (1996a) Effect of nonlinear equilibrium on the mass transfer rate of α-amylase extraction by reversed micelles, *Process Biochem.*, 31, 249–252.

Brandani, V., Di Giacomo, G., and Spera, L. (1996b) Recovery of α-amylase extracted by reverse micelles, *Process Biochem.*, 31, 125–128.

Campa, A. (1991) Biological roles of plant peroxidase: known and potential function, in *Peroxidases in Chemistry and Biology*, L. Everse, K.E. Everse, and M.B. Grisham, Eds., CRC Press, Boca Raton, FL, pp. 26–43.

Cardoso, M.M., Barradas, M.J., Kroner, K.H., and Crespo, J.G. (1999) Amino acid solubilization in cationic reversed micelles: factors affecting amino acid and water transfer, *J. Chem. Technol. Biotechnol.*, 74, 801–811.

Carlson, A. and Nagarajan, R. (1992) Release and recovery of porcine pepsin and bovine chymosin from reverse micelles: a new technique based on isopropyl alcohol addition, *Biotechnol. Progress*, 8, 85–90.

Chang, Q.L. and Chen, J.Y. (1996) Liquid–liquid reversed micellar extraction for isolating enzymes: studies on the purification of α-amylase, *Process Biochem.*, 31, 371–375.

Chang, Q.L., Chen, J.Y., Zhang, X.F., and Zhao, N.M. (1997) Effect of the cosolvent type on the extraction of α-amylase with reversed micelles: circular dichroism study, *Enzyme Microb. Technol.*, 20, 87–92.

Chaudhuri, J.B. and Spirovska, G. (1994) Recovery of proteins from reversed micelles using a novel ion-exchange material, *Biotechnol. Technol.*, 8, 909–914.

Chen, S.J., Evans, D.F., Ninham, B.W., Mitchel, D.J., Blum, F.D., and Pickup, S. (1986) Curvature as a determinant of microstructure and microemulsions, *J. Phys. Chem.*, 90, 842–847.

Childs, R.E. and Bardsley, W.G. (1975) The steady-state kinetics of peroxidase with 2,2′-azino-di-(3-ethyl-benzthiazoline-6-sulphonic acid) as chromogen, *Biochem. J.*, 145, 93–103.

Clark, S., Fletcher, P.D.I., and Ye, X. (1990) Interdroplet exchange rates of water-in-oil and oil-in-water microemulsion droplets stabilized by pentaoxyethylene monododecyl ether, *Langmuir*, 6, 1301–1309.

Clementi, F. and Palade, G.E. (1969) Intestinal capillaries I. Permeability to peroxidase and ferritin, *J. Cell Biol.*, 41, 33–58.

Contreras-Padilla, M. and Yahia, E.M. (1998) Changes in capsaicinoids during development, maturation, and senescence of chile peppers in relation with peroxidase activity, *J. Agric. Food Chem.*, 46, 2075–2079.

Dahuron, L. and Cussler, D.L. (1988) Protein extractions with hollow fibres, *AICHE J.*, 34, 130–136.

Dekker, M. (1990) Enzyme Recovery Using Reversed Micelles, Ph.D. thesis, Wagenigen Agricultural University, Holland.

Dekker, M., Van't Riet, K., Van Der Pol, J.J., Baltussen, J.W.A., Hilhorst, R., and Bijsterbosch, B.H. (1991) Effect of temperature on the reversed micellar extraction of enzymes, *Chem. Eng. J.*, 46, B69–74.

Duarte-Vázquez, M.A., García-Almendárez, B., Regalado, C., and Whitaker, J.R. (2000) Purification and partial characterization of three turnip (*Brassica napus* L. var. *esculenta* D.C.) peroxidases, *J. Agric. Food Chem.*, 48, 1574–1579.

Dunford, H.B. (1991) Horseradish peroxidase: structure and kinetic properties, in *Peroxidases in Chemistry and Biology*, Vol. II, J. Everse, K.E. Everse, and M.B. Grisham, Eds., CRC Press, Boca Raton, FL, pp. 1–24.

Dungan, S.R., Bausch, T., Hatton, T.A., Plucinsky, P., and Nitsch, W. (1991) Interfacial transport processes in the reversed micellar extraction of proteins, *J. Colloid Interface Sci.*, 145, 33–50.

Eastoe, J., Robinson, B.H., Steytler, D.C., and Thorn-Lesson, D. (1991) Structural studies of microemulsions stabilised by aerosol-OT, *Adv. Colloid Interface Sci.*, 36, 1–31.

Eicke, H.F. and Kvita, P. (1984) Reverse micelles and aqueous microphases, in *Reverse Micelles: Biological and Technological Relevance of Amphiphilic Structures in Apolar Media*, P.L. Luisi and B.E. Straub, Eds., Plenum Press, New York, pp. 21–35.

Fletcher, P.D.I., Howe, A.M., and Robinson, B.H. (1987) The kinetics of solubilisate exchange between water droplets of a water-in-oil microemulsion, *J. Chem. Soc. Faraday Trans. I*, 83, 985–1006.

Goklen, K.E. and Hatton, T.A. (1985) Protein extraction using reverse micelles, *Biotechnol. Prog.*, 1, 69–74.

Goto, M., Ono, T., Nakashio, F., and Hatton, T.A. (1997) Design of surfactants suitable for protein extraction by reverse micelles, *Biotechnol. Bioeng.*, 54, 26–32.

Gupta, R.B., Han, C.J., and Johnston, K.P. (1994) Recovery of proteins and amino acids from reverse micelles by dehydration with molecular sieves, *Biotechnol. Bioeng.*, 44, 830–836.

Hashimoto, Y., Ono, T., Goto, M., and Hatton, A. (1998) Protein refolding by reversed micelles utilizing solid–liquid extraction technique, *Biotechnol. Bioeng.*, 57, 620–623.

Hatton, T.A. (1989) Reversed micellar extraction of proteins, in *Surfactant Based Separation Processes*, J.F. Scamehorn and J.H. Harwell, Eds., Marcel Dekker, New York, pp. 55–90.

Hayes, D.G. (1997) Mechanism of protein extraction from the solid state by water-in-oil microemulsions, *Biotechnol. Bioeng.*, 53, 583–593.

Hayes, D.G. and Marchio, C. (1998) Expulsion of proteins from water-in-oil microemulsions by treatment with cosurfactant, *Biotechnol. Bioeng.*, 59, 557–566.

Hentsch, M., Menoud, P., Steiner, L., Flaschel, E., and Renken, A. (1992) Optimization of the surfactant (AOT) concentration in a reverse micellar extraction process, *Biotechnol. Technol.*, 6, 359–364.

Hossain, M.J., Takeyama, T., Hayashi, Y., Kawanishi, T., Shimizu, N., and Nakamura, R. (1999) Enzymatic activity of *Chromobacterium viscosum* lipase in an AOT/Tween 85 mixed reverse micellar system, *J. Chem. Technol. Biotechnol.*, 74, 423–428.

Hu, Y. and Gulari, E. (1996a) Extraction of aminoglycoside antibiotics with reverse micelles, *J. Chem. Technol. Biotechnol.*, 65, 45–48.

Hu, Y. and Gulari, E. (1996b) Protein extraction using the sodium *bis*(2-ethylhexyl)phosphate (NaDEHP) reverse micellar system, *Biotechnol. Bioeng.*, 50, 203–206.

Huang, S.Y. and Lee, Y.C. (1994) Separation and purification of horseradish peroxidase from *Armoracia rusticana* root using reversed micellar extraction, *Bioseparation*, 4, 1–5.

Huang, W., Li, X., Zhou, J., and Gu, T. (1996) Activity and conformation of bovine pancreatic ribonuclease A in reverse micelles formed by dodecylammonium butyrate and water in cyclohexane, *Colloids Surfaces B. Biointerfaces*, 7, 23–29.

Jarudilokkul, S., Poppenborg, L.H., and Stuckey, D.C. (1999) Backward extraction of reverse micellar encapsulated proteins using a counterionic surfactant, *Biotechnol. Bioeng.*, 62, 593–601.

Jarudilokkul, S., Paulsen, E., and Stuckey, D.C. (2000a) The effect of demulsifiers on lysozyme extraction from hen egg white using reverse micelles, *Bioseparation*, 9, 81–91.

Jarudilokkul, S., Paulsen, E., and Stuckey, D.C. (2000b) Lysozyme extraction from egg white using reverse micelles in a Graesser contactor: mass transfer characterization, *Biotechnol. Bioeng.*, 69, 618–626.

Kadam, K.L. (1986) Reverse micelles as a bioseparation tool, *Enzyme Microbiol. Technol.*, 8, 266–273.

Kawakami, L.E. and Dungan, S.R. (1996) Solubilization properties of α-lactalbumin and β-lactoglobulin in AOT-isooctane reversed micelles, *Langmuir*, 12, 4073–4083.

Krei, G.A. and Hustedt, H. (1992) Extraction of enzymes by reverse micelles, *Chem. Eng. Sci.*, 47, 99–111.

Krell, H.W. (1991) Peroxidase: an important enzyme for diagnostic test kits, in *Biochemical, Molecular and Physiological Aspects of Plant Peroxidases*, J. Lobarzewsky, H. Greppin, C. Penel, and T. Gaspar, Eds., University M. Curie, Lublin, Poland; University of Geneva, Geneva, Switzerland, pp. 469–478.

Krieger, N., Taipa, M.A., Aires-Barros, M.R., Melo, E.H.M., Lima-Filho, J.L., and Cabral, J.M.S. (1997) Purification of the *Penicillium citrinum* lipase using AOT reverse micelles, *J. Chem. Technol. Biotechnol.*, 69, 77–85.

Lazarova, Z. and Tonova, K. (1999) Integrated reversed micellar extraction and stripping α-amylase, *Biotechnol. Bioeng.*, 63, 583–592.

Leodidis, E.B. and Hatton, T.A. (1989) Specific ion effects in electrical double layers: selective solubilization of cations in aerosol-OT reversed micelles, *Langmuir*, 5, 741–753.

Leser, M.E., Mrkoci, K., and Luisi, P.L. (1993) Reverse micelles in protein separation: the use of silica for the back-transfer process, *Biotechnol. Bioeng.*, 41, 489–492.

Lu, Q., Li, K., Zhang, M., and Shi, T. (1998) Study of centrifugal extractor for protein extraction using reverse micellar solutions, *Sep. Sci. Technol.*, 33, 2397–2409.

Luisi, P.L. and Magid, L.J. (1986) Solubilization of enzymes and nucleic acids in hydrocarbon micellar solutions, *Crit. Rev. Biochem.*, 20, 409–474.

Luisi, P.L. and Steinmann-Hofmann, B. (1987) Activity and conformation of enzymes, in reverse micellar solutions, in *Methods in Enzymology*, Vol. 136, K. Mosbach, Ed., Academic Press, New York, pp. 188–219.

Luisi, P.L., Bonner, F.J., Pellegrini, A., Wiget, P., and Wolf, R. (1979) Micellar solubilization of proteins in aprotic solvents and their spectroscopic characterization, *Helv. Chim. Acta.*, 62, 740–753.

Lye, G.J., Asenjo, J.A., and Pyle, D.L. (1996) Reverse micellar mass-transfer processes: spray column extraction of lysozyme, *AICHE J.*, 42, 713–726.

Maitra, A.N. (1984) Determination of size parameters of water–aerosol OT-oil reverse micelles from their nuclear magnetic resonance data, *J. Phys. Chem.*, 88, 5122–5125.

McGuire, R.G. (1992) Reporting of objective color measurements, *Hort. Sci.*, 27, 1254–1255.

Melo, E.P., Costa, S.M.B., and Cabral, J.M.S. (1996) Denaturation of a recombinant cutinase from *Fusarium solani* in AOT-isooctane reverse micelles: a steady-state fluorescence study, *Photochem. Photobiol.*, 63, 169–175.

Melo, E.P., Carvalho, C.M.L., Aires-Barros, M.R., Costa, S.M.B., and Cabral, J.M.S. (1998) Deactivation and conformational changes of cutinase in reverse micelles, *Biotechnol. Bioeng.*, 58, 380–386.

Muñoz de Chávez, M., Chávez Villasana, A., Roldán Amaro, J.A., Ledesma Solano, J.A., Mendoza Martínez, E., Pérez-Gil Romo, F., Hernández Cordero, S.L., and Chaparro Flores, A.G. (1996) *Tablas de Valor Nutritivo de los Alimentos de Mayor Consumo en Latinoamérica*, PAX, México, pp. 74, 91, 96.

Nagayama, K., Nishimura, R., Doi, T., and Imai, M. (1999) Enhanced recovery and catalytic activity of *Rhizopus delemar* lipase in an AOT microemulsion with guanidine hydrochloride, *J. Chem. Technol. Biotechnol.*, 74, 227–230.

Naoe, K., Nishino, M., Ohsa, T., Kawagoe, M., and Imai, M. (1999) Protein extraction using sugar ester reverse micelles, *J. Chem. Technol. Biotechnol.*, 74, 221–226.

Nishii, Y., Nii, S., Takahashi, K., and Takeuchi, H. (1999) Extraction of proteins by reversed micellar solution in a packed column, *J. Chem. Eng. Jpn.*, 32, 211–216.

Nitsch, W. and Plucinsky, P. (1990) Two phase kinetics of the solubilization in reverse micelles, *J. Colloid Interface Sci.*, 136, 338–351.

Ono, T., Goto, M., Nakashio, F., and Hatton, T.A. (1996) Extraction behavior of hemoglobin using reverse micelles by dioleyl phosphoric acid, *Biotechnol. Prog.*, 12, 793–800.

Orlich, B. and Schomaecker, R. (1999) Enzymatic reduction of a less water-soluble ketone in reverse micelles with NADH regeneration, *Biotechnol. Bioeng.*, 65, 357–372.

Orlich, B., Berger, H., Lade, M., and Shoemäcker, R. (2000) Stability and activity of alcohol dehydrogenases in W/O-microemulsions: enantioselective reduction including cofactor regeneration, *Biotechnol. Bioeng.*, 70, 638–646.

Paradkar, V.M. and Dordick, J.S. (1993) Affinity-based reverse micellar extraction and separation (ARMES): a facile technique for the purification of peroxidase from soybean hulls, *Biotechnol. Prog.*, 9, 199–203.

Paul, K.G. and Stigbrand, T. (1970) Four isoperoxidases from horseradish root, *Acta Chem. Scand.*, 24, 3607 3617.

Pérez-Arvizu, O., García, B.E., and Regalado, C. (1999) Purification of peroxidase from waste frozen vegetable processing companies using reverse micelles, in *Food for Health in the Pacific Rim*, J.R. Whitaker, N.F. Haard, C.F. Shoemaker, and R.P. Singh, Eds., Food & Nutrition Press, Trumbull, CT, pp. 206–215.

Pires, M.J. and Cabral, J.M.S. (1993) Liquid–liquid extraction of a recombinant protein with a reverse micelle phase, *Biotechnol. Prog.*, 9, 647–650.

Plucinsky, P. and Nitsch, W. (1989) Two phase kinetics of the solubilization in reverse micelles extraction of lysozyme, *Ber. Bunsenges. Phys. Chem.*, 93, 994–997.

Rariy, R.V., Bec, N., Klyachko, N.L., and Levashov, A.V. (1998) Thermobarostability of α-chymotrypsin in reversed micelles of aerosol OT in octane solvated by water-glycerol mixtures, *Biotechnol. Bioeng.*, 57, 552–556.

Regalado, C., Asenjo, J.A., and Pyle, D.L. (1994) Protein extraction by reverse micelles. Studies on the recovery of horseradish peroxidase, *Biotechnol. Bioeng.*, 44, 674–681.

Regalado, C., Asenjo, J.A., and Pyle, D.L. (1996) Studies on the purification of peroxidase from horseradish roots using reverse micelles, *Enzyme Microb. Technol.*, 18, 332–339.

Regalado, C., Pérez-Arvizu, O., García-Almendárez, B.E., and Whitaker, J.R. (1999) Purification and properties of two acidic peroxidases from Brussels sprouts (*Brassica oleraceae* L.), *J. Food Biochem.*, 23, 435–450.

Robinson, B.H., Toprakcioglu, C., Dore, J.C., and Chieux, P. (1984) Small-angle neutron-scattering study of microemulsions stabilized by aerosol-OT. Part 1. Solvent and concentration variation, *J. Chem. Soc. Faraday Trans. I*, 80, 13–27.

Sadana, A. and Raju, R.R. (1990) Bioseparation and purification of proteins, *BioPharm.*, May, 53–60.

Shiomori, K., Ishimura, M., Baba, Y., Kawano, Y., Kubio, R., and Komasawa, I. (1996) Characteristics and kinetics of lipase-catalyzed hydrolysis of olive oil in a reverse micellar system, *J. Ferment. Bioeng.*, 81, 143–147.

Shiomori, K., Ebuchi, N., Kawano, Y., Kuboi, R., and Komasawa, I. (1998) Extraction characteristics of BSA using sodium *bis*(2-ethylhexyl) sulfosuccinate reverse micelles, *J. Ferment. Bioeng.*, 86, 581–587.

Smith, P.K., Krohn, R.I., Hermanson, G.T., Mallia, A.K., Gartner, F.H., Provenzano, M.D., Fujimoto, E.K., Goeke, N.M., Olson, B.J., and Klenk, D.C. (1985) Measurement of protein using bicinchoninic acid, *Anal. Biochem.*, 150, 76–85.

Spirovska, G. and Chaudhuri, J.B. (1998) Sucrose enhances the recovery and activity of ribonuclease during reversed micelles extraction, *Biotechnol. Bioeng.*, 58, 374–379.

Stobbe, H., Yunguang, X., Zihao, W., and Jufu, F. (1997) Development of a new reversed micelle liquid emulsion membrane for protein extraction, *Biotechnol. Bioeng.*, 53, 267–273.

Sun, Y., Ichikawa, S., Sugiura, S., and Furusaki, S. (1998) Affinity extraction of proteins with a reversed micellar system composed of cibacron blue-modified lecithin, *Biotechnol. Bioeng.*, 58, 58–64.

Tijssen, P. (1985) Properties and preparation of enzymes used in enzyme inmunoassays, in *Practice and Theory of Enzyme Inmunoassays*, Vol. 15, T.H. Burdon and P.H. Knippenberg, Eds., Elsevier, Amsterdam, pp. 173–219.

Tong, J. and Furusaki, S. (1997) Mass transfer performance and mathematical modeling of rotating disc contactors used for reversed micellar extraction of proteins, *J. Chem. Eng. Jpn.*, 30, 79–85.

Vasudevan, M. and Wiencek, J.M. (1997) Role of the interface in protein extractions using nonionic microemulsions, *J. Colloid Interface Sci.*, 186, 185–192.

10

Improving Biogeneration of Aroma Compounds by In Situ Product Removal

Leobardo Serrano-Carreón

CONTENTS

Introduction

Consumer preferences for natural food additives has led to an increasing demand for natural aroma compounds; however, the poor productivity of these novel biotechnological processes often prevents industrial application. Several works concerning the production of flavors by microorganisms have shown that the production rate decreases when the product concentration exceeds a toxicity threshold beyond which growth and/or production are inhibited. Attention is now being directed to extractive fermentation, which is regarded as a promising fermentation technology because it is capable of relieving end-product inhibition and increasing productivity. Also, it can recover fermentation products *in situ* and hence simplify downstream processing. Reports on *in situ* aroma extraction from the culture broth are mostly focused on two general techniques: (1) extraction into another phase, and (2) membrane separation. This chapter presents a review of the literature concerning the experimental results, primarily from the last decade, on attempts to improve biogeneration of aroma compounds by *in situ* product removal.

Downstream Processes for Aroma Compounds

Flavor and aroma chemicals used in the food, cosmetic, and pharmaceutical industries are products of great commercial significance. Increased production of processed foods is accompanied by an increasing demand for flavors; however, chemical synthesis, now used to produce aroma compounds, is markedly declining because of new regulations concerning food additives and because of consumer dislike of synthetic compounds. Existing technologies for flavor and aroma production by microorganisms or enzymatic synthesis offer an alternative to chemical synthesis (Gatfield, 1988). The industrial exploitation of microorganisms (bacteria, yeast, and fungi) for the production of flavors is in fact an extension of the biotechnology used for traditional processes. Most of the aroma compounds (terpenes, esters, ketones, lactones, alcohols, and aldehydes) are secondary metabolites.

Secondary metabolites are compounds that cells do not require for growth. They are present at low concentrations during the logarithmic growth phase, but appear in large quantities during the stationary phase. In many cases, such compounds result from detoxification processes developed by the cell to contend with unfavorable environmental conditions (e.g., when a high concentration of nutrients or metabolites is present). In aroma biosynthesis, the success of the process depends on the individual stages, namely strain screening, improvement of selected strains, process design, and downstream processing (Belin et al., 1992).

Downstream processing is the general term used to describe separation processes for the recovery of biological products at some stage in their production (Liddell, 1994). The current industrial standard in aroma biotechnology is batch fermentation with off-line distillation (or extraction), which requires the handling of large volumes of dilute aqueous solutions (Berger, 1995). The concentration of product in the bioreaction mixture is a major factor in production costs. Data on product concentration and retail price for a wide range of biological products show a strong inverse correlation between these two parameters (Humphrey, 1994). For example, for a production level of 1 g/L, annual depreciation based on a 5-year amortization period, and no discount rate, the breakeven price for a hypothetical flavor decreases from $1240/kg for production of 1000 kg/yr to around $300/kg for 10,000 kg/y, and to $202/kg for 100,000 kg/yr (Welsh, 1994). However, in many cases, the biosynthesis of aroma compounds is far from reaching such high concentrations. Several factors may limit process productivity (Berger, 1995):

1. Loss of product via the waste airstream due to volatility
2. Biochemical instability of the product in the presence of the producing cell
3. Inhibition phenomena
4. Nonstationary product concentration in conventional batch processes

In any of these cases, it is necessary to perform *in situ* extraction of the aroma compound from the culture broth. The objective of *in situ* product removal (ISPR), also known as *extractive fermentation*, is to remove the product as it is formed from the vicinity of the producing cell (and the reaction medium). While ISPR is considered primarily for the improvement of existing processes, in some cases, where product–cell interference is intensive, ISPR is essential for carrying out the envisaged process (Freeman et al., 1993).

In Situ Product Removal of Aroma Compounds

Reports on *in situ* aroma extraction from the culture broth are focused on two general techniques: extraction into another phase and membrane separation. Most of the work reported in this area deals with the use of ISPR for the production of ethanol (and other related compounds such as butanol and acetone), organic acids (lactic and propionic), lactones (6-pentyl-α-pyrone and γ-decalactone), ketones, and some aldehydes. An interesting point is that the use of various ISPR techniques has been evaluated for most of the aroma compounds cited here. That such evaluations have been performed by many different research groups reflects the tremendous effort required to find the best alternative for each case. It is possible to find general trends for the suitability of each ISPR technique, and in this review the emphasis will be on how ISPR has improved aroma productivity (in relation to conventional fermentation) and, when applicable, on the long-term performance of the process.

Extraction into Another Phase

Liquid–Liquid Extraction

The principle of liquid–liquid extraction involves an efficient contact of two liquid phases: a feed phase and an extracting solvent. The phases are put in contact by dispersing one liquid as a drop dispersion into the second liquid, which remains as a continuous phase. The chemical properties of the solvent are chosen to facilitate selective uptake of the desired product from the feed phase and to ensure rapid disengagement of the two phases after contact (Weatherley, 1996). Organic solvents used as an *in situ* extractive system can lead to physical, morphological, and biochemical changes of the microorganism. It is accepted that low-polarity/high-molecular-weight solvents are less toxic for cells (Brink and Tramper, 1985). In general, solvent biocompatibility (low or no toxic effect) can be related to the logarithm of the partition coefficient of the solvent in a standard octane–water (1:1 v/v) system, called log P. It has been reported that minor toxic effects of solvents are found for log P > 4 (Laane et al., 1987; Yabannavar and Wang, 1991). The use of anion-exchange liquid membranes for the extraction of carboxylic acids and amino acids provides high permeability of the products while preventing the extractant from contaminating the product and the feed (Eyal and Bressler, 1993). The most recent work on ISPR by a two-phase, liquid–liquid process is briefly discussed below.

Ethanol production from lactose by liquid–liquid extractive fermentation has been reported (Jones et al., 1993). The authors reported a critical log P value for ethanol production by *Candida pseudotropicalis* of 5.2, and the use of Adol 85 F (log P = 8) showed a 60% improvement in lactose yield and ethanol production, as well as a 75% higher volumetric productivity when compared with control cultures.

Like other weak organic acids, propionic acid is known to inhibit cell growth, substrate consumption, and acid production as a result of its antimicrobial activity (Lueck, 1980). The maximal propionic acid volumetric productivity (3 g/L/h) can be obtained if propionic acid concentration is maintained below 3 g/L in a *Propionibacterium thoenii* fermentation (Gu et al., 1998).

In situ product removal of lactic acid has been studied extensively. The use of secondary and tertiary amines as extraction solvents for lactic acid by *Lactobacillus delbrueckii* has been reported (Honda et al., 1995). The use of *Alamine 336* and oleyl alcohol as extractant and back-extractant, respectively, improved the total lactic acid productivity by 40% in relation to the control cultures. Planas et al. (1996) reported the use of an aqueous two-phase system (ATPS) of extractive fermentation for enhanced production of lactic acid by *Lactobacillus lactis*. The productivity of lactic acid increased from 2.5 to 3 m*M*/h when a copolymer of ethylene oxide and propylene oxide was used with hydroxypropyl starch as the ATPS. When the ATPS was replaced with a fresh top phase, the productivity reached a value of 4.7 m*M*/h, which is still far from the 7 m*M*/h required for commercial fermentation (Planas et al., 1997). The use of a poly(ethyleneimine)/hydroxyethyl cellulose ATPS extractive fermentation allowed only slight improvements in the lactic acid productivity in the continous cultivation of *Lactococcus lactis* (Kwon et al., 1996).

The authors reported maximal productivities of 9.3 and 10.2 g/L/h for the control (D = 0.54 h) and ATPS extraction (D = 0.3 h), respectively.

Production of 6-pentyl-α-pyrone (6PP) by *Trichoderma* spp. has been limited by product inhibition, as biomass growth inhibition occurs at 6PP concentrations as low as 100 mg/L (Prapulla et al., 1992). The potential of a polyethylene glycol/phosphate ATPS for the recovery of 6PP has been reported (Rito-Palomares et al., 2000). The use of octane as an extraction solvent in combination with a surface culture of *Trichoderma viride* over wooden rods yielded 432 mg/L of 6PP (Tekin et al., 1995); however, the main bottleneck of this methodology is the scaling-up of the process.

The production of ketones in aqueous-decane two-phase systems by microencapsulated spores of *Penicillium roquefortii* was investigated (Park et al., 2000). The biotransformation of hexanoic and octanoic acid or their alkyl esters to 2 alkanones was only possible in the decane-based two-phase system. Indeed, the bioconversion of fatty acids into methyl ketones by spores of *Penicillium roquefortii* was improved by the use of an isoparaffin solvent, Hydrosol IP 230 (Creuly et al., 1992). The batch-fed bioconversion in the Hydrosol (log P = 7)-water system resulted in 21, 73, and 57 g/L of solvent for 2-pentanone, 2-heptanone, and 2-nonanone, respectively.

The main advantage of product recovery by solvent extraction is the low cost, as no infrastructure investment is required. Solvent can be recycled and downstream costs are reduced as the product is concentrated in the solvent (Mathys et al., 1999); however, the use of organic solvents may exert toxic effects on the producing cells. This is of crucial importance in the production of ethanol, butanol, acetone, and organic acids, as the better the extraction solvents (affinity) are, unfortunately, the more toxic they are to the cells. Therefore, the results obtained are modest when liquid–liquid extraction is used for the recovery of hydrophilic products. This also could be the primary reason for improved performance in liquid–liquid extraction when hydrophobic aroma compounds are extracted, as in the case of methyl ketones and 6PP.

Solid-Phase Extraction

This process can be defined as the adsorption of a product in an inorganic or organic polymeric solid. Synthetic adsorbents are organic polymeric materials with a hydrophobic surface, which can be exploited in downstream processes. They are now widely used in the field of biotechnology and for natural product isolation from laboratory scale to industrial production. The use of synthetic adsorbents has several advantages over conventional solvent-extraction processes (Takayanagi et al., 1996):

1. They are nontoxic for the cells.
2. Synthetic adsorbents have a molecular sieving function based on their pore structure.

3. Adsorbents can be used either in a contained form (column) or in batch mode (in suspension).

4. Adsorbents can be used repeatedly, as the adsorbed compounds can be eluted under mild conditions.

The use of adsorbents in ISPR during aroma production has been widely reported. With the objective of harvesting aroma compounds from fermentation broths, Krings et al. (1993) evaluated 31 different adsorptive materials and their adsorption properties. The best adsorbent materials turned out to be activated carbons, which adsorbed more than 86% of the compounds (model solution of 12 aroma compounds at 33 ppm each); however, these adsorbents exhibited poor desorption capabilities when common organic solvents were use as desorbents.

The most suitable adsorbents were found to be styrene-divinybenzene resine and zeolite, which exhibited adsoption rates similar to activated carbons and better desorption properties. Very few examples can be found regarding the use of adsorbents for improving aroma compound production, probably because ISPR by adsorption of aromas has attained only modest results; however, the development of new synthetic resins in the area of chromatography could lead to the appearance of better materials with higher selectivity, high achievable loadings, and long-term performance.

Production of 6-pentyl α-pyrone by *Trichoderma viride* was improved using Amberlite XAD-2 (Prapulla et al., 1992). When this adsorbent was added (13.3 g/L) at the beginning of the fermentation, 6PP production reached 248 ppm, 2.75 times greater than the control.

When a packed column with an anion-exchange resin (Amberlite IRA-400) was used to improve lactic acid production by *Lactobacillus delbreuckii*, the lactic acid yield was improved from 0.83 to 0.93 g/g substrate, while the productivity went from 0.31 to 1.67 g/L/h in relation to conventional batch mode (Srivastava et al., 1992).

Evaluation of the most suitable adsorbents for extractive fermentation of benzaldehyde and γ-decalactone resulted in the selection of Lewatit 1064 (styrene-divinyl-benzene resin) due to the high achievable loading, even at a low solute concentration (Krings and Berger, 1995). Adsorption of γ-decalactone for the bioconversion of methyl ricinoleate by *Sporidiobolus salmonicolor* has been studied (Souchon et al., 1998). However, under the experimental conditions tested, the production of γ-decalactone was not improved when polystyrene-type polymers (Porapaq Q, Chromosorb 105, or Resin SM4) were used, probably because methyl ricinoleate (precursor) was also adsorbed. In another example, the production of thiopenes by hairy root cultures of *Tagetes petula* increased up to 40% when XAD-7 was used in an elicitation–extractive fermentation system (Buitelaar et al., 1993).

Gas Extraction

Supercritical fluids — fluids at temperatures and pressures slightly above the critical points (e.g., 31°C and 7.38 MPa for CO_2) — exhibit unique combined properties: liquid-like density (and hence solvent power), high compressibility, very low viscosity, and high diffusivity. The first two properties make the solvent power of supercritical fluids easy to control by changing pressure and/or temperature, while low viscosity and high diffusivity markedly enhance mass transport phenomena and hence extraction process kinetics (Jarzebski and Malinowski, 1995). Extraction of aroma compounds with supercritical carbon dioxide offers an ideal process for general use in the food and flavor industries because of the high physiological compatibility of CO_2, the absence of residues in the final product, low temperature of separation, and protection from oxidizing atmospheres during processing. A highly relevant feature of this technique is the ability to separate and select specific aroma fractions within individual products and thus provide a high degree of control over final product specification (Carbonell, 1991). Despite the fact that this technique requires high-cost investment, the use of supercritical CO_2 in extractive fermentation has been reported.

The production of ethanol by *Saccharomyces cerevisiae* and *Saccharomyces rouxii* under hyperbaric conditions has been evaluated (L'Italien et al., 1989). Using high cell concentrations it was possible to obtain a volumetric productivity of 10.9 g/L/h under 7-MPa pressure of CO_2. This performance can be further improved by combining a high-cell-concentration fermentation under atmospheric pressure (which gave a maximum volumetric productivity of 29.7 g/L/h) with a period of hyperbaric conditions for the rapid recovery of ethanol by supercritical CO_2.

Direct supercritical CO_2 extraction of 2-phenylethyl alcohol from a culture of *Kluyveromyces marxianus* was not possible due to the drastic effect of CO_2 on yeast viability (Fabre et al., 1999). Indeed, the use of a cell-broth separation step (ultrafiltration) coupled with two depressurization steps at the outflow of the extraction vessel resulted in 97% of aroma extracted with a mass purity of 91%.

Aroma extraction with supercritical fluids (mainly CO_2) seems to be restricted to the recovery of these compounds from plant extracts. Its use in a fermentation process (one aroma compound at the time) is still far from being economically feasible.

Membrane Separation

Membrane separation covers a wide range of different processes, ranging from microfiltration to electrodialysis. The common factor linking this wide range of separation operations is the physical arrangement of the process. Separation occurs between two fluids that are separated by a thin physical barrier, or interface. This interface constitutes the membrane and allows some materials to pass through the membrane while others are retained (Bell and Cousins, 1994). Typical membrane processes applied for downstreaming of aroma compounds are hollow-fiber membranes and pervaporation.

Hollow-Fiber Membranes

Hollow-fibers are capillary tubes installed in a shell-and-tube arrangement and provide a very high surface area for a given volume unit. Because of the small pore of the fibers (<0.5 mm), the feed must be free of solids. For ultra-filtration, the high pressure is on the side of the fibers; for reverse osmosis, the high pressure is on the shell side (Bell and Cousins, 1994). Some examples of ISPR by hollow-fiber membranes are described below. The most promising results have been reported for the production of organic acids by the combined use of hollow-fiber membrane separation coupled with solvent extraction.

The production of acetone, butanol, and ethanol (ABE) by *Clostridium acetobutylicum* was improved by the use of membrane extraction (Grobben et al., 1993). The use of a polypropylene-based, hollow-fiber membrane module, in combination with a mixture of olelyl alcohol and decane (50/50 v/v) as extraction solvents, increased the total ABE production to 100% and the product yield by 60%, when compared with the control system.

The extractive bioconversion of isovaleraldehyde by isoamyl alcohol oxidation by *Gluconobacter oxydans* has been studied in a hollow-fiber membrane bioreactor (Molinari et al., 1997). This bioconversion was possible when the aqueous bioconversion phase (isoamyl alcohol and cells) and the organic phase (isooctane) were maintained apart with the aid of a hydrophobic membrane (made of microporous polypropylene). The process was performed in a continuous mode, and the selective *in situ* removal of the aldehyde allowed higher conversions (72 to 90%) and improved overall productivity (2 to 3 g/L/h); however, cell inactivation was observed after only 10 to 12 h of fermentation.

Enhanced propionic acid production was obtained by *Propionibacterium acidipropionici* using an amine extractant and a hollow-fiber membrane extractor (Jin and Yang, 1998). Compared with conventional batch fermentation, the extractive fermentation exhibited a much higher productivity (1 g/L/h, a fivefold increase), higher propionate yield (20% higher), higher final product concentration (75 g/L or higher), and higher purity (up to 90%). The process gave consistent long-term performance over the 1.5-month period of the study. Similar results were obtained in a 5-m^3 bioreactor by the use of cell recycling (microfiltration), where 30 to 40 g/L of propionic acid at 1.2 g/L/h were produced by *Propionibacterium acidipropionici* (Colomban et al., 1993). The process has shown long-term stability (over 900 h).

The extraction of lactic acid from fermented broth based on anion exchange reaction with tri-*n*-octylmethylammonium chloride and microporous hollow fiber has been studied (Tong et al., 1998). No adverse effect of yeast extract was observed over the membrane performance. The results, obtained after batch fermentation, demonstrated the feasibility of integrating membrane extraction with lactic acid fermentation.

Pervaporation

Pervaporation is a separation technique based on a selective transport through a dense layer (generally composed of a polymer) associated with an evaporation technique (Baudot et al., 1999). Pervaporation has two methods of operation: (1) the liquid to be separated circulates on one side of the membrane and a vacuum is applied to the other, or (2) the vacuum is replaced by an inert carrier gas that transports the permeate away from the membrane, thus maintaining a low permeate pressure (Bell and Cousins, 1994). The membrane employed in the pervaporation process and the total permeate pressure determine process selectivity and hence the application (Baudot et al., 1999). During pervaporation organics from dilute organic–water mixtures will be enriched as they are transported from the liquid feed to the vapor permeate. Basically, the mass transfer through the membrane occurs in three consecutive steps: (1) selective absorption of permeates at the feed side, (2) selective diffusion through the membrane, and (3) desorption to the vapor permeate (Karlsson and Trägårdh, 1997). This technique has shown important development in the last decade, and some interesting applications are described here.

Bengtson et al. (1992) investigated 6-pentyl-α-pyrone recovery from a cell-free broth by pervaporation using hydrophobic membranes, reaching an eightfold enrichment in the membrane compartment; however, only modest results were obtained when fermentation and pervaporation were coupled due to the low production rate of the molecule.

Fermentation coupled with a complete cell-recycled pervaporation system allowed an increase in the ethanol productivity of *Saccharomyces cerevisiae* (Wei et al., 1995). By continuous concentration and removal of the ethanol from the membrane bioreactor (silicone/polysulphone), the ethanol concentration in the broth was kept to about 20 g/L, while ethanol concentration in the permeate was around 100 g/L. Under such conditions, the ethanol productivity was increased from 1.74 to 3.25 g/L/h over 72 h. Better results were obtained when a continuous fermentation/membrane pervaporation system was used (O'Brien and Craig, 1996). During continuous operation of the pervaporation module (polydimethylsiloxane membranes) and with broth cell densities of 15 to 23 g/L, ethanol productivities of 4.9 to 7.8 g/L/h were achieved. However, the long-term operation of an industrial continuous fermentation/pervaporation system would require the incorporation of a dilution stream to counteract the inhibitory effects of the build-up of minor fermentation products (i.e., organic acids) in the fermentation broth.

In a recent work, various membranes were studied for the removal of acetone, butanol, and ethanol (ABE) from model systems (Kawedia et al., 2000). It was found from the overall performance (selectivity and flux) that the styrene butadiene rubber membrane was the best for the quaternary mixture (acetone, butanol, ethanol, and water) studied. A silicone membrane was used to study butanol separation from fermentation broth in an ABE fermentation of a hyperproducing mutant of *Clostridium beijerinckii* strain (Qureshi and Blaschek, 1999). The integrated process of ABE fermentation–recovery allowed an increase of 100% of solvent productivity in comparison with control batch fermentation productivity (from 24.2 to 51.5 g/L of total solvents).

Production of benzaldehyde by *Bjerkandera adusta* has been improved by the use of polydimethylsiloxane membranes in a fermentation/pervaporation system (Lamer et al., 1996). During the continuous recovery, the benzaldehyde concentration in the fermenter remained below the inhibitory concentration (300 ppm), while in the pervaporate the benzaldehyde concentration reached concentrations between 3.2 and 6.2 g/L. Because of the presence of fungal biomass, a preliminary separation step was necessary to maintain biomass in the fermenter and to prevent membrane fouling.

Recovery of the aroma profile (48 selected aroma compounds) from commercial wine-must fermentation by pervaporation was recently reported (Schäfer et al., 1999). The process is especially challenging because it requires the extraction of each aroma compound to the same degree (keeping the molar relationships among them), because the profile might otherwise be altered. Using a polyoctylmethylsiloxane on polyetherimide membrane, the optimum time span for recovery was 3 to 5 days of fermentation. During this period, the complete muscatel aroma profile could be recovered as a concentrate because the enrichment of aroma compounds remained almost constant (up to 90-fold).

Pervaporation has been shown to be a powerful extraction technique in the case of high-volatile/low-molecular-weight aroma compounds, such as in the production of ethanol, butanol, and acetone. Further developments of this technique should lead to better improvements and applications, as in the case of the extraction of a complex aroma profile in wine-must fermentation.

Conclusion

In situ product removal (ISPR) for improving aroma compound biogeneration has focused on two aroma compound categories: (1) bulk products such as solvents and organic acids where an economically feasible process is based on high yield and productivity, and (2) aroma compounds causing biocatalyst product inhibition, as in the case of lactones, ketones, and some aldehydes, where ISPR is essential not only to increase productivity but, in some cases, simply to make the process possible.

Regarding ISPR techniques, the high occurrence of reports confirms that pervaporation is the most promising technique. The main reasons include low investment, the use of mild process conditions, and problems related to solvent toxicity. Nowadays, the use of supercritical CO_2 seems to be limited to high-value products (pharmaceutical); however, it represents a powerful technique due to the possibility of performing one-step extraction and purification of the aroma compounds. It can be expected that research in the field of aroma biogeneration will increase significantly over the next several years.

Acknowledgments

The financial support of DGAPA (IN209799) and IFS (E/2548-2) is acknowledged. The critical reviewing of the manuscript by Dr. Enrique Galindo is acknowledged with thanks.

References

Baudot, A., Souchon, I., and Marin, M. (1999) Total pressure influence on the selectivity of the pervaporation of aroma compounds, *J. Membrane Sci.*, 158, 167–185.

Belin, J.M., Bensoussan, M., and Serrano-Carreón, L. (1992) Microbial biosynthesis for the production of food flavours, *Trends Food Sci. Technol.*, 3, 11–14.

Bell, G. and Cousins, R.B. (1994) Membrane separation processes, in *Engineering Processes for Bioseparations*, Laurence R. Weatherley, Ed., Butterworth-Heinermann, Oxford, UK, pp. 135–165.

Bengston, G., Böddeker, K.W., Hanssen, H.P., and Urbasch, I. (1992) Recovery of 6-pentyl-alpha-pyrone from *Trichoderma viride* culture medium by pervaporation, *Biotechnol. Techniques*, 6, 23–26.

Berger, R.G. (1995) Bioprocess technology, in *Aroma Biotechnology*, R.G. Berger, Ed., Springer-Verlag, Berlin, pp. 139–148.

Brink, L.E.S. and Tramper, J. (1985) Optimization of organic solvent in multiphase catalysis, *Biotechnol. Bioeng.*, 27, 1258–1269.

Buitelaar, R.M., Leenen, E.J.T.M., Geurtsen, G., De Groot, E., and Tramper, J. (1993) Effects of the addition of XAD-7 and elicitor treatment on growth, thiopene production, and excretion by hairy roots of *Tagetes patula*, *Enzyme Microbiol. Technol.*, 15, 670–676.

Carbonell, E.S. (1991) Extraction of flavours with supercritical carbon dioxide, *Cereal Foods World*, 36, 935–937.

Colomban, A., Roger, L., and Boyaval, P. (1993) Production of propionic acid from whey permeate by sequential fermentation, ultrafiltration and cell recycling, *Biotechnol. Bioeng.*, 42, 1091–1098.

Creuly, C., Larroche, C., and Gros, J.B. (1992) Bioconversion of fatty acids into methyl ketones by spores of *Penicillum roquefortii* in a water-organic solvent, two-phase system, *Enzyme Microbiol. Technol.*, 14, 669–678.

Eyal, A.M. and Bressler, E. (1993) Industrial separation of carboxylic and amino acids by liquid membranes: applicability, process considerations and potential advantages, *Biotechnol. Bioeng.*, 41, 287–295.

Fabre, C.E., Condoret, J.S., and Marty, A. (1999) Extractive fermentation of aroma with supercritical CO_2, *Biotechnol. Bioeng.*, 64, 392–400.

Freeman, A., Woodley, J.M., and Lilly, M.D. (1993) *In situ* product removal as a tool for bioprocessing, *BioTechnology*, 11, 1007–1012.

Gatfield, I.L. (1988) Production of flavour and aroma compound by biotechnology, *Food Technol.*, 42, 111–169.

Grobben, N.G., Eggink, G., Cuperus, P., and Huizing, H.K. (1993) Production of acetone, butanol, and ethanol (ABE) from potato wastes: fermentation with integrated membrane extraction, *Appl. Microbiol. Biotechnol.*, 39, 494–498.

Gu, Z., Glatz, B.A., and Glatz, C.E. (1998) Effects of propionic acid on propionibacteria fermentation, *Enzyme Microbiol. Technol.*, 22, 13–18.

Honda, H., Toyama, Y., Takahashi, H., Nakazeko, T., and Kobayashi, T. (1995) Effective lactic production by two-stage extractive fermentation, *J. Ferment. Bioeng.*, 79, 589–593.

Humphrey, A.E. (1994) Plant cells as chemical factories: control and recovery of valuable products, in *Advances in Bioprocess Engineering*, E. Galindo and O.T. Ramírez, Eds., Kluwer Academic, Dordrecht, The Netherlands, pp. 103–108.

Jarzebski, A.B. and Malinowski, J.J. (1995) Potentials and prospects for application of supercritical fluid technology in bioprocessing, *Process Biochem.*, 30, 343–352.

Jin, Z. and Yang, S.T. (1998) Extractive fermentation for enhanced propionic acid production from lactose by *Propionibacterium acidipropionici*, *Biotechnol. Prog.*, 14, 457–465.

Jones, T.D., Havard, J.M., and Daugulis, A.J. (1993) Ethanol production from lactose by extractive fermentation, *Biotechnol. Lett.*, 15, 871–876.

Karlsson, H.O.E. and Trägårdh, G. (1997) Aroma recovery during beverage processing, *J. Food Eng.*, 34, 159–178.

Kawedia, J.D., Vishwas, G.P., and Niranjan, K. (2000) Pervaporative stripping of acetone, butanol and ethanol to improve ABE fermentation, *Bioseparation*, 9, 145–154.

Krings, U. and Berger, R.G. (1995) Porous polymers for fixed bed adsorption of aroma compounds in fermentation processes, *Biotechnol. Techniques*, 9, 19–24.

Krings, U., Kelch, M., and Berger, R.G. (1993) Adsorbents for the recovery of aroma compounds in fermentation processes, *J. Chem. Technol. Biotechnol.*, 58, 293–299.

Kwon, Y.J., Kaul, R., and Mattiasson, B. (1996) Extractive lactic acid fermentation in poly(ethyleneimine)-based aqueous two-phase system, *Biotechnol. Bioeng.*, 50, 280–290.

Laane, C., Boeren, S., Vos, K., and Veeger, C. (1987) Rules for optimization of biocatalysis in organic solvents, *Biotechnol. Bioeng.*, 30, 81–87.

Lamer, T., Spinnler, H.E., Souchon, I., and Voilley, A. (1996) Extraction of benzaldehyde from fermentation broth by pervaporation, *Process Biochem.*, 31, 533–542.

Liddell, J.M. (1994) Introduction to downstream processing, in *Engineering Processes for Bioseparations*, L.R. Weatherley, Ed., Butterworth-Heinermann, Oxford, UK, pp. 5–34.

L'Italien, Y., Thibault, J., and LeDuy, A. (1989) Improvement of ethanol fermentation under hyperbaric conditions, *Biotechnol. Bioeng.*, 33, 471–476.

Lueck, E. (1980) Propionic acid, in *Microbial Food Additives: Characteristics, Uses, Effects*, E. Lueck, Ed., Springer-Verlag, New York, pp. 175–182.

Mathys, R.G., Schmid, A., and Witholt, B. (1999) Integrated two-liquid phase bioconversion and product recovery processes for the oxidation of alkanes: process
design and economic evaluation, *Biotechnol. Bioeng.*, 64, 459–477.

Molinari, F., Aragozzini, F., Cabral, J.M.S., and Prazeres, D.M.F. (1997) Continous
production of isovaleraldehyde through extractive bioconversion in a hollow-
fiber membrane bioreactor, *Enzyme Microbiol. Technol.*, 20, 604–611.

O'Brien, D.J. and Craig, Jr., J.C. (1996) Ethanol production in a continuous fermen-
tation/membrane pervaporation system, *Appl. Microbiol. Biotechnol.*, 44,
699–704.

Park, O.J., Holland, H.I., Khan, J.A., and Vulfson, E.N. (2000) Production of flavour
ketones in aqueous-organic two-phase systems by using free and microencapsulated fungal spores as biocatalysts, *Enzyme Microbiol. Technol.*, 26, 235–242.

Planas, J., Rådström, P., Tjerneld, F., and Hahn-Hågerdal, B. (1996) Enhanced production of lactic acid through the use of a novel aqueous two-phase system as
an extractive fermentation system, *Appl. Microbiol. Biotechnol.*, 45, 737–743.

Planas, J., Lefebvre, D., Tjerneld, F., and Hahn-Hågerdal, B. (1997) Analysis of phase
composition in aqueous two-phase systems using a two-column chromatographic method: application to lactic acid production by extractive fermentation, *Biotechnol. Bioeng.*, 54, 303–311.

Prapulla, S.G., Karanth, N.G., Engel, K.H., and Tressl, L. (1992) Production of 6-
pentyl-alpha-pyrone by *Trichoderma viride*, *Flavour Fragance J.*, 7, 231–233.

Qureshi, N. and Blaschek, H.P. (1999) Production of acetone-butanol-ethanol by a
hyper-producing mutant strain of *Clostridium beijerinckii* BA101 and recovery
by pervaporation, *Biotechnol. Prog.*, 15, 594–602.

Rito-Palomares, M., Negrete, A., Galindo, E., and Serrano-Carreón, L. (2000) Aroma
compounds recovery from mycelial cultures in aqueous two-phase processes,
J. Chromatogr. B, 743, 403–408.

Schäfer, T., Bengston, G., Pingel, H., Böddeker, K.W., and Crespo, J.P.S.G. (1999)
Recovery of aroma compounds from a wine-must fermentation by organophilic
pervaporation, *Biotechnol. Bioeng.*, 62, 412–421.

Souchon, I., Spinnler, H.E., Dufossé, L., and Voilley, A. (1998) Trapping of γ-decalactone by adsorption on hydrophobic sorbents: application to the bioconversion
of methyl ricinoleate by the yeast *Sporidiobolus salmonicolor*, *Biotechnol. Techniques*, 12, 109–113.

Srivastava, A., Roychoudhury, P.K., and Sahai, V. (1992) Extractive lactic acid fermentation using ion-exchange resin, *Biotechnol. Bioeng.*, 39, 607–613.

Takayanagi, H., Fukuda, J., and Miyata, E. (1996) Non-ionic adsorbents in separation
processes, in *Downstream Processing of Natural Products: A Practical Handbook*,
M. Verral, Ed., John Wiley & Sons, West Sussex, U.K., pp. 159–178.

Tekin, A.R., Oner, M.D., and Kaya, A. (1995) Production of coconut-like aroma by
Trichoderma viride in aqueous and two phase fermentation, *Trans. J. Eng. Environ.
Sci.*, 19, 247–251.

Tong, Y., Hirata, M., Takanashi, H., Hano, T., Kubota, F., Goto, M., Nakashino, F., and
Matsumoto, M. (1998) Extraction of lactic acid from fermented broth with
microporous hollow fiber membranes, *J. Membrane Sci.*, 143, 81–91.

Weatherley, L.R. (1996) Solvent extraction of fermentation broth, in *Downstream Processing of Natural Products: A Practical Handbook*, M. Verral, Ed., John Wiley &
Sons, West Sussex, U.K., pp. 71–91.

Wei, Z., Xingju, Y., and Quan, Y. (1995) Ethanol fermentation coupled with complete cell recycle pervaporation system: dependence of glucose concentration, *Biotechnol. Techniques*, 9, 299–304.

Welsh, F.W. (1994) Overview of bioprocess flavor and fragance production, in *Bioprocess Production of Flavor, Fragance, and Color Ingredients*, A. Gabelman, Ed., John Wiley & Sons, New York, pp. 1–16.

Yabannavar, V.M. and Wang, D.I.C. (1991) Strategies for reducing solvent toxicity in extractive *fermentations, Biotechnol. Bioeng.*, 37, 716–722.

11

Lupines: An Alternative for Debittering and Utilization in Foods

Cristian Jiménez-Martínez, Humberto Hernández-Sánchez, and Gloria Dávila-Ortíz

CONTENTS

Introduction

In studies on legume seed protein spanning more than a century, the soy bean has been for many decades the only legume crop on which significant research has been undertaken. Besides its traditional uses, an extremely wide area of use has been developed for this crop in animal feeds, human foods, and other industrial applications; however, soy beans are not adapted to many climatic conditions. On the other hand, lupines are able to grow under climatic conditions on soils that soy beans would never tolerate; in fact, the soil would be improved because the lupine roots fix nitrogen, and crop rotation can provide better yields (Feldheim, 1999). Cultivation of lupines is still limited and has never exceeded 7000 ha per year; the potential planting area, however, is estimated at 1 million ha (Imane and Al Faïz, 1999). About 90 species have been reported in México, but these wild lupines have not been exploited at a commercial level.

Many nutritional studies (Yañez et al., 1979; Shoenenberger et al., 1982; Groos et al., 1983; Feldheim, 1999) in animals and humans have shown that lupines compare very favorably with soy beans. *Lupinus campestris* seed, as well as other *Lupinus* species have high protein (44%) and oil (13%) contents. Nutritionally, lupines offer advantages in comparison with soy beans, as they contain only small amounts of trypsin inhibitors, tannins, phytate, saponins (Kyle, 1995; García et al., 1999), and other antinutritional factors. The main limitation for wider use of lupines is their high content of quinolizidine alkaloids; furthermore, the seeds have relatively high levels of the raffinose family of α–galactosides (Trugo et al., 1988), which have been reported as flatulence promoters as they are not digested by humans (Calloway et al., 1971). Cooking and other methods facilitate the elimination of antinutritional factors to improve the nutritive value of legumes. Debittering is an ancient procedure widely used by the inhabitants of the Andean Highlands in order to wash out the bitter components of the lupine seeds (Groos et al., 1983).

Chemical Composition of *Lupinus campestris*

The chemical composition of the *L. campestris* seed is presented in Table 11.1. The protein content is high (440 g/kg) and similar to that of the *L. luteus* seed, which has been reported to be 415 g/kg (Yañez et al., 1979) and 435 g/kg (Feldheim, 1994). This value is higher than the protein contents of others legumes such as the lentil and the common bean, which are in the range of 60 to 250 g/kg (Bourges, 1987). The lipid content is also similar to that of the *L. albus* seed (Yañez et al., 1979) but higher than that of *L. luteus* (49 g/kg), according to Ballester et al. (1980); however, it is below the value of *L. mutabilis*, which has a lipid content of 170 to 210 g/kg (Shoenenberger et al., 1982). An improved variety has reached a lipid content of 250 g/kg (Gross and Von Baer, 1977). The crude fiber content of *L. campestris* is 147 g/kg, similar to that found in *L. angustifolius*, *L. luteus*, and *L. notarius* (130 to 190 g/kg) (Aguilera and Trier, 1978).

TABLE 11.1

Chemical Composition of Four Species of *Lupinus* (g/kg)

Species	Protein (N × 6.25)	Ether Extract	Fiber Crude	Carbohydrates (by difference)	Ash
L. luteus	390	47	168	234	35
L. albus	344	109	117	268	32
L. mutabilis	465	166	68	262	39
L campestris	440	131	147	247	35

Debittering of Seeds

As a consequence of aqueous and alkaline thermal treatments for debittering, the *L. campestris* seeds absorb more than twice their original weight in water. In Figure 11.1, this behavior can be observed as a function of the time. After 6 h of thermal processing, all of the seeds have hydrated completely (Figure 11.1). The weight gain (Figure 11.2 is higher for the alkaline treatment, as the percentage of hydrated grains is greater than that for the aqueous treatment; however, the behavior is similar in either case.

In addition to hydrating the seed, the debittering process helped eliminate alkaloids, destroy their germinative power, inactivate cell enzymes such as lipase and lipoxygenase, eliminate microorganisms adhered to the seed

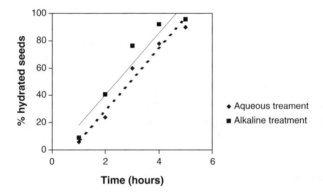

FIGURE 11.1
L. campestris hydratation during debittering.

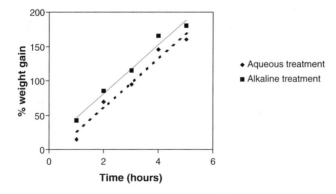

FIGURE 11.2
Weight gain (%) during the hydratation of *Lupinus campestris* seeds.

(which could produce toxins), and increase the cell wall permeability, facilitating the extraction of alkaloid, oligosaccharides, or other antinutritional factors. Also, as a consequence of the high temperature, proteins are denatured and their loss to the water is prevented (Torres et al., 1980). The hydration time for soy beans is 14 h at room temperature (Bianchi et al., 1983). It is possible that this difference is due to the fact that *L. campestris* is a wild species that requires genetic improvement to eliminate the disadvantage (relative hull impermeability) with respect to other legumes with short hydration times, varying from 40 to 180 min using thermal treatment (Adewusi and Falade, 1996) or 6 to 12 h soaking at room temperature after a short boiling time (20, 30, or 45 min) or using a pressure cooker (121°C for 15 min) (Onayemi et al., 1986; Jood et al., 1985).

During the debittering process, the different seed components modify their proportions, showing in some cases a great variation (Table 11.2). The protein content increased from 440 to 500 and 528 g/kg when aqueous and alkaline treatments, respectively, were used. Torres et al. (1980) found that during the process of debittering, 120 to 270 g protein per kg solids were lost, and that this depended on the kind of treatment used. This same trend was observed for fat content, which increased from the 131 g/kg originally present in the seed to approximately 210 and 220 g/kg. Under these conditions, lupine could be considered as an oilseed (Groos et al., 1983). The final protein content showed by the *L. campestris* seeds is slightly smaller (523 g/kg) than that presented by the *L. mutabilis* seed (Duque, 1995) after debittering (593 g/kg).

Total Carbohydrates and Oligosaccharides

The results obtained for carbohydrate content in the *L. campestris* seeds are shown in Table 11.3 (Jiménez-Martínez et al., 2001). It can be observed that in the sample without treatment the content was 169.2 g/kg, consisting of monosaccharides, disaccharides, and oligosaccharides. It can also be observed that the monosaccharide concentration was higher than that

TABLE 11.2

Variation in the *Lupinus campestris* Seed Chemical Composition During Aqueous and Alkaline Thermal Treatments (g/kg)

Analyses	Original Seed	Aqueous Treatment	Alkaline Treatment
Protein (N ×6.25)	440 ± 2.0	500 ± 5.0	528 ± 1.3
Lipids	131 ± 2.0	212 ± 5.0	220 ± 4.5
Crude fiber	147 ± 1.1	102 ± 1.7	102 ± 0.6
Carbohydrates	247 ± 1.3	156 ± 1.9	120 ± 1.8
Ash	35 ± 0.1	30 ± 0.3	30 ± 0.6

Note: Values shown are mean ± standard deviation.

Source: From Jiménez-Martínez et al., *J. Sci. Food Agric.*, 81, 421–428, 2001. With permission.

TABLE 11.3

Effect of Aqueous and Alkaline Thermal Treatments on the Carbohydrate Content of the *Lupinus campestris* Seed (g/kg)

Treatment	Monosaccharides	Sucrose	Raffinose	Stachyose	Verbascose	Others	Total
None	50.6 ± 0.2	28.4 ± 0.2	8.9 ± 0.1	35.9 ± 2.6	13.2 ± 0.2	32.3 ± 0.3	169.2
			Alkaline				
1 h	37.2 ± 2.6	28.1 ± 0.3	8.6 ± 0.1	35.0 ± 9.4	10.7 ± 0.2	29.6 ± 0.1	149.9
3 h	23.6 ± 0.5	22.2 ± 0.5	1.7 ± 0.8	15.5 ± 2.3	4.0 ± 0.3	18.2 ± 0.2	71.6
6 h	8.6 ± 0.1	0.8 ± 0.1	0.2 ± 0.03	0.2 ± 0.05	0.2 ± 0.05	6.0 ± 0.1	17.7
			Aqueous				
1 h	47.9 ± 1.3	28.2 ± 0.2	8.7 ± 1.2	35.8 ± 2.3	12.0 ± 2.2	31.1 ± 8.9	163.6
3 h	39.9 ± 2.1	18.0 ± 1.1	7.1 ± 0.3	12.1 ± 1.7	8.6 ± 0.3	29.8 ± 0.3	115.5
6 h	10.6 ± 0.8	7.5 ± 2.1	1.8 ± 0.6	7.7 ± 0.7	2.7 ± 0.09	20.9 ± 0.1	51.2

Note: Values shown are mean ± standard deviation.

Source: From Jiménez-Martínez et al., *J. Sci. Food Agric.*, 81, 421-428, 2001. With permission.

found in other varieties, such as *L. albus* (12.0 g/kg) (Sosulski et al., 1982), or in other legumes such as peanut (23.4 g/kg) (Cegla and Bell, 1976) and soy beans (7.4 g/kg) (Sosulski et al., 1982).

Jiménez-Martínez et al. (2001) detected a sucrose concentration of 28 g/kg which constitutes 16.8% of the total carbohydrates. Quemmener (1988) reported a slightly greater percentage for varieties such as *L. angustifolius* seed (37.6 g/kg), *L. albus* seed (37.4 g/kg), and *L. kiew* mutant seed (28 g/kg). In bean seeds (Cegla and Bell, 1976), the values are in the range of 16.5 to 24.5 g/kg, while in sunflower seeds the content is 22.9 g/kg.

With respect to the α-galactosides content, it was observed that stachyose was the most abundant of the three oligosaccharides present in *L. campestris* seed. The 35.9 g/kg concentration is similar to that determined for other species such as *L. albus* seed (35.0 to 41.1 g/kg) (Sosulski et al., 1982) and *L. angustifolius* (35 to 48 g/kg) (Quemmener, 1988) and in other legumes such as lentils (24.4%) (Carlsson et al., 1992) or soy beans (29.6 to 41.4 g/kg) (Kennedy et al., 1985). However, some soy bean varieties contain a greater percentage of these oligosaccharides (78 g/kg) (Cegla and Bell, 1976). Species having higher stachyose concentrations include *L. albus kali* seed, with 71 g/kg (Eskin et al., 1980); *L. consentinii* seed, with 63 g/kg; *L. hispanicus* seed, with 19 g/kg (Saini and Lymbery, 1983); and *L. luteus* seed, with 118 g/kg (Matterson and Saini, 1977). These differences can be attributed to the various methods used for elimination of these sugars.

The results obtained by aqueous and alkaline treatments are shown in Table 11.3. It can be observed that, after 6 h, reductions in the total original carbohydrates of 70 and 90% were achieved by the aqueous and alkaline treatments, respectively. This decrease was beneficial, as stachyose, raffinose, and verbascose were included among these carbohydrates. The elimination of these oligosaccharides is necessary because all of them cause flatulence in monogastric animals and humans due to their lack of the α-galactosidase enzyme that helps to unfold these sugars (Saini and Lymbery, 1983).

Elimination of 70% of carbohydrates achieved by the aqueous treatment is considered efficient, especially if compared with the results obtained by Reddy and Salunke (1980), who reported a 25% oligosaccharide reduction in the case of black gram (*Phaseolus mungo*) after 40 min of boiling at 116°C in aqueous solution. Bianchi et al. (1983) found a very similar decrease in a Brazilian soy bean variety after 30 min of boiling and a decrease of more than 80% when the process was carried out at 90°C. Silva and Leite (1982) reported that boiling for 60 min reduced the content of total carbohydrates in beans by 20 to 45%; however, Rao and Belavady (1978) found an increase in carbohydrate levels after boiling for 15 min in the case of red gram (*Cajanus cajan*), chick-pea (*Cicer arietum*), black gram (*Phaseolus mungo*), and green gram (*Phaseolus aureus*).

Oligosaccharide reduction by means of alkaline treatment was similar in our study (Jiménez-Martínez et al., 2001) to that reported by different authors (Bianchi et al., 1983; Calloway et al., 1971) and superior to that reported by Molnar-Perl et al. (1984), who obtained a decrease of 26% in total carbohydrates present in soy beans using acid treatment (HCl) and a boiling temperature of 100°C for 5 min. Jood et al. (1985) eliminated 70% of the total carbohydrates in different legumes by soaking for 6 h and boiling in a $NaHCO_3$ solution for 45 min; 86% was removed when soaking was extended to 12 h and a pressure cooker (121°C for 20 min) was used. The oligosaccharide content after both treatments is shown in Table 11.3 (Jiménez-Martínez et al., 2001). It can be observed that with the alkaline thermal treatment sucrose was also reduced to 3% of its initial value. The percentage of reduction is similar to that obtained by Molnar-Perl et al. (1984), who achieved a 96% carbohydrate decrease by applying an acidified solution. Other authors have reported a maximum decrease of 54 to 84% by soaking for 6 to 12 h and boiling for 45 or 30 min (Silva and Leite, 1982; Jood et al., 1985; Adewusi and Falade, 1996).

In our study, stachyose (the most abundant oligosaccharide) was reduced 94.7%, with respect to the initial concentration, a greater reduction than that achieved by Jood et al. (1985), who applied soaking and boiling in $NaHCO_3$ and treatment in a pressure cooker at 121°C for 20 min. Verbascose and raffinose reductions from original content were 98.7% and 98%, respectively. Jood et al. (1985) reduced the content of these oligosaccharides by 83 to 96% and 77 to 80%, respectively, with autoclaving treatment and achieved reductions of 73.4% for sucrose, 79.7% for raffinose, 78% for stachyose, and 77.5% for verbascose with aqueous thermal treatment. These reductions were superior to those obtained by Silva and Leite (1982) and Adewusi and Falade (1996), which were in the range of 20 to 60%. However, Jood et al. (1985) obtained a greater reduction by applying soaking for 12 h and treatment in a pressure cooker at 121°C for 20 min, and total elimination by means of germination. From these results, it can be concluded that the processes discussed are effective in reducing the oligosaccharide levels, increasing the protein and oil concentrations, and possibly resulting in nutritional improvement.

Quinolizidine Alkaloids

The results obtained in the determination of the *L. campestris* total quinolizidine alkaloids (TQAs) are presented in Table 11.4 (Jiménez-Martínez et al., 2001). An initial concentration of 2.74% was determined, comparable with the concentrations found in other varieties of bitter lupines (*Lupinus* spp.) (Muzquiz et al., 1994) and slightly smaller than that found by Hatzold et al. (1983) for *L. mutabilis* seed (3.1%). Because the content of TQAs in *L. campestris* seeds is between 0.3 and 3.0%, it can be classified as a bitter variety (Wink, 1998). This is of importance from the toxicologic point of view, as it is well known that quinolizidine alkaloids have high pharmacological activity, and the *L. campestris* seed is no exception.

TABLE 11.4

Effect of Aqueous and Alkaline Thermal Treatment on the Quinolizidine Alkaloids Content of *Lupinus campestris* Seed (g/kg)

Retention Time (min)	Quinolizidine Alkaloids	None	Treatment Aqueous	Alkaline
11.94	Lupanine	0.02	0	0
12.11	UD	0.01	0	0
12.71	Dehydro-oxo-sparteine	0.08	0	0
13.78	Hydroxyaphylline	11.03	0.20	0.015
14.54	Hydroxyaphyllidine	9.4	0.04	0.011
15.09	UD	5.37	0.05	0.003
15.55	UD	0.34	0	0.002
15.89	UD	0.34	0	0
18.06	UD	0.20	0	0
18.70	UD	0.56	0	0
Total		27.37	0.27	0.021
% residual alkaloids		100	1.03	0.092

Note: UD = unidentified.

Source: From Jiménez-Martínez et al., *J. Sci Food Agric.*, 81, 421–428, 2001. With permission.

During alkaloid separation, nine components were isolated in our study and the major alkaloid was identified as hydroxyaphylline with a 1.1% concentration, constituting 40.3% of the TQAs. Hydroxyaphyllidine was identified as having a 0.95% concentration (34.54% of TQA). Also, an unidentified alkaloid with a concentration of 0.4% (19.61% of TQA) was detected. These three alkaloids constitute 94.5% of the TQA. It is important to note that lupanine, a common alkaloid in lupines, had a concentration of only 0.002%.

The alkaloids present in seeds of varieties such as *L. albus* are lupanine, hydroxyaphylline, albine, and multiflorine; the major alkaloid is lupanine, comprising 50 to 80% of the TQAs (Wink, 1998). In *L. mutabilis* seeds, the main quinolizidine alkaloids are lupanine, 13-hydroxylupanine, and 4-hydroxylupanine (Hatzold et al., 1983). *L. latifolius* and *L. hispanicus* have anagyrine and lupanine, respectively, as their main alkaloids (Meeker and Kilgore, 1987). The alkaloid found in the smallest proportion and those that are indicated in Table 11.4 as being unidentified (UD) represent 0.15% and are comprised of five different alkaloids, tentatively differentiated by their characteristic ions and molecular ions presented in the spectrum bulk.

With respect to the toxicity of quinolizidine alkaloids, those belonging to the sparteine and lupanine types are relatively toxic when injected, but if orally ingested the toxic effect is lower. Lupanine has a moderate effect in vertebrates, while alkaloids such as α-piridone, cytisine, and anagyrine are highly poisonous (Wink, 1994). It is important to note that none of these alkaloids has been found in *L. campestris* seeds.

Some pharmacological studies on the acute toxicity (oral LD_{50}) of sparteine and lupanine determined in mice showed values of 350 to 510 and 410 mg/kg (Yovo et al., 1984; Wink, 1998), while the oxidized derivatives, with the exception of the hydroxylated ones (such as lupanine), presented a smaller toxicity than lupanine (Petterson et al., 1987; Wink, 1998). This is important, as few studies exist related to quinolizidine alkaloid metabolism. It is known that sparteine is transformed into 2-dehydro-oxo-sparteine in rats and rabbits through the neurosomal oxidation system with the participation of cytochrome P_{450}. In relation to this finding, it has been established that in humans it is possible to differentiate a genetic polymorphism with two phenotypes: the active individuals and the weak sparteine metabolizers (Ohnheus et al., 1985).

The clinical effect that manifests quinolizidine alkaloid toxicity in domestic animals is of a neurological type, for which the principal symptoms are shaking, convulsions, and breathing problems. It is known that quinolizidine alkaloids act by inhibiting the ganglionar transmission impulse of the sympathetic nervous system (Yovo et al., 1984).

In Table 11.4, a gradual decrease in the content of QA when the seed was submitted to debittering can be observed. In the case of aqueous thermal treatment, a 56% decrease was obtained in the first 3 h, while the most important decrease (76.5%) was obtained with alkaline thermal treatment. This behavior could be due to the fact that in the alkaline environment seed hydration was greater than in the aqueous environment (78% and 62%, respectively). After 5 h of alkaline treatment, the TQA concentration in the seed was 0.03%, which means that about 99% of the total alkaloids were eliminated. However, in the case of the aqueous treatment, this level was obtained only after 6 h. The final concentration of TQAs in debittered seeds with the alkaline thermal treatment was 0.002%, reflecting a 99.9% TQA elimination.

Torres et al. (1980) found that the application of aqueous thermal treatment (using an unspecified alkali) reduced 98.6% of the TQAs. Other authors have reported that the use of alkaline media resulted in a reduction of 70 to 80% in the content of TQAs (Ortiz and Mukherjee, 1982).

In our study, the final concentrations were (as determined by high-performance liquid chromatography, or HPLC): 0.0005% for hydroxyaphylline, 0.0011% for hydroxyaphyllidine, and 0.0003% for the unidentified alkaloid. The relative abundance of quinolizidine alkaloids was compared with that of the internal standard (caffeine), which had a 0.0002% concentration. It was also observed that, in the seeds that underwent the aqueous thermal treatment, the main quinolizidine alkaloids were more abundant than in the seeds with alkaline thermal treatment, in which the elimination was almost total.

Even with the aqueous debittering, the final alkaloid concentration found was well below that of the toxicity safety limit, which is 0.04 to 0.05% for animal and human consumption (Muzquiz, 1989). It is well established that a quinolizidine alkaloid content above 0.03% results in a decrease in nutrient ingestion and consequently a decrease in animal growth (Lucas and Sotelo, 1995).

As a special case, the domestic animal most tolerant to quinolizidine alkaloids is the rabbit; some studies have reported that these animals can be fed with rations containing up to 50% of *Lupinus* flour seed as the only protein source with only a slight decrease in growth observed. In the case of pigs, the effect of quinolizidine alkaloids depends more on the alkaloid type than on the TQA content. Some alkaloids, such as sparteine, are considerably more toxic than others (Yovo et al., 1984). As stated previously, this last alkaloid was not found in the *L. campestris* seed.

Few studies have been conducted with respect to quinolizidine alkaloid chronic toxicity. It is assumed that because these alkaloids are soluble in water it is relatively easy to eliminate them from the animal or human body with no accumulative toxic effect. In this regard, one study indicated that rats fed with *L. albus* seed flour (0.025% lupanine) did not show any harmful effect after two generations. Furthermore, the rats surviving a lethal dose completely recovered, without showing any clinical abnormality signs, and were able to reach a normal weight as well as physiological maturity *per se* (Petterson et al., 1987).

Tannins

Table 11.5 shows the tannin content in *L. campestris* seed. The original content was 0.31%, which is 33% less than in soy bean seed and soy milk powder. Chavan et al. (1979) found a variety of soy bean (IS-2825) with a tannin content of 3.4%, a quantity ten times higher than that of the common varieties. Varieties with contents of 1.18 to 2.4% (Hoff and Singleton, 1977) and 0.40 to 0.46% (Price et al., 1980) can also be found. The low content of these compounds in *L. campestris* (0.32%) indicates that the seed could be considered nutritionally good in terms of its mineral bioavailability and protein digestibility (Adewusi and Falade, 1996). The minimal quantity in the diet to produce a negative growth response has not been established yet. Price et al. (1980) found that 0.1% of tannic acid (a hydrolyzable form) in diets given to chickens did not cause any harmful effect; however, at levels of 0.5 and 2.0% growth ceased after 7 weeks in 3% and 32%, of the population respectively. At a 5% tannin level, 70% of the chicken population died. The effects of tannins in the human diet are unknown; however, some epidemiological considerations suggest a possible relationship between the presence of condensed tannins and esophageal cancer (Price et al., 1980).

The boiling process generally decreases the quantity of the tannins in legumes. In *L. campestris* a reduction of more than 70% during swelling and 6-h boiling was obtained. When alkaline treatment (in a solution of 0.5% $NaHCO_3$) was used, the decrease was 77%, which is slightly higher than with the aqueous treatment (71%). Ziena et al. (1991) reported a reduction of 10% of the broad bean *(Vicia faba)* tannin content by means of soaking,

TABLE 11.5

Tannin Content in *Lupinus campestris* Seed during the Debittering Process (g/kg)

Time (hr)	Aqueous	Alkaline
0	3.187 ± 0.13	3.187 ± 0.24
3	2.707 ± 0.16	1.810 ± 0.27
6	0.970 ± 0.12	0.800 ± 0.33
Soy flour[a]		5.188 ± 0.11
Soy milk[a]		3.935 ± 0.17

[a]Without treatment.

Note: Values shown are mean ± standard deviation.

Source: From Jiménez-Martínez et al., *J. Sci Food Agric.*, 81, 421–428, 2001. With permission.

but Ologhobo and Fetuga (1982) showed that boiling cowpeas (*Vigna unguiculata*) reduced the concentration in more than 47%.

Tannin reduction can improve the nutritional quality of the seeds, as it increases the bioavailability of minerals. Adewusi and Falade (1996) reported that boiling cowpeas (*Vigna unguiculata*) reduced the tannin level between 23 and 68%. In African jam beans (*Stenostylis stecocarpa*) and rajma beans (*Phaseolus vulgaris*), reductions of 57 to 94% and 72 to 92%, respectively, were obtained. These values were similar to the one obtained in this work.

The use of alkali allows a greater tannin extraction, as lupines treated with this solution have presented a concentration of 0.08% in the seed after the treatment, while in the case of aqueous debittering the value was 0.097%. In either case, the reduction was good (more than 70%), but it was better in the case of alkaline treatment. In this study, the final concentration was well below the value that is considered risky for condensed tannins (0.1%) (Price et al., 1980).

Lupine Milk Elaboration

Figure 11.3 shows a generalized flowchart developed in the author's laboratory for the production of lupine milk from cleaned seeds and includes a material balance for *L. campestris*. Table 11.6 shows the results of the proximate analyses of cow, soy, and lupine milks (Jiménez-Martínez et al., 2001). It can be observed that the lupine milk had a higher protein concentration than the other milks. This is due to the fact that during the debittering process some carbohydrates and other compounds were eliminated, allowing the relative proportion of protein to increase. This is important, as protein is one of the main elements contributing to the nutritional value of any food.

The concentration of protein in the lupine milk is slightly higher than the value reported by Camacho (1989), but similar to the values for a lupine milk obtained from dehulled seeds (Camacho et al., 1988). Johnson and Snyder (1978) obtained soy milk with a protein concentration of 28 to 33 g/L using

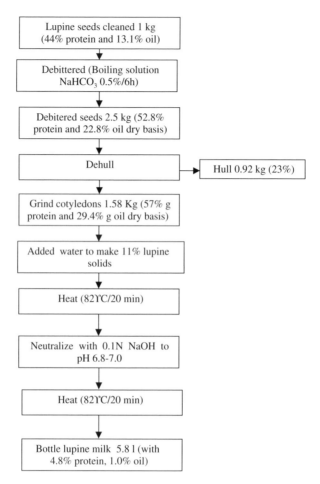

FIGURE 11.3
Preparation of lupine milk from *Lupinus campestris*.

TABLE 11.6

Proximate Analysis of Cow, Soy, and *L. campestris* Milk (g/kg dry basis)

Milk	Protein (N×6.38)	Fat	Fiber	Carbohydrates[a]	Ash
Lupine	580 ± 4.24	294 ± 4.24	0.75 ± 0.21	91.8 ± 0.42	33.45 ± 0.64
Soy	391 ± 0.85	70.5 ± 0.71	1.00 ± 0.14	476 ± 0.49	61.45 ± 2.19
Cow	262 ± 1.41	134 ± 5.73	0.9 ± 0.14	547 ± 0.57	56.25 ± 3.89

[a] By difference.

Note: Values shown are mean ± standard deviation.

Source: From Jiménez-Martínez et al., *J. Sci Food Agric.*, 81, 421–428, 2001. With permission.

alkaline thermal treatment and 42 to 44 g/L using the aqueous processing method. In the author's study, the milk obtained by alkaline thermal treatment had a protein concentration even higher than any of the soy milks obtained by Johnson and Snyder (1978). On the other hand, the fat content was higher in the lupine milk than in the soy milk, which had a content of 26 to 42 g/L (Johnson and Snyder, 1978). Camacho (1989) reported a fat content of 12 g/L for a lupine milk prepared from *L. albus-multolupa*, a value similar to the one used in our study. The *L. campestris* milk yield obtained in the process developed in our work was 5.8 L/kg seed. The concentration of bitter compounds (alkaloids) decreased through the process because they were partially eliminated with the hull or were leached out during the alkaline heat treatment.

Fermentation Profiles

Figures 11.4 and 11.5 show the results obtained for lactic acid production, pH profile, and bacterial counts, respectively, plotted against fermentation time. The graphs were obtained from cow and lupine milks fermented with a commercial yogurt starter culture. Figure 11.3 shows the kinetics of lactic acid production for both substrates, and it can be seen that the pattern is similar during the first two hours of fermentation. After this time, the culture developed more acidity in the cow milk. However, after 8 h, the lupine milk reached an acidity of 0.87%, which is a value in the range of commercial cow milk yogurt (0.8 to 1.2%) (García, 1986). In the case of a yogurt made from peanut milk (Beuchat and Nail, 1978), the final acidity was in the range of 0.3 to 0.53% after a fermentation time of 48 to 72 h. Shirai et al. (1992) reported a value of 0.75% for a yogurt made from a mixed substrate composed of soy milk, oat flour, and whey. This value, although similar, is still lower than the one obtained in this study, as are all the above cited studies with legume milks. Some of the more common problems in legume-based yogurts are low acidity and a beany flavor (Karleskind et al., 1991); however, in the case of the lupine yogurt, the acidity and flavor were found to be acceptable by sensorial evaluation.

Figure 11.4 shows the profiles of pH during the fermentation of cow and lupine milks. The pattern is very similar, both types of milk having a value of around 4.0 after a period of 8 h at 45°C. A pH value of 4.7 or less is important in this product, as it has been related to a good body (texture), flavor, aroma, and stability (Karlensind et al., 1991). In soy yogurt, pH values ranged from 3.9 to 4.3 (Ankenman and Morr, 1996; Cheng et al., 1990; Shirai et al., 1992) and in peanut milk yogurt the pH values were in the range of 4.43 to 4.73 (Beuchat and Nail, 1978). The lupine yogurt had a pH value in the range of most of the legume milk yogurts. Figure 11.6 shows the profiles of the lactic acid bacteria counts during the fermentation of cow and lupine milk. It can be observed that after 8 h the count was

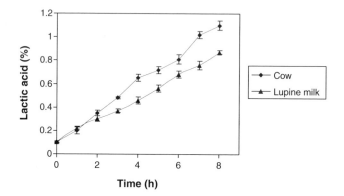

FIGURE 11.4
Profile of acid production during the fermentation of cow and lupine milk by *S. thermophilus* and *L. delbrueckii sp. bulgaricus.*

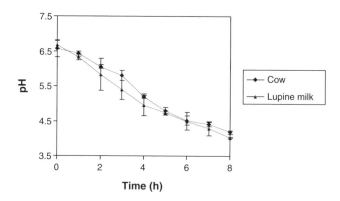

FIGURE 11.5
Profile of pH during the fermentation of cow and lupine milk by *S. thermophilus* and *L. delbrueckii sp. bulgaricus.*

higher in the cow milk yogurt (2×10^9 CFU/mL) than in the lupine yogurt (3.2×10^8 CFU/mL). The results were similar to those obtained by Shirai et al. (1992) and can be explained in terms of the better proportion of nutrients in the cow milk. Mital and Steinkraus (1974) reported a count of 6.8×10^7 CFU/mL for soy milk after a fermentation time of 16 to 18 h. In peanut milk yogurt, Beuchat and Nail (1978) reported a count of 7.24×10^8 CFU/mL after 72 h of fermentation. Again, it can be said that the counts found in this study for the lupine yogurt are in the range of the results obtained by other authors.

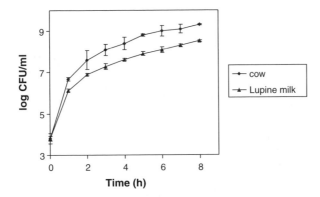

FIGURE 11.6
Lactic acid bacteria count during the fermentation of cow and lupine milk.

Amino Acid Composition

The amino acid composition of the *L. campestris* seed is presented in Table 11.6. *L. campestris* seed was found, as for other *Lupinus* varieties, to be deficient in sulfur-containing amino acids with respect to FAO/WHO standards (1985); however, the concentration was higher than that reported for *L. albus* and *L. mutabilis* (Ballester et al., 1980; Duque, 1995).

Isoleucine, leucine, lysine, and threonine were in lower proportion than that recommended by FAO/WHO (1985) for 2-year-old children; however, almost all amino acids were above the standard recommendations for adults, as is true for other legumes. The protein of *L. campestris* seed presented a higher concentration of isoleucine, phenylalanine, tyrosine, threonine, valine, and histidine than *L. albus* (Ballester et al., 1980; Gross and Von Baer, 1977) and *L. mutabilis* seeds (Duque, 1995); however, it showed a smaller amount of lysine in comparison with other lupines varieties.

Table 11.7 shows amino acid analyses of the samples of lupine seeds, milk, and yogurt. It can be observed that the lupine protein contains all the amino acids in the concentrations recommended by FAO/WHO for adults in 1985, with the exception of sulfur-containing amino acids, which are limiting in all the legume seeds and beverages (Taha et al., 1986; Ivanovic, 1983). During preparation of the lupine yogurt, the concentration of some amino acids increased slightly (serine, glycine, threonine, and anagirine). Similarly, the concentration of others decreased (isoleucine, valine, methionine, and glutamic acid). All these changes may be a result of the use of proteins by the microorganism during the fermentation process. The amino acid composition in lupine milk and yogurt was very similar to that of soy milk and yogurt (Lee et al., 1990).

TABLE 11.7

Amino Acid Content in the Proteins of the Milk and Yogurt of *Lupinus campestris*[a]

Amino Acid	*L. campestris*	Lupine Milk	Yogurt	FAO/WHO (1985)		
				Children < 2 Years Old	Children 2 to 5 Years Old	Adults
Ile	4.15	4.35	3.98	4.6	2.8	1.3
Leu	5.75	6.62	6.13	9.3	6.6	1.9
Lys	3.33	3.92	3.11	6.6	5.8	1.6
Met + Cys[b]	1.27	1.17	0.96	4.2	2.5	1.7
Phe + Tyr	9.18	2.78	2.51	7.2	6.3	1.9
Thr	4.46	4.41	4.84	4.3	3.4	0.9
Val	4.50	4.56	4.16	5.5	3.5	1.3
Hys	4.33	3.71	4.02			
Ala	4.10	3.61	4.00			
Arg	7.25	8.39	9.96			
Asp	9.70	9.11	9.05			
Glu	19.93	20.44	18.36			
Gly	3.81	3.72	4.37			
Pro	4.48	5.45	5.36			
Ser	5.56	5.45	6.03			

[a] Grams amino acids per 16 g N.

[b] Not determined.

Note: Values are the mean of triplicate determinations.

References

Adewusi, S.R.A. and Falade O.S. (1996) The effects of cooking on the extractable tannins, phytate, sugar and mineral solubility in some improved Nigerian legume seeds, *Food Sci. Technol. Int.*, 2, 231–239.

Aguilera, J.M. and Trier, A. (1978) The revival of lupin, *Food Technol.*, 8, 70–76.

Ankenman, G.L. and Morr, C.V. (1996) Improved acid, flavor and volatile compound production in high protein and fiber soymilk yogurt-like product, *J. Food Sci.*, 61, 331–336.

Ballester, D., Yañez, E., Garcia, R., Erazo, S., López, F., Haardt, E., Cornejo, S., López, A., Pokniak, J., and Chichester, C. (1980) Chemical composition, nutritive value and toxicological evaluation of two species of sweet lupine (*L. albus* and *L. luteus*), *J. Agric. Food Chem.*, 28, 402–405.

Beuchat, L.R. and Nail, B.J. (1978) Fermentation of peanut milk with *Lactobacillus bulgaricus* and *L. acidophilus*, *J. Food Sci.*, 43, 1109–112.

Bianchi, M.L.P., Silva, H.C., and Campos, M.A.D. (1983) Effect of several treatments on the oligosaccharides content of a Brazilian soybean variety, *J. Agric. Food Chem.*, 31, 1364–1366.

Bourges, R.H. (1987) Las leguminosas en la alimentación humana (parte 1), *Cuadernos de Nutrición*, 10(1), 17–32.

Calloway, D., Hickey, C.A., and Murphy E.L. (1971) Reduction of intestinal gas forming properties of legumes by traditional and experimental food processing methods, *J. Food Sci.*, 36, 251.

Camacho, L. (1989) Comparative nutritional and quality studies of drum-drying and spray drying of lupin imitation milk, *Nutr. Rep. Int.*, 39, 539–604.

Camacho, L., Vásquez, M., Leiva, M., and Vargas, E. (1988) Effect of processing and methionine addition on the sensory quality and nutritive values of spray-dried lupin milk, *Int. J. Food Sci. Technol.*, 23, 233–240.

Carlsson, N.G., Karlson, H., and Sardhberg, A. (1992) Determination de oligosaccharides in food, diets, and intestinal contents by high-temperature gas chromatography and gas chromatography/mass spectrometry, *J. Agric. Food Chem.*, 40, 2404–2412.

Cegla, G.F. and Bell, K.R. (1976) High pressure liquid chromatography for analysis of soluble carbohydrates in defatted oilseed flours, *J. Am. Oil. Chem. Soc.*, 54, 150–152.

Chavan, J.K., Kadam, S.S., Ghonsikar, C.P., and Salunke, D.K. (1979) Removal of tannins and improvement of *in vitro* protein digestibility of sorghum seed by soaking in alkali, *J. Food Sci.*, 44, 1319–1321.

Chen, S. (1984) *Principios de la Producción de Leche de Soya*, American Soybean Association México, D.F. México.

Cheng, Y.J., Thompson, L.D., and Brittin, H.C. (1990) Sogurt, a yoghurt-like soybean product: development and properties, *J. Food Sci.*, 55, 1178–1179.

Duque, R.L. (1995) Evaluación química, biológica y teratogénica de *L. mutabilis*, Tesis de Maestría IPN-ENCB, D.F. México.

Eskin, N.A., Johnson, M.S., Vaisey-Genser, M., and McDonald, B.E. (1980) A study of oligosaccharides in a select group of legumes, *Can. Inst. Food Sci. Technol. J.*, 13(1), 40–42.

FAO/WHO (1985) Energy and protein requirements, report of a joint FAO/WHO meeting, Tech. Rep. Ser. 724, World Health Organization, Geneva.

Feldheim, W. (1994) Fermentation of lupin fibre, *Proc. 7th International Lupin Conference*, Evora, Portugal, 18–23, April 1993, 445–450.

Feldheim, W. (1999) The use of lupins in human nutrition, in *Lupin: An Ancient Crop for the New Millennium*, Proceedings of the 9th International Lupin Conference, Klink/Müritz, Germany, 20–24, June 1999.

García, G.M. (1986) Yoghurt, aspectos microbianos y de elaboración, *Tecnol. Alim.* (Mex), 21, 5–13.

García, L.P.M., Muzquiz, M., Zamora, N.J.F., Burbano, C., Pedrosa, M.M., and Cuadrado C. (1999) Chemical composition, phytate and galactoside content of wild Mexican lupines, in *Lupine: An Ancient Crop for the New Millennium*, Proceedings of the 9th International Lupin Conference, Klink/Müritz, Germany, 20–24, June 1999.

Gross, R. and Von Baer, E. (1977) Posibilidades del *Lupinus mutabilis* y *L. albus* en los países andinos, *Arch. Latinoam. Nutr.*, 27, 451–471.

Groos, U., Godomar, G.R., and Schoenenberger, H. (1983) The development and acceptability of lupine (*L. mutabilis*) products, *Qual. Plant Food Hum. Nutr.*, 32(2), 155–164.

Hatzold, T., Ibrahim, E., Gross, R., Wink, M., Hartmann, H., and Witte, L. (1983) Quinolizidine alkaloids in seed of *Lupin mutabilis*, *J. Agric. Food Chem.*, 31, 934–938.

Hoff, J.E. and Singleton, K.I. (1977) A method for determination of tannins in foods by means of immobilised protein, *J. Sci. Food Agric.*, 42(6), 1566–1569.

Imane, T.A. and Al Faiz, C. (1999) Lupine use and research in Morocco, in *Lupine: An Ancient Crop for the New Millennium*, Proceedings of the 9th International Lupin Conference, Klink/Müritz, Germany, 20–24, June 1999.

Ivanovic, D. (1983) Formulación y valor nutritivo de un sustituto lácteo en base a lupinus dulce, *Arch. Latinoam Nutr.*, 33, 620–632.

Jiménez-Martínez, C., Hernández-Sánchez, H., Alvarez-Manilla, G., Robledo-Quintos, N., Martínez-Herrera, J., and Dávila-Ortiz, G. (2001) Effect of aqueous and alkaline thermal treatments on chemical composition and oligosaccharide, alkaloid and tannin contents of *Lupinus campestris* seeds, *J. Sci. Food Agric.*, 81, 421–428.

Jood, S., Metha, U., Randhir, S., and Charajit, M.B. (1985) Effects of processing on flatus-producing factors in legumes, *J. Agric. Food Chem.*, 33, 268–271.

Johnson, K.W. and Snyder, H.E. (1978) Soy milk: a comparison of processing methods on yields and composition, *J. Food Sci.*, 43, 349–353.

Karleskind, I., Laye, E., Halpin, E., and Morr, C.V. (1991), Improving acid production in soy-based yogurt by adding cheese whey proteins and mineral salts, *J. Food Sci.*, 56, 999–1001.

Kennedy, I.R., Mwandemele, O.D., and McWhirter, K.S. (1985) Estimation of sucrose, raffinose and stachyose in soybean seed, *Food Chem.*, 17, 85–93.

Kyle, W.S.A. (1995) *The Current and Potential Uses of Lupins for Human Food*, Department of Food Technology, Victoria University.

Lee, S.Y., Morr, C.V., and Seo, R. (1990) A comparison of milk-based and soy milk-based yogurt, *J. Food Sci.*, 55, 532–536.

Lucas, F.B. and Sotelo, L.A. (1995) Aspectos toxicológicos de lupinus, *Plantas: Biotecnología, Agronomía y Nutrición*, COFAA-IPN, México, pp. 31–34.

Macrae, R. and Zand-Moghaddam, A. (1978) The determination of the component oligosaccharides of lupin seed by high pressure liquid chromatography, *J. Sci. Food Agr.*, 29, 1083–1086.

Matterson, N.K. and Saini, H.S. (1977) Polysaccharides and oligosaccharides changes in germination in lupin cotyledons, *Phytochemistry*, 16, 59–66.

Meeker, J.E. and Kilgore, W.W. (1987) Identification and quantification of the alkaloids of *Lupinus angustifolius*, *J. Agric. Food Chem.*, 36, 125–128.

Mital, B.K. and Steinkraus, K.H. (1974) Growth of lactic acid bacteria in soy milk, *J. Food Sci.*, 39, 1018–1022.

Molnar-Perl, I., Pinter-Szakcs, K., and Petroczy, J. (1984) Gas-Liquid chromatographic determination of the raffinose family of oligosaccharides and their metabolites present in soy beans, *J. Chrom.*, A (295), 433–443.

Muzquiz, M.A. (1989) Chemical study of *L. hispanicus* seed-nutritional components, *J. Sci. Food Agric.*, 47,197–204.

Muzquiz, M.A., Cuadrado, C., Ayet, G., Cuadra, C., Burbano, C., and Osagie, A. (1994) Variation of alkaloid components of lupin seed in 49 genotypes of *Lupinus albus* from different countries and locations, *J. Agric. Food Chem.*, 42, 1447–1450.

Ohnheus, E.E., McManus, M.E., Schuarz, D.M., and Thoegerson, S.S. (1985) Kinetics of spartein metabolism in rat and rabbit liver microsomes, *Biochem. Pharm.*, 34, 439–440.

Ologhobo, A.D. and Fetuga, B.L. (1982) Polyphenols, phytic acid and other phosphorus compounds of lima beans (*Phaseolus lunatus*), *Nutr. Rep. Int.*, 26, 605–611.

Onayemi, O., Osibogun, A., and Obembe, O. (1986) Effect of different storage and cooking methods on some biochemical, nutritional and sensory characteristics of cowpea (*Vigna ungiculata l. walp*), *J. Food Sci.*, 51(1), 153–156.

Ortiz, J.G.F. and Mukherjee, K.D. (1982) Extraction of alkaloids and oil from bitter lupine seed, *J. Am. Oil Chem. Soc.*, 59(5), 241–244.

Petterson, D.S., Ellis, Z.L., Harris, D.J., and Spadek, Z.E. (1987) Acute toxicity of the major alkaloids of cultivated *Lupinus angustifolius* seed to rats, *J. Appl. Toxicol.*, 7, 51–53.

Price, M.L., Hagerman, A.E., and Buttler, L.G. (1980) Tannin content of cow pea, chick peas, pigeon peas and mung beans, *J. Agric. Food Chem.*, 28, 459–461.

Quemmener, B. (1988) Improved in the high-pressure liquid chromatographic determination of amino sugar and α-galactosides in faba bean, lupine and pea, *J. Agric. Food Chem.*, 36, 754–759.

Rao, U.P. and Belavady, B. (1978) Oligosaccharides in pulses: variety differences and effects of cooking and germination, *J. Agric. Food Chem.*, 26(2), 316–319.

Reddy, N.R. and Salunke, D.K. (1980) Changes in oligosaccharides during germination and cooking of black gram and fermentation of black gram/rice blend, *Cereal Chem.*, 57, 356–360.

Saini, H.S. and Lymbery, J. (1983) Soluble carbohydrates of developing lupine seed, *Phytochemistry*, 22(6), 1367–1370.

Shirai, K., Gutierrez-Durán, M., Marshall, V.M.E., Revah-Moiseev, S., and García Garibay, M. (1992) Production of yoghurt luke product from plant foodstuff and whey: sustrate, preparation and fermentation, *J. Sci. Food Agric.*, 59, 199–204.

Shoenenberger, H., Gross, R., Cremer, H.D., and Elmadfa, I. (1982) Composition and protein quality of *Lupinus mutabilis*, *J. Nutr.*, 112(1), 70–76.

Silva, H.C. and Leite, B.G. (1982) Effect of soaking and cooking on the oligosaccharides content of dry beans (*Phaseolus vulgaris L.*), *J. Food Sci.*, 47, 924–925.

Singleton, V.L. and Roos, J.A. (1965) Colorimetry of total phenolics with phosphomolybdic-phosphotungtic acid reagents, *Am. J. Enol. Viticult.*, 16, 144.

Sosulski, F.W., Elkowicz, L., and Reichert, R.D. (1982) Oligosaccharides in eleven legumes and their air-classified protein and starch fractions, *J. Food Sci.*, 47, 498–502.

Taha, F.S., Mohamed, S.S., and El-Nocrashy, A.S. (1986) The use of soy bean, sunflower and lupin seed in the preparation of protein bases for nutritious beverages, *J. Sci. Food Agric.*, 37, 1209–1216.

Torres, T.F., Nagata, A., and Dreifua, W.S. (1980) Métodos de eliminación de alcaloides en la semilla de *L. mutabilis*, *Arch. Latin. Nutr.*, 30(2), 200–207.

Trugo, C.L., Almeida, D.C.F., and Gross, R. (1988) Oligosaccharides content in the seed of cultivated Lupins, *J. Sci. Food Agric.*, 45, 21–24.

Wink, M. (1994) Biological activities and potential application of lupin alkaloids, *Proc. 7th International Lupin Conference*, Evora, Portugal, 18–23, April, 1993, 161–178.

Wink, M., Ed. (1998) *Alkaloids: Biochemistry, Ecology and Medicinal Applications*, Plenum Press, New York.

Yañez, E., Gattas, V., and Ballester, D. (1979) Valor nutritivo del lupinus y su potencial como alimento humano, *Arch. Latinoam. Nutr.*, 29, 510–520.

Yovo, K., Huget, F., Pothier, J., Durend, D., Bretau, M., and Narcisse, G. (1984) Comparative pharmacological study of sparteine and ketonic derivative lupanine seed of *Lupinus albus*, *Planta Med.*, 50, 420–424.

Ziena, H.M., Youssef, M.M., El Mahady, A.R. (1991) Amino acid composition and some antinutritional factors of cooked faba beans (*Medammis*). Effects of cooking temperature and time, *J. Food Sci.*, 56, 1347–1352.

12

Recent Development in the Application of Emulsifiers: An Overview

Victor T. Huang

CONTENTS

1-56676-892-6/03/$0.00+$1.50
© 2003 by CRC Press LLC

Introduction

Natural emulsifiers, such as proteins and phospholipids, have been consumed by humans since ancient times, while synthetic emulsifiers have been used in the pharmaceutical, chemical, cosmetics, and food industries since the 1930s. Both types of emulsifiers play critical roles in stabilizing emulsions such as milk, cream, butter, margarine, salad dressing, mayonnaise, whipped topping, and ice cream. Emulsifiers also interact with other components in various food systems to deliver specific functional properties such as the interaction of emulsifiers with gluten and starch resulting in dough with modified rheological properties and baked goods with extended shelf life. Emulsifiers can also modify the crystallization behavior of fats.

Over the last 30 years, development in this field has been carried out in the pharmaceutical, cosmetic, and chemical industries, taking advantage of the interactions among emulsifiers, water, and lipids to formulate emulsifier-based products with unique properties. Some examples of mesophases are found in floor cleaners, pharmaceutical preparations, and agrochemical formulations. The food industry has been slow to progress in this area, perhaps due to the complexity of typical food systems and to the relatively low investment in food research.

In the last decade, an increased number of patents have been granted to food companies for the utilization of lyotropic liquid crystals, liposomes, and microemulsions for developing unique ingredient delivery systems and products with improved stability or functionality. Procter & Gamble, Nestle, Unilever, and Nabisco have been active in patenting in these areas. Recently, quite a few review articles have described the preparation methods and physical properties of those emulsifier-based mesophases (Paul and Mouliuk, 1997; Engström and Larsson, 1999; Bergenstahl, 1997). This chapter reviews the application of emulsifiers as reported in recent food science journals and patent literature.

The Mesophases of Emulsifier–Water Binary System

Due to the presence of both hydrophilic and hydrophobic moieties in emulsifier molecules, emulsifiers are surface active and can partition to the interfacial region of unmixable phases such as oil and water. Emulsifiers at the interface serve to lower surface tension between the two phases and thus stabilize the product against phase separation. Their surface-active nature also contributes to the formation of unique mesophases in water.

Liquid Crystalline Phases

Structure

Mesophases resulting from the association of emulsifiers with water have long-range order, albeit having liquid-like disorder in atomic scale, so they are called *liquid crystalline phases* (Eliasson and Larsson, 1993). Emulsifiers can form various mesophases in water with different structures and viscosities, depending on temperature, concentration, ionic strength, other solutes present in the system, and chemical composition of emulsifiers (Bos et al., 1997). The length of the fatty acid chain and the number and configuration of double bonds in the emulsifier structures are critical. These mesophases include lamellar (L_a), hexagonal (H_i), reversed hexagonal (H_{ii}), cubic, fluid isotropic, and gel phases. Physical analytical techniques such as polarized light microscopy, x-ray diffraction, differential scanning calorimetry (DSC), and nuclear magnetic resonance (NMR) are commonly used to identify the mesophases.

Lamellar Phase

Some emulsifiers, such as glycerol monosterate, exist as beta crystals at room temperature. In the presence of water, at a temperature above the Kraft temperature, the hydrocarbon chain will transform into a disordered liquid state, while water penetrates the polar region forming the L_a phase. The L_a phase consists of stacked infinite lipid bilayers separated by water and having long-range order in one dimension. The L_a phase is usually less transparent and less viscous than other mesophases. The water layer thickness in the polar region can swell almost indefinitely under optimum conditions when small amounts of ionic emulsifiers are present, due to electric repulsion effect. However, the swelling capacity of the L_a phase decreases greatly when sodium chloride concentration is increased. At a concentration greater than 0.3% sodium chloride in water, the benefit of adding ionic emulsifiers to the L_a phase disappears (Krog and Lauridsen, 1976).

Gel Phase

When the L_a phase is cooled below the Kraft point, the hydrocarbon chains crystallize and arrange themselves in a regular lattice. Water may still be present between the polar groups, and an *alpha-crystalline* gel is formed. Such gels are normally metastable and will eventually change due to a decrease in the water layer thickness. When all the water is expelled, the gel phase is transformed into a *beta-crystalline* coagel, which is a microcrystalline suspension of the emulsifier in water. The stability of alpha-crystalline gel can be enhanced by adding ionic emulsifiers, by keeping the pH closer to neutral, by minimizing salt concentration (Krog and Lauridsen, 1976), or by a wide distribution of the fatty-acid-chain composition (Bergenstahl, 1997).

Hexagonal Phase

Hexagonal phases, which have long-range order in two dimensions, consist of infinite cylinders having either a hydrocarbon core (H_i) or a water core (H_{ii}). The H_{ii} phase is quite commonly present in food emulsifier–water systems, while H_i exists only in the aqueous systems of very hydrophilic emulsifiers such as polysorbates (Krog and Lauridsen, 1976).

Cubic Phase

The cubic liquid crystalline phase, which has long-range order in three dimensions, is continuous with respect to both water and emulsifier. It is based on curved, nonintersecting lipid bilayers, which are organized to form two unconnected continuous systems of water channel (Eliasson and Larsson, 1993). The cubic phase often occurs in the temperature and composition ranges between H_{ii} and L_a, between L_a and H_i, and between H_{ii} and the fluid isotropic phase. The cubic phase is isotropic and thus completely transparent. It is usually very viscous, thus sometimes can be problematic during processing. For example, some sugars used in frozen dessert mixes can form viscous cubic phase particles that might be left behind on the screen prior to homogenization and heat treatment, resulting in lower effective concentration of the emulsifier in the final product.

Fluid Isotropic Phase

The fluid isotropic phase is a binary mixture of a small amount of water in an emulsifier continuous matrix; otherwise, it is similar to L_2 microemulsion in a water–oil–emulsifier ternary system, which will be discussed in the next section.

Effect of Molecular Structure of Emulsifiers

Most phospholipids tend to form the L_a phase, except phosphatidylethanolamine, which is relatively less hydrophilic and tends to form the H_{ii} phase. However, the hydophilicity of phosphatidylethanolamine can be increased either by partial hydrolysis to form lysolecithin or by acetylation (Bergenstahl, 1997).

Effect of Temperature, Concentration, and Other Solutes

At relatively high temperatures the disorder along the hydrocarbon chain increases, which means that the shape will become more wedge-like. The transitions with temperature will therefore follow the sequence lamellar → cubic → reversed hexagonal. A similar trend holds true when water is added to nonionic emulsifiers; however, this trend is reversed for ionic emulsifiers due to the increased lateral repulsion between the polar heads at higher water contents, resulting in a reduced wedge shape (Eliasson and Larsson, 1993).

Low-molecular-weight carbohydrates also affect the formation of mesophases. For example, disaccharides such as trehalose and sucrose enhance H_{ii} phase formation while destroying the L_a phase in phosphatidylethanolamine–water systems. These sugars strongly decrease the transition temperature from L_a to H_{ii}. Thus, a phase transition occurs when sugar is added to a monoglyceride–water system. These phase transitions result in an increased average wedge shape of the liquid molecule. Crowe et al. (1987) found that trehalose stabilizes the L_a phase over the H_{ii} phase.

Application

The L_a phase has been found to be the most active physical state for the interactions of emulsifiers with other food components, such as starch and protein (Krog and Lauridsen, 1976). Beta-crystalline coagel, which is prepared from the L_a phase, is being used extensively as a prehydrated emulsifier in the baking industry for modifying dough rheology or for extending the shelf life of baked goods. The prehydrated emulsifier comes in various emulsifier concentrations. The prehydrated monoglyceride can be more homogeneously distributed in the dough during mixing, and, due to the small crystal size and the resulting high surface area, it can be quickly transformed into the L_a phase for interacting with free amylose in the aqueous phase during the early stage of baking (Krog and Lauridsen, 1976). Emulsifiers that tend to form the L_a phase are positive for baking performance of breads (Eliasson and Larsson, 1993). The mesophase structure of emulsifiers is also important in cake baking. Six cakes were prepared by adding 0.4% of monoglycerides to water at varying temperatures to yield six different mesophases and then combining with the remaining ingredients. Only the lamellar dispersion and alpha-crystalline gel phase produced satisfactory batter and cake volumes, while all others produced unacceptable products (Lauridsen, 1976). The L_a phase can also stabilize emulsion by forming interfacial multilayer structures.

These mesophases have been patented as fat replacers, consistency control agents, moisture retention agents, and encapsulation agents. El-Nokaly (1992) used low-molecular-weight hydrophobic cellulose derivatives as emulsifiers to form a polymeric L_a phase that functions as a reduced-calorie, fiber-containing fat substitute. Polysaccharides are added for controlling rheology. Miller et al. (2000) used blends of ionic and nonionic emulsifiers to form a stable L_a phase for stabilizing emulsions and dispersions in low-fat and fat-free products. El-Nokaly (1993, 1997) proposed the use of polymeric liquid crystals based on low-molecular-weight hydroxypropyl cellulose as encapsulating agents for delivering nutrients and flavors in foods.

Liposomes

Structure

The L_a phase can be diluted with an excess of water under stirring conditions to produce liposomes. Liposomes, also known as *dispersions* or *globular vesicles*, are essentially closed lipid bilayer membranes in the form of sacs containing an entrapped aqueous core. Liposomes may be unilamellar or multilamellar emulsifier vesicles enclosing a three-dimensional space. Liposomes can be up to several micrometers in diameter, with the innermost layer having a minimal diameter of a few hundred angstroms. The structure of the liposome provides a unique and convenient carrier for various components entrapped in the internal aqueous layer, which is separated from the external aqueous environment. Similar to emulsions, liposomes are kinetically stable. The rate of release of the encapsulant depends on the composition of the lipid bilayer, the concentration and type of the encapsulant, and the phase-transition temperature of the emulsifier. Phospholipids have been used extensively for liposome preparations. The temperature of liposomes should be carefully controlled to avoid breakdown of the lipid bilayer structure to other mesophases. On the other hand, in some applications, temperature is being used as a release mechanism.

Application

Liposomes have been researched widely in the medical and pharmaceutical industries due to their potential uses as target carriers of drugs, including bioactive macromolecules. Recently, in the food industry, liposomes have been suggested as a fat replacer, carrier, encapsulating agent, moisturizing agent, and depanner. Koide and Karel (1987) reported the use of lecithin liposome for encapsulating enzymes. Enzyme-encapsulated liposomes may improve flavor development — for example, in the field of accelerated cheese ripening (Law and King, 1985; Gripon, 1986; Kirby et al., 1987; Piard et al. 1994). The liposome-encapsulated protease was added along with rennet and lactic starter to the milk. The protease remained in the liposomes during the curd formation process, preventing the premature degradation of casein matrix. The protease was not released until the curd began to undergo ripening. This approach also gave more uniform distribution than spreading a powdered mixture of salt/enzyme over the milled curd. Lecithin liposome has also been used as a research vehicle for studying antioxidant activity of grape extracts (Yi et al., 1997).

In the early 1990s, Nabisco received four liposome patents, two of which were for the encapsulation of oxidizable, lipophilic substances, such as omega-3 fatty acid, flavoring, preservatives, or antioxidants to stabilize against oxidation and rancidity. Liposomes were prepared by first dissolving fish oil in a phospholipid, followed by water addition and sonication. Liposome-encapsulated fish oil has 3 to 4 times longer shelf life. Lengerich (1991) showed the use of liposomes for encapsulating vitamins, flavor, antioxidants, coloring agents, enzymes, and acidulants and for reducing sugar or non-nutritive sweetener in extrusion-baked cookies. One example is the encapsulation of browning agents at high pH so they will survive dough mixing and post-extrusion baking and will not release until storage and/or baking. Haynes et al. (1992b) used liposomes to encapsulate lysine at high pH for browning during microwave heating. Silva et al. (1992) reported the use of high (>35%) phosphatidylcholine containing phospholipid as a liposome-forming ingredient.

Typical liposome stabilizers are animal and plant sterols, triglycerides, diglycerides, monoglycerides, phenolics, and sucrose esters of long-chain fatty acids. Mechansho et al. (1998) reported the use of cholesterol-stabilized, lecithin-based liposomes for encapsulating divalent mineral salts and vitamins to prevent discoloration, off flavor, and astringency.

The Mesophases of an Emulsifier–Water–Oil Ternary System

In addition to L_a, H_i, H_{ii}, and cubic phases, water–oil–emulsifier systems also form microemulsions. Microemulsions are single, thermodynamically stable, optically isotropic, and nonviscous liquid solutions. Microemulsions have three or more components: a hydrophobic component, a hydrophilic component, and at least one emulsifier. Microemulsions can form spontaneously without mechanical energy input, with a droplet size of less than 0.1 μm. Two types of microemulsions exist. In oil/water (O/W) microemulsions (L_1), oil (O) is inside the droplets, while water (W) is inside the droplet in W/O microemulsions (L_2).

Microemulsions have been widely used for non-food applications, such as cosmetics, drycleaning fluids, paint technology, tertiary oil recovery, precious metal recovery, photochemical and polymerization reactions, advanced fuel technology, drug delivery, and biomedical applications (Paul and Mouliuk, 1997; Shad and Shechter, 1977). The biggest challenge for food application is the limited type of oils and emulsifiers allowed. The surfactants and co-solvents also must be food grade and of good sensory quality. Triglycerides, especially long-chain triglycerides (mainly C16 and C18), are much more difficult to solubilize into microemulsions than hydrocarbons or alkyl esters. Thus, they tend to form liquid crystalline mesophases (Alander and Warenheim, 1989a,b). For more hydrophilic oils, the regions for both the L_1 and L_2 phases are extended (Tokuoka et al, 1993).

O/W Microemulsion (L_1)

Structure

Triglycerides containing unsaturated or short-chain fatty acids have better tendency to form L_1 phase compared to triglycerides with saturated or long-chain fatty acids. Water-soluble co-solvents, such as sucrose and alcohol, when used alone or in combination, can destabilize the liquid crystalline mesophases and promote the formation of the L_1 phase, as confirmed by x-ray diffraction and Polaroid microscopy. Sucrose enhances the formation of the L_1 phase while destroying the L_2 phase (Joubran et al. 1994).

Application

The first food O/W microemulsion patent was based on using a high amount of alcohol (>25%) as a co-solvent and 1 to 30% of high hydrophile–lipophile balance (HLB) emulsifiers (Wolf and Hauskotta, 1989). The alcohol phase includes ethanol, propylene glycol, glycerol, sugar, and sugar alcohol. The microemulsion concentrate is stable at 70 to 75°F for at least a year and can be added up to 0.2% in beverages or up to 50% in salad dressings. Chung et al. (1994a,b) reported L_1-type microemulsions for delivering a small amount of hydrophobic spearmint mouthwash and fragrance oil without using alcohol. Gaonkar (1994) showed L_1 phases comprising only those oils that cannot be formed into a microemulsion in a matrix of water and alcohol specifically for aromatized coffee oil and oil-soluble egg flavor. Chmiel et al. (1996) reported a soluble preconcentrated emulsion of hydrolyzed coffee oil. Upon heating of the food product above the melting point of the hydrolyzed fat, the preconcentrated emulsion spontaneously forms an L_1 phase for rapid release of the flavor. The advantages of these procedure are that there are no unappealing oil slicks on the coffee surface and no exceedingly high levels of emulsifier, because the hydrolyzed coffee oil contains 75 to 85% free fatty acids, which behave as emulsifiers. Chmiel et al. (1997) further extended the same utility from hydrolyzed coffee oil to hydrolyzed fat. Up to 2% can be used in frozen or chilled food products to release flavor upon heating. Merabet (1999) reported that an L_1 phase has microwave absorption characteristics that make it highly suitable as a crisping and browning agent when added onto the surface of a food product. This L_1 phase absorbs microwave energy in a thin layer at the surface of the food product and thus heats the

dispersed oil droplets up to about 200°C or higher, while the water continuous phase quickly evaporates. This characteristic gives a high microwave
heating rate and a small microwave penetration depth. Logan and Porzio
(2000) reported the method of making flavored vinegar by diluting a flavored
L_1 phase concentrate containing 25 to 70% vinegar and a relatively high level
of ethanol at 5 to 35%.

W/O Microemulsion (L_2)

Structure

A few triglyceride-based L_2 systems have been reported in the food science
literature (Friberg and Ridhay, 1971; Schwab et al., 1983; Hernqvist, 1986;
Engstroem, 1990; El-Nokaly et al., 1991; Dunn et al., 1992, 1993). Lindstrom
et al. (1981) characterized small areas of L_2 systems containing triglycerides,
monoglycerides, and an aqueous phase. Joubran et al. (1993) reported that
triglycerides easily formed an L_2 phase at a 3:1 ratio of ethoxylated
monoglycerides to monoglycerides. The L_2 phase can be used for protecting
fat from oxidation. Ascorbic acid and tocopherol can inhibit oxidation of oil
in an L_2 system (Moberger et al., 1987; Jakobsson and Sivik, 1994). Another
use of the L_2 phase is as a novel reaction medium, eliminating the insolubility
problem frequently encountered with triglycerides and other lipophilic substances. It has been used for enzymatic preparation of monoglycerides
(Stamatis et al., 1993), for lipase-catalyzed transesterification of unsaturated
lipids with stearic acid (Osterberg et al., 1989), for lipase-catalyzed interesterification of butterfat (Piard et al, 1994), for lipase-catalyzed hydrolysis of
oils (Chandrasekharam and Basu, 1994), and for phospholipase-catalyzed
synthesis of phosphatidylcholine with ω-3 fatty acids (Na et al., 1990). Other
uses of the L_2 phase include controlling the availability of water for microbiological activities (Pfammatter et al., 1992), freezing (Garti et al., 2000), and
as a separation matrix for proteins (Ayala et al., 1992; Hayes, 1997). Dungan
(1994) used the L_2 phase for separating various fractions of whey protein by
changing pH and salt concentration during protein extraction from the L_2
phase.

Application

The L_2 phase can be used to disperse water-soluble nutrients, vitamins,
flavor, and flavor precursors in oils. El-Nokaly (1991b) reported an L_2 phase
containing up to 90% oil and 9% mono- and di-unsaturated C18 ester of di-
and triglycerol as a delivering system for up to 5% water-soluble nutrients
and flavors in foods. Leser (1995) showed an L_2 phase formed from up to
33% water, up to 30% phospholipid, and diacylglyceride esters from nitrate,

tartrate, lactylates, and up to 30% monoglycerides. Alander et al. (1996) used L_2 microemulsion to deliver up to 2% of nanometer-sized water droplets uniformly into a chocolate matrix to make it heat resistant. Merabet (2000) took advantage of the high microwave absorption of super-cooled water in the L_2 phase to thaw frozen foods uniformly. The microemulsion has been applied either onto or into the frozen food, so that it absorbs microwave energy to create a uniform blanket of heat that thaws the frozen food evenly. The L_2 phase is based upon water, medium-chain triglycerides, and diglycerol monooleate. Typically, water of up to 15% of total weight of the L_2 phase is supercooled to at least $-8°C$ (Garti, 2000). Thus, water also has a higher dielectric loss factor at frozen temperatures (Senatra et al, 1985).

Conclusions

More food companies are learning from the emulsifier-based mesophase technologies developed in the chemical and pharmaceutical industries. The regulations on emulsifier usage level and the sensory quality of emulsifiers have limited the applicability of those different mesophases in commercial food production. A new generation of emulsifiers with better flavor quality is needed before this technology can be widely applied.

References

Alander, J. and Warenheim, T. (1989a) Model microemulsions containing vegetable oils. Part I. Nonionic surfactant systems, *JAOCS*, 66(11), 1656–1660.

Alander, J. and Warenheim, T. (1989b) Model microemulsions containing vegetable oils. Part II. Ionic surfactant systems, *JAOCS*, 66(11), 1661–1665.

Alander, J., Warnheim, T., and Luhti, E. (1996) Heat-Resistant Chocolate Composition and Process for the Preparation Thereof, U.S. Patent No. 5486376.

Ayala, C.A., Kamat, S., and Beckman, A.J. (1992) Protein extraction and activity in reverse micelles of a nonionic detergent, *Biotechnol. Bioeng.*, 39(8), 806–814.

Bergenstahl, B. (1997) Physicochemical aspects of an emulsifier functionality, in *Food Emulsifiers and Their Applications*, H.L. Hasenhuettl and R.W. Hartel, Eds., Chapman & Hall, New York, pp. 149–172.

Bos, M., Nylander, T., Arnebrant, T., and Clark, D. (1997) Protein/emulsifier interactions, in *Food Emulsifiers and Their Applications*, H.L. Hasenhuettl and R.W. Hartel, Eds., Chapman & Hall, New York, pp. 127.

Chandrasekharam, C.V. and Basu, A.K. (1994) Lipase catalyzed hydrolysis of oils in microemulsion medium, *J. Oil Technologists' Association of India*, 26(2), 53–58.

Chmiel, O., Traitler, H., and Voepel, K. (1997) Food Microemulsion Formulations, U.S. Patent No. 5674549.

Chmiel, O., Traitler, H., Watzke, H., and Westfall, S.A. (1996) Coffee Aroma Emulsion Formulation, U.S. Patent No. 5576044.

Chung, S.L., Tan, C.T., Tuhill, I.M., and Scharpf, L.G. (1994a) Transparent Oil-in-Water Microemulsion Flavor Concentrate, U.S. Patent No. 5320863.

Chung, S.L., Tan, C.T., Tuhill, I.M., and Scharpf, L.G. (1994b) Transparent Oil-in-Water Microemulsion Flavor or Fragrance Concentrate, Process for Preparing Same, Mouthwash or Perfume Composition Containing Said Transparent Microemulsion Concentrate, and Process for Preparing Same, U.S. Patent No. 5283056.

Crowe, J.H., Crowe, L.M., Carpenter, J.F., and Aurell-Wistrom, C., (1987) Stabilization of dry phospholipid bilayers and proteins by sugars, *Biochem. J.*, 242, 1.

Dungan, S. (1994) Purification of milk proteins using reversed micelle systems, *California Dairy Beat*, 2, 3–5.

Dunn, R.O., Schwab, A., and Bagby, M. (1992) Physical property and phase studies of nonaqueous triglyceride/unsaturated long chain fatty alcohol/methanol systems, *J. Dispersion Sci. Technol.*, 13(1), 77–93.

Dunn, R.O., Schwab, A., and Bagby, M. (1993) Solubilization and related phenomena in nonaqueous triglyceride/unsaturated long chain fatty alcohol/alcohol/methanol systems, *J. Dispersion Sci. Technol.*, 14(1), 1–16.

Eliasson, A. and Larsson, K., Eds. (1993) *Cereals in Breadmaking: A Molecular Colloidal Approach*, Marcel Dekker, New York, pp. 11–16.

El-Nokaly, M. (1992) Food Products Containing Reduced Calorie, Fiber Containing Fat Substitute, U.S. Patent No. 5106644.

El-Nokaly, M. (1993) Encapsulated Cosmetic Compositions, U.S. Patent No. 5215757.

El-Nokaly, M. (1997) Encapsulated Materials, U.S. Patent No. 5599555.

El-Nokaly, M., Hiler, G.D., and McGrady, J. (1991a) Food Microemulsion, U.S. Patent No. 5045337.

El-Nokaly, M., Hiler, G., and McGrady, J. (1991b) Solubilization of water and water-soluble compounds in triglycerides, in *Microemulsions and Emulsions in Foods*, A. Magda and C. Donald, Eds., ACS Symposium Series #448, American Chemical Society, Washington, D.C., pp. 26–43.

Engstroem, L. (1990) Aggregation and structural changes in the L_2-phase in the system of water/soybean oil/sunflower oil monoglycerides, *J. Dispersion Sci. Technol.*, 11(5), 479–489.

Engström, S. and Larsson K. (1999) Microemulsions in foods, in *Handbook of Microemulsion Science and Technology*, K.L. Mittal and P. Kumar, Eds., Marcel Dekker, New York.

Friberg, E. and Rydhag, L. (1971) Solubilization of triglycerides by hydrotropic interactions: liquid crystalline phases, *JAOCS*, 48, 113–115.

Gaonkar, A.G. (1994) Microemulsion of Oil and Water, U.S. Patent No. 5376397.

Garti, N. (2000) The properties of water in W/O microemulsion at subzero temperatures, 8th International Symposium on the Properties of Water (ISOPOW) meeting abstract, Zichron Yaakov, Israel, Sept. 16-21, p. 17.

Garti, N., Clement, V., Fanun, M., and Leser, M.E. (2000) Some characteristics of sugar ester nonionic microemulsions in view of possible food applications, *J. Agri. Food Chem.*, 48, 3945–3956.

Gripon, J.C. (1986) Acceleration of cheese ripening with liposome-entrapped proteinase, *Biotech Lett.*, 8, 241–246.

Hayes, D.G. (1997) Mechanism of protein extraction from the solid state by water-in-oil microemulsions, 53(6), 583–593.

Haynes, L.C., Levine, H., and Finley, J.W. (1991) Liposome Composition for the Stabilization of Oxidizable Substances, U.S. Patent No. 5015483.

Haynes, L.C., Levine, H., and Finley, J.W. (1992a) Method and Liposome Composition for the Stabilization of Oxidizable Substances, U.S. Patent No. 5139803.

Haynes, L.C., Levine, H., Otterburn, M.S., and Mathewson, P. (1992b) Microwave Browning Composition, U.S. Patent No. 5089278.

Hernqvist, L. (1986) An electron microscopy study of the L_2 phase (microemulsion) in a ternary system: triglyceride/monoglyceride/water, in *Food Emulsion and Foam*, E. Dickinson, Ed., The Royal Society of Chemistry, Leeds, England, pp. 158–169.

Jakobsson, M. and Sivik, B. (1994) Oxidative stability of fish oil included in a microemulsion, *J. Dispersion Sci. Technol.*, 15(5), 611–619.

Joubran, R.F., Cornell, D., and Parris, N. (1993) Microemulsion of triglycerides and nonionic surfactant: effect of temperature and aqueous phase composition, *Colloids Surf.*, 80, 153–160.

Joubran, R.F., Parris, N., Lu, D., and Trevino, S. (1994) Synergetic effect of sucrose and ethanol on formation of triglyceride microemulsions, *J. Dispersion Sci. Technol.*, 15(6), 687–704.

Kirby, C.J., Brooker, B.E., and Law, B.A. (1987) Accelerated ripening of cheese using liposome-encapsulated enzyme, *Int. J. Food Sci. Technol.*, 22, 355–375.

Koide, K. and Karel, M. (1987) Encapsulation and stimulated release of enzymes using lecithin liposomes, *Int. J. Food Sci. Technol.*, 22, 707–723.

Krog, N. and Lauridsen, J.B. (1976) Food emulsifiers, in *Food Emulsions*, S. Friberg, Ed., Marcel Dekker, New York, pp. 67–114.

Lauridsen, J.B. (1976) Food emulsifiers: surface activity, edibility, manufacture, composition, and application, *J. Am. Oil Chem. Soc.*, 53, 795–802.

Law, B.A. and King, J.S. (1985) Use of liposomes for proteinase addition to cheddar cheese, *J. Dairy Res.*, 52, 183–188.

Lengerich, B., Haynes, L.C., Levine, H., Otterburn, M.S., Mathewson, P., and Finley, J. (1991) Extrusion Baking of Cookies Having Liposome Encapsulated Ingredients, U.S. Patent No. 4999208.

Leser, M.E. (1995) Thermodynamically Stable Transparent Edible Water-in-Oil Microemulsion Comprises Oil, Oil-Soluble Surfactants and Polar Ingredients, European Patent No. 657104.

Lindstrom, M., Lusberg-Wahren, H., and Larsson, K. (1981) Aqueous lipid phase of relevance to intestinal fat digestion and absorption, *Lipids*, 16(10), 749–754.

Logan, S.S. and Porzio, M.A. (2000) Flavored Oil-in-Vinegar Microemulsion Concentrates, Method for Preparing the Same, and Flavored Vinegars Prepared from the Same, U.S. Patent No. 6077559.

Mechansho, H., Mellican, R.I., and Trinh, T. (1998) Use of Bilayer Forming Emulsifiers in Nutritional Compositions Comprising Divalent Mineral Salts to Minimize Off-Tastes and Interactions with Other Dietary Components, U.S. Patent No. 5707670.

Merabet, M. (1999) Edible Micro-Emulsion and Method of Preparing a Food Product Treated with the Micro-Emulsion, U.S. Patent No. 5891490.

Merabet, M. (2000) Microwave Thawing Using Micro-Emulsions, U.S. Patent No. 6149954.

Miller, M., Akashe, A., and Das, D. (2000) Mesophase-Stabilized Emulsions and Dispersions for Use in Low-Fat and Fat-Free Food Products, U.S. Patent No. 6068876.

Moberger, L., Larson, K., Buchheim, W., and Timmen, H. (1987) A study of fat oxidation in a microemulsion system, *J. Dispersion Sci. Technol.*, 8(3), 207–215.

Na, A., Eriksson, S.G., Osterberg, E., and Holmberg, K. (1990) Synthesis of phospho-tidylcholine with w-3 fatty acids by phopholipase A2 in microemulsion, *JAOCS*, 67(11), 766–770.

Osterberg, E., Blomstrom, A.C., and Holmberg, K. (1989) Lipase-catalyzed transes-terification of unsaturated lipids in a microemulsion, *JAOCS*, 66(9), 1330–1333.

Paul, B.K. and Mouliuk, S.P. (1997) Microemulsions: an overview, *J. Disp. Sci.*, 18(4), 301–367.

Pfammatter, N., Hochkoeppler, A., and Luisi, P.L. (1992) Solubilization and growth of Candida pseudotropicalis in water-in-oil microemulsions, *Biotech. Bioeng.*, 10(1), 167–172.

Piard, J.C., El Soda, M., Alkhalaf, W., Rousseau, M., Desmazeaud, M., Vassal, L., Safari, M., and Kermasha, S. (1994) Interesterification of butterfat by commer-cial microbial lipase in a cosurfactant-free microemulsion systems, *JAOCS*, 71(9), 969–973.

Schwab, A., Nielsen, H., Brooks, D., and Ryde, E. (1983) Triglycerides/aqueous eth-anol/1-butanol microemulsions, *J. Dispersion Sci. Technol.*, 4(1), 1–17.

Senatra, D., Guarini, G.T., Gabrielli, G., and Zoppi, M. (1985) Thermal and dielectric behavior of free and interfacial water in water-in-oil microemulsions, in *Macro-and Microemulsions: Theory and Practice*, D.O. Shah, Ed., ACS Symposium Series #272, American Chemical Society, Washington, D.C., pp. 133–148.

Shad, D.O. and Shechter, R.S., Eds. (1977) *Improved Oil Recovery by Surfactant and Polymer Flooding*, Academic Press, New York.

Silva, R.S., Fierro, J., Buccino, J., and Jodlbauer, H. (1992) Food Composition and Method, U.S. Patent No. 5120561.

Stamatis, H., Xenakis, A., Menge, U., and Kolisis, F.N. (1993) Kinetic study of lipase-catalyzed esterification reactions in water-in-oil microemulsions, *Biotechnol. Bioeng.*, 42(8), 931–937.

Tokuoka, Y., Uchiyama, H., and Abe, M. (1993) Phase diagrams of surfactant/water/synthetic perfume ternary systems, *Colloid Polymer Sci.*, 272, 317–323.

Wolf, P.A. and Hauskotta, M.J. (1989) Microemulsion of Oil in Water and Alcohol, U.S. Patent No. 4835002.

Yi, O.S., Meyer, A.S., and Frankel, E.N. (1997) Antioxidant activity of grape extracts in a lecithin liposome system, *JAOCS*, 74(10), 1301–1307.

13

Drying of Biotechnological Products: Current Status and New Developments

Arun S. Mujumdar

CONTENTS

Introduction

Drying, by definition, involves removal of a liquid (generally water, but in many bioprocessing applications it could be an organic solvent or an aqueous mixture) from a solid, semi-solid, or liquid material to produce a solid product by supplying thermal energy to cause a phase change, which converts the liquid to vapor.

In the exceptional case of freeze-drying, the liquid is first solidified and then sublimed. Bioproducts are produced by microbial action and are related to living organisms. Bioproducts are a subset of a broader generic definition of biomaterials which includes wood, coal, biomass, foods (biopolymers), vegetables, fruits, etc. This overview is limited to such bioproducts as whole cells (e.g., baker's yeast, bacteria, vaccines), fermented foods (e.g., yogurt, cheese), synthetic products of both low molecular weight (e.g., amino acids, citric acid) and high molecular weight (e.g., antibiotics, xanthene), and enzymes.

All of these products are characterized by their high thermal sensitivity; they are damaged or denatured and inactivated by exposure to certain temperatures specific to the product. Some are inactivated by mechanical stress or damage caused to the cell walls during the drying operation. These products are often produced in smaller qualities in batch mode. Further, they are typically high-value products so that the cost of drying is often secondary to quality constraints. It is therefore not unusual to use more expensive drying techniques (e.g., freeze-drying, vacuum-drying etc.) even when less expensive techniques such as heat-pump drying could be applied successfully. Of course, some biotechnology products are produced in bulk in continuous operation using conventional drying technologies such as spray- or fluid-bed drying.

The activity of water in a bioproduct is determined by the state of water in it. So-called *free water* represents the intracellular water in which nutrients needed by the living cells are in solution. *Bound water* is built into cells or the biopolymer structures. It is more strongly held to the solid matrix and is also resistant to freezing. The ratio of the vapor pressure expected by the water in the product to the equilibrium vapor pressure of pure water at the same temperature is referred to as the *water activity*. For safe storage, the objective of a drying process is to reduce the product moisture content so as to lower its activity below a threshold value safe for storage.

Effect of Drying on Bioproduct Quality

As noted earlier, quality of the dried product is of paramount concern in selecting a dryer and its operating conditions for thermolabile bioproducts. Numerous and varied undesirable changes can occur in the product during drying; in the worst case scenario, one may obtain a dry but totally inactivated product. Table 13.1 summarizes such changes for various biomaterials and their effects on product quality.

Various indices are used to quantify changes in quality as a result of drying, and their choice clearly must depend on the product, but it is beyond the scope of this brief overview to discuss this important issue. Briefly, typical quality criteria may be as follows:

- For food biopolymers, criteria include color, texture, organoleptic properties, nutritional value (vitamin content), taste, and flavor.
- For "live" products (e.g., bacteria, yeast) or products such as enzymes or proteins that are thermally destabilized or inactivated, quality index A may be represented in terms of the degradation kinetics: $\dfrac{dA}{dt} = f(C_i X_i)$ where the C_i represent moisture or temperature and X_i are process variables.

TABLE 13.1

Possible Quality Changes during Bio-Material Drying

Material	Change Type	Effect
Yeast	Biochemical	Atrophy of cells
Bacteria	Biochemical	Atrophy of cells
Molds	Biochemical	Atrophy of cells
Enzymes	Enzymatic	Loss of activity
Vitamins	Enzymatic	Loss of activity
Proteins, fats, carbohydrates, antibiotics	Chemical	Loss of activity, nutritive contents
Other	Physical/chemical/ biochemical	Solubility, rehydration, loss of aroma, shrinkage

A usual simplification that works satisfactorily within engineering accuracy involves assumption of first-order kinetics:

$$\frac{dA}{dt} = -k_d A$$

where $k_d = k_\infty \exp(-\Delta E/RT)$; R = 8.314 J/mol/K; ΔE = activation energy (J/ mol); T = absolute temperature (K). Small changes in temperature can have a dramatic effect on the degradation rate constant, k_d, which can increase three- to eight-fold over a temperature rise of 60 to 80 K. Kudra and Strumillo (1998) have given values of k_d for selected bioproducts.

As an example of the diverse quality criteria employed in practice for a bioproduct, Table 13.2 lists quality indices often used to define suitability of dried proteins or protein-containing compounds. Not all of these criteria are used for a given product, however.

For biomaterials such as various foods, fruits, and vegetables, numerous other quality parameters apply (refer to Krokida and Maroulis [2000] for a detailed review). Physical properties such as shrinkage, puffing, porosity, and texture are also important in these applications.

TABLE 13.2

Quality Changes: Drying of Protein-Containing Compounds

Quality Indices
Nitrogen solubility index (NSI)
Protein dispersibility index (PDI)
Water dispersed protein (WDP)
Water-soluble protein (WSP)
Nitrogen solubility curve (NSC)
Protein precipitate curve (PPC)

Note: For fruits, vegetables, and other foods, other criteria apply, including color, texture, taste, flavor, nutrition, organoleptic properties, etc.

Commonly Used Dryers

Often, the wet bioproduct to be dried is in the form of a wet solid, sludge, filter cake, a suspension, or solution. Mujumdar (1995) and Devahastin (2000) presented a classification scheme for the numerous dryer types and their selection criteria in a general way, and the reader is referred to these basic references for details. Suffice it to say here that the choice of dryers for bioproducts is constrained mainly by the ability of the dryer to handle the material physically, while the choice of the operating conditions is determined by the thermal sensitivity of the material. Table 13.3 lists some of the conventional dryers, as well as some emerging drying techniques for heat-sensitive bioproducts, many of which are already commercialized but not commonly offered by vendors yet. Table 13.4 summarizes the key restrictive criteria that determine suitability of a given drying technology for biotechnology products. Note that aside from heat, such products may be damaged by the presence of oxygen. Some products may have to be stabilized by additives such as sugars or salts, as in the case of drying of some enzymes. Certain cryoprotective chemicals are used when freeze-drying live cells to avoid rupture of the cell walls. The rate of drying may have a direct or indirect effect on the quality as well as on the physical handling of the product.

Spray-drying and freeze-drying are some of the most common drying technologies used for drying of bioproducts, although fluid-bed, batch- and continuous-tray, spin-flash, and vacuum dryers are also common. Pilosof and Terebiznik (2000) reviewed the literature on the drying of enzymes using spray- and freeze-drying.

In recent years, we have seen a considerable rise in applications of enzymes as industrial catalysts, as pharmaceutical products, as clinical diagnostic chemicals, and in molecular biology. Because most enzymes are not stable in water, dehydration is used to stabilize them. Multistage drying systems (e.g., spray dryer to remove surface moisture followed by a fluidized or vibrated bed to remove internal moisture at milder drying conditions over an extended period) are often used to speed up the overall drying process while maintaining product quality. Low-pressure fluid-bed drying can be used to achieve drying of particulate solids at lower temperatures, although it is not a commonly used process.

Freeze-drying (lyophilization) is used extensively in the industry to dry ultra-heat-sensitive biomaterials (e.g., some pharmaceuticals). Some $200 billion worth of pharmaceutical products are freeze-dried worldwide each year. It is a very expensive dehydration process, justified by the high value of the product. Liapis and Bruttini (1995) have given an excellent account of freeze-drying technologies.

TABLE 13.3

Commonly Used Dryers and Emerging Drying Technologies Suitable for Biotech Products

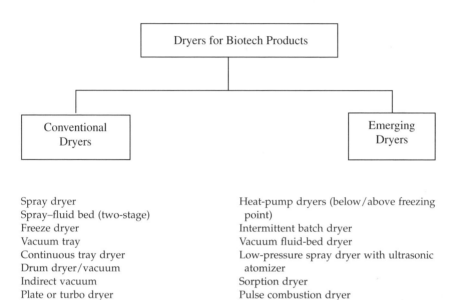

Conventional Dryers	Emerging Dryers
Spray dryer	Heat-pump dryers (below/above freezing point)
Spray–fluid bed (two-stage)	
Freeze dryer	Intermittent batch dryer
Vacuum tray	Vacuum fluid-bed dryer
Continuous tray dryer	Low-pressure spray dryer with ultrasonic atomizer
Drum dryer/vacuum	
Indirect vacuum	Sorption dryer
Plate or turbo dryer	Pulse combustion dryer
	Cyclic pressure/vacuum dryer
	High electric field (HEF) dryer
	Superheated steam dryer at low pressures

Some Emerging Drying Technologies

Numerous new drying techniques proposed and tested over the past decade have potential for application to biotech products. Extensive discussion of the basic principles, advantages, and limitations of each of these is beyond the scope of this brief review. Table 13.3 lists some such emerging technologies that have potential for bioproducts. For detailed discussion of most of these, the reader is referred to Mujumdar (1995), Mujumdar and Suvachittanont (1999, 2000), and Kudra and Mujumdar (2001), among others.

For recent advances in heat-pump drying (HPD) of heat-sensitive products, the reader is referred to Chua et al. (2000), who provide a comprehensive overview of the numerous variants possible, including those involving multistage heat pumps; multistage dryers; drying below the freezing point; HPD with supplemental heat input by conduction, radiation, or dielectric (MW or RF) fields; and use of cyclical variation of the drying air temperature for batch drying of heat-sensitive products. Figure 13.1 is a schematic of the

TABLE 13.4

Choice of Dryers and Drying Conditions for Biotech Products Depending on Specific Constraints

Restrictive Criterion When Drying Biotech Products	Possible Dryer/Drying Conditions
Highly heat-sensitive; thermally inactivated; or damaged	Dehumidified air drying (heat pump or adsorption dehumidifier) at low temperatures Vacuum drying with indirect heat supply Intermittent batch drying Freeze drying
Damaged by oxidation	Convective drying in N_2 or CO_2 Vacuum drying Freeze drying
Product subject to destabilization (e.g., enzymes)	Addition of sugars, maltodextrin, salts, etc. to stabilize some enzymes Control of pH change during drying
Product affected by physical processing	Use of gentle drying (e.g., packed bed or continuous tray as opposed to fluid bed) Better drying of some products in one type of dryer than others (e.g., yeast in spouted bed vs. fluid bed)

wide assortment of HPDs possible; not all of these have been tested at laboratory or pilot scales. Several of these are of interest when drying highly heat-sensitive biotechnology products, as they are more cost effective than freeze dryers. The two-stage heat-pump dryer designed by Alves-Filho and Strommen (1996), in which the first stage is a fluid-bed freezer/freeze dryer at atmospheric pressure and the second stage is a fluid-bed dryer operated with dehumidified air but above the freezing point, can successfully compete with the freeze dryer for certain products. It yields dried product properties that resemble those obtained by the much more expensive freeze-drying process.

When the biomaterial to be dried is very sticky due to the presence of proteins, fats, or sugars, special drying techniques may be necessary. Such problems must be solved on an individual product basis, however.

Sadykov et al. (1997) have proposed an interesting technique to dry bio-active materials. It involves cycling the operating pressure in a batch mode. Heat is supplied convectively at atmospheric pressure for a certain length of time and then the moisture is flashed off in a subsequent cycle when vacuum is applied to the chamber for a given, but different, length of time. This process may be repeated several times.

Some French researchers recently proposed a similar idea, suggesting that heat be supplied indirectly by conduction through the chamber walls. For heat-sensitive products, intermittent application of high-pressure and low-pressure environments can achieve drying at low product temperatures. The process must be operated in batch mode, however. More research and development are suggested to test this interesting new concept.

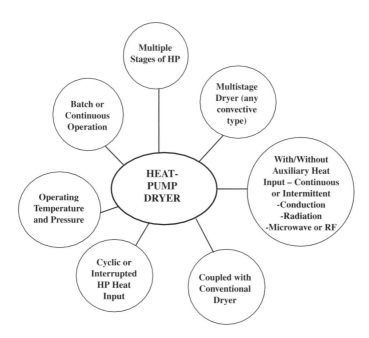

FIGURE 13.1
A classification scheme for heat-pump dryers.

An early Russian doctoral thesis also had a similar idea but implemented it differently, placing the drying material in a cylindrical chamber for which volume (and pressure) could be altered cyclically at a desired frequency (or cycle time) by a tight-fitting reciprocating piston.

The idea of pulse combustion drying has been proposed and revisited several times over the past two decades with limited success. In principle, even highly heat-sensitive products such as vitamins, enzymes, and yeasts can be dried by direct injection into the highly turbulent pulse combustor exhaust tailpipe; despite the ultra-high temperatures of the exhaust, the rapid heat and mass transfer rates and fine atomization of the feed (slurry or dilute paste) by the highly turbulent flow allow drying in a fraction of a second and without thermal degradation. The process has not been a commercial success yet, possibly due to problems of noise, scale-up, and capital cost.

Energy consumption for thermal dehydration depends to a great extent on the dryer or drying system chosen and on the wet feed and properties of the dried product to some extent. Sometimes a lower thermal efficiency dryer is chosen for a given application, as the alternative higher efficiency dryers yield a lower quality product. For example, fish meal dried in a direct rotary dryer gives a 10% better yield of salmon weight than that dried in a thermally more efficient steam tube rotary dryer (Flesland et al., 2000). The drying technique used can affect such properties of fish meal as protein digestibility, feed utilization, and the growth rate of the fish that eat it. Indirect drying, both atmospheric and vacuum, is found to produce a fibrous

TABLE 13.5

Comparison of Specific Energy Consumption of Various Drying Technologies

Drying Technology (with Indirect Rotary Dryer as Predryer)	Specific Energy Consumption (kWh/kg)
Superheated steam dryer	0.75
Indirectly heated steam dryer	1.00
Hot-air rotary dryer	1.15
Mechanical vapor recompression (MVR)	0.04
evaporator	0.30
Three-stage evaporator	0.20
MVR-superheated steam dryer	0.20

product with low flowability and hence a homogeneously extruded feed product.

It is interesting to compare the energy consumption figures for commercial-scale drying of fish meal provided by Flesland et al. (2000), who considered the case of a 42 T/h unit for drying fish meal. About 19.2 T/h of water is removed in the evaporator. Table 13.5 gives the estimated specific energy consumption figures (kWh/kg water removed) for alternative drying technologies.

If a superheated steam dryer could replace the entire drying capacity without use of a predryer, the potential energy savings is about 50%, but with a loss of quality. Depending on whether or not a predryer is used, different process layouts yield different specific energy consumption figures, as listed below (in kWh/kg water evaporated):

	With Predryer	Without Predryer
Hot-air drying/three-stage evaporation	0.82	0.83
Hot-air drying/mechanical vapor recompression (MVR)	0.65	0.66
Superheated steam drying/MVR	0.58	0.55
Superheated steam drying/ MVR + vapor reuse)	0.47	0.44
MVR dryer/MVR evaporator	0.46	0.33

Superheated steam drying is a concept that has been around for over a century, although commercial products appeared on the market only two decades ago for such products as pulp, waste sludge, hog fuel in the paper industry, beet pulp, etc. For heat-sensitive materials that are damaged in an atmosphere containing oxygen, superheated drying is possible only at low operating pressures. This technique has been shown by the author to be successful for drying of silkworm cocoons. The resulting silk is also found to be stronger and brighter. More recent laboratory studies have focused on drying of vegetables but the results are tentative. No work has been reported on biotech product drying to date. Due to the fact that most biotech products are made in small quantities and in batch mode, it is unlikely that super-heated steam drying will be a major contender in this application area.

Nitrogen could be used to provide the inert medium when oxygen must be excluded.

For batch drying, intermittent supply of energy is an especially interesting concept if the bulk of the drying takes place in the falling rate period. Jumah et al. (1996) explained the principle with application to a novel intermittently spouted and intermittently heated spouted-bed dryer for grains. It was shown that appreciable reductions in energy and air consumption could be made while enhancing product quality due to the lower product temperature attained, as well as reduced mechanical handling of the grain due to intermittent spouting. This idea has been extended to fluidized beds as well. Again, no direct biotech applications have been reported, but the concept is fundamentally sound and is expected to find new applications.

When a product to be dried is used in a mixture, one of the components of the mixture can be used as a carrier. This drying method is known as contact-sorption drying. The carrier can have different roles:

- If the product is a liquid suspension, the particulate-form carrier disperses it, thus providing a large interfacial area for evaporation of the moisture while producing a granulated product.
- The presence of the carrier effectively reduces the hygroscopicity of the material.
- Dispersion of the liquid on a "dry" substrate makes the mixture easier to handle (e.g., fluidize, convey, feed), thus permitting the use of a number of conventional dryers.

Sorption dryers of various designs have been reported in the literature, ranging from single- or multistage fluidized bed dryers to cocurrent spray dryers in which the carrier is dispersed in the zone with the drying air in the atomizer zone.

High electric field (HEF) drying is a relatively new application for a well-known technique. Kulacki (1982) discussed the fundamental principles of electrohydrodynamics and the effect of electrical field on heat and mass transfer. In the HEF technique, wet materials can be dried at ambient temperature and pressure (or at lower temperatures and pressures) using an AC high electric field (Hashinaga et al., 1999). Unlike MW or RF, heat is not generated in the material, so no loss of color, nutrients, or texture occurs during drying. The apparatus is very simple, consisting of point and plate electrodes. The main cost is that of electrical power consumption. Bajgai and Hashinaga (2000) showed the high quality attained in HEF drying in a field of 430 KV/m of chopped spinach. The drying rates were very low, but the dried product quality was very high. Although not tested for biotech products, this technique could have potential for drying smaller batches of materials. Further research is needed to evaluate and compare the techno-economics of this technique with competing drying methods.

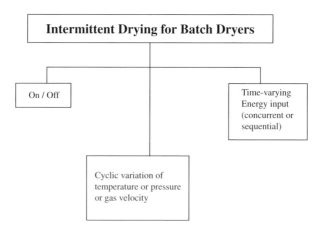

FIGURE 13.2
A schematic of the various types of intermittent batch dryer operations. Time-varying energy input may be by convection, conduction, radiation, or MW or RF; several heat transfer modes can be used concurrently.

Figure 13.2 summarizes some other ideas for intermittent or cyclically operated batch dryers. This figure is not all inclusive; rather, it is intended to give the reader ideas to consider when developing new drying systems for a new product or even when designing a new facility for an existing product line. Chua et al. (2000) and Chou et al. (2001) demonstrated experimentally and by mathematical modeling the superior performance of intermittent drying of heat-sensitive fruits in terms of the quality parameters such as color and ascorbic acid content. The drying time may be increased marginally.

Closing Remarks

Clearly, it is impossible to provide an overview all of the emerging drying technologies that are relevant to drying of the diverse and ever increasing numbers of bioproducts. The reader is referred to recent books by Mujumdar (2000) and Mujumdar and Suvachittanont (1999, 2000), as well as Kudra and Mujumdar (2001) for detailed discussions of some of the new technologies listed in this chapter. Descriptions of conventional drying technologies as well as most of the new ones can also be found in Mujumdar (1995) and Strumillo and Kudra (1998). The proceedings of the biennial International Drying Symposium (IDS) series initiated in 1978 at McGill University in Canada by the author now represent a gold mine of technical literature providing the latest information on emerging drying techniques and research and development. Researchers interested in drying will find these proceedings invaluable for their work. The website (www.geocities.com/drying_guru/) gives ready access to all the key sources of literature and their availability. The proceedings of the recently initiated Inter-American Drying Conferences (1997–), Asia-Oceania Drying Conferences (1999–), and Nordic Dewatering Conferences (2001–) provide additional sources of new technical information on drying technology. For archival information on the subject, *Drying Technology — An International Journal* (Marcel Dekker) remains the premier periodical for both academic and industrial practitioners.

References

Alves-Filho, O. and Strommen, I. (1996) Performance and improvements in heat pump dryers, *Drying '96*, C. Strumillo and Z. Pakowski, Eds., Krakow, Poland, pp. 405–415.

Bajgai, T.R. and Hashinaga, F. (2000) High electric field drying of spinach, in *Proc. 12th Int. Drying Symp.*, Amsterdam, The Netherlands.

Cao, C.W. and Liu, X.D. (2000) Experimental study on impinging stream drying of particulate materials, in *Proc. 12th Int. Drying Symp.*, Amsterdam, The Netherlands.

Chou, S.K., Chua, K.J., Mujumdar, A.S., Ho, J.C., and Hawlader, M.N.A. (2001) On intermittent drying of an agricultural product, *Trans. Instn. Chem. Eng. (London)* (in press).

Chua, K.J., Mujumdar, A.S., Chou, S.K., Hawlader, M.N.A., and Ho, J.C. (2000) Convective drying of banana, guava and potato pieces: effect of cyclical variations of air temperature or drying kinetics and color change, *Drying Technol.*, 18(5), 907–936.

Devahastin, S. (2000) *Mujumdar's Practical Guide to Industrial Drying Technology*, Exergex, Canada (mujumdar_guide@hotmail.com).

Flesland, O., Hostmark, O., Samuelsen, T.A., and Oterhals, A. (2000) Selecting drying technology for production of fish meal, in *Proc. IDS '2000*, P.J.A.M. Kerkhof, W.J. Coumans, and G.D. Mooiweer, Eds., Elsevier, Amsterdam.

Jumah, R., Mujumdar, A.S., and Raghavan, G.S.V. (1996) Batch drying of corn in a novel rotating jet spouted bed, *Can. J. Chem. Eng.*, 74, 479–486.

Krokida, M. and Maroulis, Z. (2000) Quality changes during drying of food materials, *Develop. Drying*, II, 149–195.

Kudra, T. and Mujumdar, A.S. (1989) Impingement stream drying for particles and pastes, *Drying Technol.*, 7(2), 219–266.

Kudra, T. and Mujumdar, A.S. (2001) *Advanced Drying Technologies*, Marcel Dekker, New York, 472 pp.

Kulacki, F.A. (1982) Electrohydrodynamically enhanced heat transfer, in *Advances in Transport Processes*, A.S. Mujumdar and R.A. Mashelkar, Eds., Wiley, New York.

Mujumdar, A.S., Ed. (1995) *Handbook of Industrial Drying*, 2nd ed., Marcel Dekker, New York, 1440 pp.

Mujumdar, A.S., Ed. (2000) *Drying Technology in Agriculture and Food Science*, SicPub, New York, and Oxford/IBH, New Delhi.

Mujumdar, A.S. and Suvachittanont, S., Eds. (1999) *Developments in Drying*. Vol. I. *Food Dehydration*, Kasetsart University, Bangkok, Thailand.

Mujumdar, A.S. and Suvachittanont, S., Eds. (2000) *Developments in Drying*. Vol II. *Drying of Food and Agro-Products*, Kasetsart University, Bangkok, Thailand.

Pilosof, A.M.R. and Terebiznik, V.R. (2000) Spray and freeze drying of enzymes, in *Developments in Drying*. Vol. II, A.S. Mujumdar and S. Suvachittanont, Eds., Kasetsart University, Bangkok, Thailand, pp. 71–94.

Sadykov, R.A., Pobedimsky, D.G., and Bakhtiyarov, F.R. (1998) Drying of bioactive products: inactivation kinetics, *Drying Technol.*, 15(10), 2401–2420.

Strumillo, C. and Kudra, T. (1998) *Thermal Processing of Bioproducts*, Gordon and Breach, London.

14

An Update on Some Key Alternative Food Processing Technologies: Microwave, Pulsed Electric Field, High Hydrostatic Pressure, Irradiation, and Ultrasound

José J. Rodríguez, Gustavo V. Barbosa-Cánovas, Gustavo Fidel Gutiérrez-López, Lidia Dorantes-Alvárez, Hye Won Yeom, and Q. Howard Zhang

CONTENTS

Introduction

For decades, various technologies have been used to preserve the quality and microbial safety of foods. Traditional preservation methods involve the use of heat (commercial sterilization, pasteurization, and blanching), preservatives (antimicrobials), and changes in the microorganism's environment, such as pH (fermentation), water availability (dehydration, concentration), or temperature (cooling and freezing). These technologies involve one or more of these mechanisms: (1) preventing access of microbes to foods, (2) inactivating microbes, or (3) slowing the growth of microbes (Gould, 1995). Heat is by far the most widely utilized technology to inactivate microbes in foods.

Despite the effectiveness of traditional technologies from a microbial safety standpoint, such technologies may also cause nutritional and sensorial food deterioration. Although food fortification can overcome certain nutritional degradation attributes, sensorial attributes such as flavor, aroma, texture, and appearance are difficult to retain in traditional heat treatments. The consumer's increasing demand for fresher, more natural foods has promoted the search for new food preservation technologies that are capable of inactivating foodborne pathogens while minimizing deterioration in food quality, from both nutritional and sensory points of view.

New technologies base their antimicrobial action on physicochemical principles that tend to reduce food quality deterioration during processing. In traditional thermal processing, heat is transferred to food by conduction or convection. This energy not only inactivates microorganisms by disrupting the chemical bonds of cellular components such as nucleic acids, structural proteins, and enzymes (Farkas, 1997) but also affects desirable food components that are responsible for flavor, aroma, texture, and appearance. Most new technologies are considered nonthermal, as their action does not imply food temperature increases (e.g., pulsed electric field, high hydrostatic pressure, irradiation, and ultrasound). Those new technologies that do imply food temperature increases (e.g., microwave heating, radiofrequency, and ohmic heating) use more efficient heat transfer modes than do traditional thermal techniques, thus allowing shorter heating times and minimizing food quality deterioration.

The purpose of this chapter is to describe five of these new technologies (microwave heating, pulsed electric field, high pressure, irradiation, and ultrasound), to explain their mechanisms of action on microbial inactivation, and to illustrate their interaction with food systems.

Microwave Processing

Description of the Technology

Microwave heating refers to the use of electromagnetic waves of certain frequencies to generate heat in a material. The Federal Communications Commission (FCC) regulates the use of electromagnetic waves in the U.S. (Curnutte, 1980). The FCC has allocated four frequencies for food processing and medical applications; however, industrial food microwaves use only 2450 and 915 MHz.

Equipment and Engineering Principles

In microwave heating, continuous electromagnetic waves are produced in the magnetron and transmitted through a hollow metallic tube into a resonant cavity where the food is processed (Van Zante, 1973). In contrast with conventional thermal processing, where heat is applied to the outside of the food, microwave processing involves heat generated from within the food through molecular vibrations (Cross and Fung, 1982). Foods are heated because of molecular friction caused by alternating polarization of molecules promoted by a time-varying electric field resulting from electromagnetic wave propagation.

Foods absorb microwave energy in the form of orientational and ionic polarization (Decareau and Peterson, 1986). Orientational polarization results from dipolar molecules, such as water, which tend to align according to the applied electric field. The electric field oscillates at 2450 or 915 million times per second (MHz), making the dipolar molecules rotate, thus promoting molecular friction, which in turn results in heat dissipation (Curnutte, 1980). Ionic polarization occurs when dissolved salts are present. Positive and negative ions tend to migrate to opposite-charged regions, colliding with other ions and converting kinetic energy into heat. Dipole rotation is more important than ionic polarization as a microwave heating mechanism (Decareau and Peterson, 1986).

The rate of heat generation per unit volume P_0 (watts/cm^3) absorbed by a substance can be expressed as (Lewis, 1987):

$$P_0 = 55.61 \times 10^{-14} f\, E_f^2 \varepsilon''$$

where f (Hz) is the frequency of radiation, E_f (V/cm) is the field strength, and ε'' is the dielectric loss factor. Therefore, the amount of power absorbed by a food will increase as the frequency, field strength, and loss factor increase. The power generated from a microwave operating at 2450 MHz is more than twice the power generated by a microwave operating at 915 MHz, provided the electric field and dielectric properties of the heated food are about the same. The field strength is another effective way to increase the rate of heating, although it is limited by the voltage breakdown strength in the microwave cavity (Lund, 1975).

Typically, the field strength, frequency, and load of materials are fixed in a microwave system; only the dielectric properties of the food will affect the amount of radiation absorbed (Lewis, 1987). Moisture and salt content are the two main food components that most affect food dielectric properties (Swami and Mudgett, 1981).

Another important parameter in microwave heating is the microwave penetration depth. This distance is defined as (Fryer, 1997):

$$d = \frac{\lambda}{2\pi \sqrt{\varepsilon_r \tan \delta}}$$

where d is the depth at which the intensity decays to 36.8% of its surface value, λ is the wavelength of the microwave radiation, and ε_r is the relative permittivity. The degree of microwave penetration depth increases with wavelength. The shorter the wavelength (higher frequencies), the greater the extent to which the energy will be absorbed at the surface of the product instead of within the product (Mudgett, 1989).

Effects of Microwaves on Microbial Inactivation

Microwave heating offers the opportunity of shortening the time required in conventional heat treatments to achieve the desired food-processing temperature (Datta and Hu, 1992; Meredith, 1998); therefore, microwave pasteurization or sterilization can potentially improve the product quality of traditional heat processes (Stenstrom, 1974). Conventional high-temperature/short-time (HTST) processing of solid foods is limited by slow heat transfer via conduction which increases the time required to transfer the heat to the cold spot, causing overheating at the surface of the solid (Ramesh, 1999). Microwave heating offers the possibility of overcoming such limitations of conduction and convection heating modes (Meredith, 1998). Heat-sensitive nutrients such as vitamins and flavor constituents can be retained better through rapid heating than by conventional heat treatments (Mudgett and Schwartzberg, 1982); however, little conclusive evidence exists for any real flavor differences between many conventionally and microwave-heated foods (Lorenz and Decareau, 1976).

Most studies have concluded that microwave energy inactivates microbes via conventional thermal mechanisms, including thermal irreversible denaturation of enzymes, proteins, and nucleic acids (Heddleson and Doores, 1994). A few studies have proposed athermal microwave mechanisms, such as production of toxic compounds that could lead to microbial inactivation (Dreyfuss and Chipley, 1980; Khalil and Villota, 1986, 1989). It is very unlikely that microwaves could induce the production of toxic compounds via an "athermal" mechanism, as their quantum energy is significantly lower than the energy required to break covalent bonds (Pomeranz and Meloan, 1987; Mudgett, 1989).

As in traditional heat processing, the time–temperature history of the coldest point determines the microbiological safety of the microwave process. The accumulated microbial lethality (F_0) in a given period of time (t) can be calculated as (Datta and Hu, 1992):

$$F_0 = \int_0^t 10^{\frac{T-T_R}{z}} \, dt$$

where T is the temperature at a specific location, T_R is the reference temperature, and z is the temperature dependency of the reaction rate. In microwave heating, it is more difficult to determine the location and history of the coldest point than in traditional heat processing. Complex models must be used and validated to locate the point of lowest integrated time history (Burfoot et al., 1988; Ramaswamy and Pillet-Will, 1992; Fleishman, 1996). Heat migration occurs from the initial, hottest locations in the interior to the surface, making it much more difficult to utilize simpler procedures such as the Ball formula (Ball and Olson, 1957). Commercial software that models electromagnetic and heat transfer is available to assess process parameters in a more efficient manner (Dibben, 2000).

Current Limitations and Status

Applications of microwave heating are found for most of the heat treatment operations in the food-processing industries. While tempering and bacon cooking account for hundreds of operating systems, most of the remaining applications are single consumer installations (Schiffmann, 2001). Sterilization using microwaves has been investigated for many years, but commercial introduction has only occurred in the past few years in Europe and Japan. Microwave pasteurization and sterilization promise fast heat processing, which should lead to small quality changes due to thermal treatments according to the HTST principle (Ohlsson, 2000). However, it turns out that, in order to fulfill these quality advantages, microwave heating requires achieving heating uniformity (Burfoot et al., 1988). The nonuniformity of microwave heating can be attributed to several factors, such as localized microwave absorption due to heterogeneity of dielectric properties and heat capacity among food components, variations in field intensity, and differences in shape, size, and placement of food (Schiffmann, 1986, 1990; Ruello, 1987; Stanford, 1990; Keefer and Ball, 1992). Besides heating uniformity, the cost of microwave processing is another limitation to its industrial use on a larger scale. The most likely future for microwave food processing, then, is the continued development of unique single systems that overcome these limitations. Examples of such systems are sausage and breakfast cereal pre-cookers. Also, Dorantes et al. (2000) suggested that some commodities such as avocado might be blanched using microwaves. Avocado develops an off-flavor when heated conventionally. When heated with microwaves, quality (including flavor) is improved, as rapid heating is achieved. It must be pointed out that the dielectric properties of avocado favor uniform heating. The rate of heating when using microwaves is the main advantage of micro-wave blanching. When heated rapidly, the quality of fruits and vegetables, such as flavor, texture, color, and vitamin content, is better kept. Thus, blanching by microwaving could be used as a pretreatment prior to canning, evaporating, frying, or freezing.

Pulsed Electric Fields Technology

Description of the Technology

Pulsed electric field (PEF) processing is based on the application of short pulses of high voltage (typically 20 to 80 kV/cm) to food placed between two electrodes. PEF is considered a nonthermal process, as foods are treated at room temperature or below for only a few microseconds, minimizing the energy loss caused by heating (Barbosa-Cánovas et al., 1999).

Equipment and Engineering Principles

In PEF technology, the energy derived from a high-voltage power supply is stored in one or several capacitors and discharged through a food material to generate the necessary electric field (Barbosa-Cánovas et al., 1999). The energy stored in one capacitor (Q [J/m^3]) is given by:

$$Q = 0.5\, C_0\, V^2$$

where C_0 is the capacitance, and V is the charging voltage.

The energy stored in the capacitors can be discharged almost instanta-neously (in a millionth of a second) at very high levels of power (Vega-Mercado et al., 1999). The discharge occurs in a treatment chamber in which the food is placed or circulates through a small gap between two electrodes (Barbosa-Cánovas et al., 1999). When a trigger signal is activated, a high-voltage switch is closed and the charge stored in the capacitor flows through the food in the treatment chamber (Zhang et al., 1995; Barsotti et al., 1999). In order to avoid undesirable thermal effects, cold water is recirculated through the electrodes to dissipate the heat generated by the electric current passing through the food (Barbosa-Cánovas et al., 1999).

Varying arrangements of capacitors, inductors, and resistors produce different types of pulses. Pulse polarity can be constant or alternating, and pulse waveform can be of exponential decay or square shape, among others (Figure 14.1) (Barbosa-Cánovas et al., 1999; Barsotti et al., 1999). Square pulses are more effective on microbial inactivation, as they maintain a peak voltage for a longer period than do exponential decay pulses. Exponential decay pulses have a long tail with a low electric field, during which excess heat is generated in the food without bactericidal effects (Zhang et al., 1995).

In the case of a square wave pulse, the electrical energy W' (J) dissipated in one pulse is given by:

$$W' = \frac{E^2 \, V \, \tau}{\rho}$$

where E is the electric field (V/cm), V is the volume of food between the two electrodes (cm³), τ is the pulse duration (sec), and ρ is the electrical resistivity of the food sample (ohms · cm).

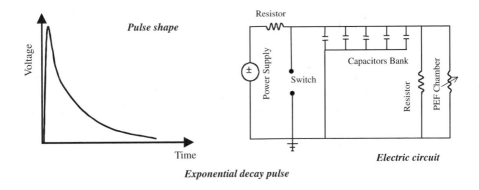

Exponential decay pulse

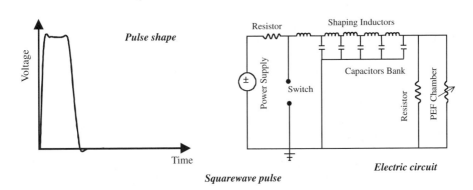

Squarewave pulse

FIGURE 14.1
Representative pulse-wave shapes and corresponding schematic electric diagrams.

Treatment chambers are designed to hold the food between two electrodes during PEF application. The chambers can be designed to work in static or continuous modes. In the static mode, food is held between two parallel electrodes, whereas in the continuous mode food circulates between the electrodes during PEF treatment. Two designs more commonly used in a continuous mode are coaxial (cylindrical or conical) and cofield (Bushnell et al., 1991; Zhang et al., 1995; Yin et al., 1997; Barbosa-Cánovas et al., 1999). Static chambers are suitable for laboratory use, whereas continuous chambers are required for large-scale operations (Qin et al., 1995, 1998; Zhang et al., 1995). Laboratory-scale PEF systems have been developed by the University of Guelph (Ho et al., 1995), Washington State University (Qin et al., 1996; Barbosa-Cánovas et al., 1997), Ohio State University (Sensoy et al., 1997; Reina et al., 1998), and PurePulse Technologies, Inc. (McDonald et al., 2000) (Figures 14.2 and 14.3).

Electric field strength and treatment time are the two most important parameters influencing microbial inactivation (Jeyamkondan et al., 1999). In a coaxial treatment chamber design, the electric field changes with position and is given by

$$E_{co} = \frac{V}{r \ln \frac{R_1}{R_2}}$$

where r is the position in which the electric field is measured, and R_1 and R_2 are the radii of the inner and outer electrodes surfaces, respectively (Zhang et al., 1995). The treatment time depends on the pulse width and the number of treatment pulses. The number of pulses (n) is given by:

$$n = \frac{f V}{\dot{V}}$$

where f is the pulse repetition rate (Hz), V is the chamber volume (cm^3), and $[\dot{V}]$ is the treated volume rate (cm^3/sec) (Zhang et al., 1995). The flow velocity profile of food is important in the determination of total treatment time, as it affects the residence time of food in the PEF chamber. Therefore, flow rate and viscosity of food should be considered in the determination of PEF processing conditions (Ruhlman, 1999).

Effects of Pulsed Electric Fields on Microbial Inactivation

Pulsed electric field (PEF) technology is used in the areas of genetic engineering and biotechnology to promote cell membrane reversible electroporation and cell electrofusion, respectively (Chang et al., 1992). The same principle is applied to PEF technology for microbial inactivation during food

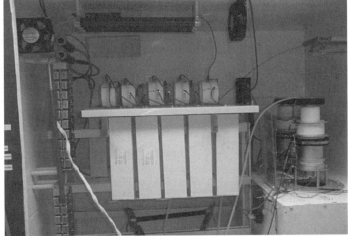

FIGURE 14.2
Pilot plant PEF generator (OSU-2C): bipolar, 50 kV/20 kW.

FIGURE 14.3
OSU-2C PEF pilot plant treatment chambers and tubular heat exchangers.

processing. The duration and intensity of the treatment are increased to make the membrane disruption an irreversible phenomenon (Jeyamkondan et al., 1999). Electroporation is the most widely accepted concept used to describe the phenomenon of cell membrane discharge (reversible electroporation) and cell membrane breakdown (irreversible electroporation) with the application of short pulses.

Sale and Hamilton (1967a) conducted one of the earliest works in PEF microbial inactivation. These authors reported the lethal effect on several bacteria and yeasts of high voltages up to 25 kV/cm applied as direct current pulses. They concluded that the products from electrolysis did not cause the inactivation nor did temperature. Their explanation for microbial inactivation was that the electric fields possibly caused an irreversible loss in the function of the membrane as a semipermeable barrier between the bacterial cell and its environment (Hamilton and Sale, 1967). According to this theory, electric fields induce a transmembrane electrical potential (TMP) that results from membrane polarization. Membrane polarization results from the fact that the cell membrane has a dielectric constant much lower than most food products; thus, the applied electric field induces the accumulation of negative and positive charges within the cell at areas closest to the cathode and anode, respectively (Zimmermann, 1986). When the TMP reaches a critical value of about 1 V (called the *breakdown* TMP), membrane integrity is lost and cells lyse (Hamilton and Sale, 1967).

The exact mechanism of electroporation is still not clear (Ho and Mittal, 1996). One proposed mechanism states that the membrane could be assumed as an ideal (linear) elastic material. Accumulated charges with different signs located at both sides of the membrane create electromechanical forces that lead to elastic strain forces on the membrane. When large compressions exceed the elastic mechanical region, the membrane is permanently ruptured and cells lyse (Croweley, 1973; Coster and Zimmermann 1975). Another theory suggests that large populations of pores are already present in cells when the electric field strength is zero (Weaver and Powel, 1989). Pores increase in size under the effect of a transmembrane electric field due to changes in both the lipid bilayer and the protein channels (Tsong, 1991). Ultimately, cell inactivation is caused by the osmotic imbalance across the cell membrane induced by the electroporation (Tsong, 1990). It has also been suggested that the structure and dynamics of electroporation may vary from cell type to cell type (Chang et al., 1992). Researchers agree that electroporation is a sequential and complex phenomenon where more than one pathway could occur simultaneously.

The PEF sensitivity of microorganisms varies with the type of microorganism. Vegetative cells are more sensitive than spores (Yonemoto et al., 1993; Vega-Mercado et al., 1996a; Raso et al., 1998a); however, depending on experimental conditions, the irreversible inactivation of bacterial spores after PEF treatment has been reported (Marquez et al., 1997). The size of microorganisms also affects the effectiveness of microbial inactivation induced by PEF. Larger yeast cells are easier to inactivate than smaller bacteria (Qin et al., 1998; Aronsson et al., 2001). Up to a 7-log reduction of yeast cells in PEF-treated orange juice was reported by Yeom et al. (2000). Gram-positive bacteria are more resistant than Gram-negative types (Mackey et al., 1994), and cells harvested in the logarithmic growth phase are more sensitive than those in the stationary growth phase (Hülsheger et al., 1983; Pothakamury et al., 1996).

Microbial inactivation by PEF also depends on the treatment medium. Lowering the medium pH increased the effectiveness against *Escherichia coli* (Vega-Mercado et al., 1996b). The combined effect of PEF and pH was attributed to a possible decrease in the cytoplasmic pH enhanced by an increase in the transfer rate of protons through the membranes when electroporated (Vega-Mercado et al., 1996b). Lowering the ionic strength also increased PEF inactivation of *E. coli* (Vega-Mercado et al., 1996b). The reduced inactivation rate on high ionic strength solutions was attributed to the stability of the cell membrane when exposed to a medium with several ions (Vega-Mercado et al., 1996b). Conductivity of the medium has been shown to influence the antimicrobial effectiveness of PEF on *Lactobacillus brevis* (Jayaham et al., 1992). As the conductivity of the medium was increased, the resistance of the treatment chamber was reduced, which in turn reduced the treatment time by reducing the pulse width. Consequently, the higher the conductivity of the treated medium, the lower the microbial inactivation attained (Jayaham et al., 1992). Treatment temperature has been shown to increase inactivation of *Escherichia coli* (Zhang et al., 1995), *Lactobacillus brevis* (Jayaham et al., 1992), and *Listeria monocytogenes* (Reina et al., 1998). The combined effect of PEF and temperature has been attributed to a decrease in the electrical breakdown potential of the membrane, due to changes in the membrane components induced by temperature (Reina et al., 1998). Nisin has shown a synergistic effect on microbial inactivation when combined with PEF (Calderón-Miranda et al., 1999; Dutreux et al., 2000; Pol et al., 2000). The site of action for both nisin and PEF is the cell membrane. This could explain why their combined effect on microbial inactivation is synergic (Calderón-Miranda et al., 1999; Dutreux et al., 2000).

The degree of microbial inactivation is known to depend on two main processing parameters: the electric field strength and the treatment time. Several mathematical models have been used to model microbial inactivation by PEF. Hülsheger et al. (1981) proposed the first mathematical model relating the survival fraction with electric field strength and treatment time:

$$s = \left(\frac{t}{t_c}\right)^{\frac{[-(E-Ec)]}{K}}$$

where s is the survival fraction, t is the treatment time (µsec) and the product between the number of pulses and the pulse width, E is the electric field strength (kV/cm), k is a specific constant for each microorganism, and t_c and E_c are the critical treatment time (µsec) and critical electric field strength (kV/cm), respectively (two threshold values above which inactivation occurs).

Peleg (1995) suggested that a model based on Fermi's equation could better describe the phenomenon. The model proposed considers a gradual transition from marginal effectiveness (or none at all) at weak electric fields to effective lethality under strong fields, instead of lethality occurring during abrupt changes in the destruction kinetic at the E_c threshold value:

$$S(V) = \frac{100}{\left(1 + \exp\left(\dfrac{V - V_c(n)}{a(n)}\right)\right)}$$

where $S(V)$ is the percent of surviving microorganisms, V is the field strength (kV/cm), $V_c(n)$ is a critical level of V where the survival level is 50%, and $a(n)$ (kV/cm) is a parameter indicating the steepness of the survival curve around V_c. Both V_c and a are functions of the number of pulses (n).

Raso et al. (2000) reported that *Salmonella senftenberg* inactivation did not follow first-order or second-order kinetics. They found that the best model to relate microbial inactivation and treatment time was a nonlinear log-logistic model initially described by Cole et al. (1993). The nonlinearity observed could originate from the different sensitivities to PEF within the microbial population (Cole et al., 1993; Raso et al., 2000).

Current Limitations and Status

Pulsed electric field technology has not yet been industrially implemented. The slow commercialization of new preservation technologies has been partially attributed to the difficulty in demonstrating the equivalency of nonthermal methods to existing thermal methods. It is foreseen that the first commercial applications of PEF technology could include acid foods, such as fruit juices, where Food and Drug Administration regulation requires a 5-log cycle reduction of specified pathogenic microorganisms (Barbosa-Cánovas et al., 1999). Some particular requirements for the industrial implementation of PEF technology include the design of PEF systems capable of assuring uniform treatment. The most important aspects of successful design include a treatment chamber capable of producing a homogeneous electric field (Qin et al., 1998), along with a good system of measurement and control. Other aspects of a successful design should account for minimizing the occurrence of dielectric breakdown produced by electric tracking along the surfaces of particulates (Barbosa-Cánovas et al., 1999).

High Hydrostatic Pressure Processing

Description of the Technology

High hydrostatic pressure (HHP) refers to the exposure of foods within vessels to high pressures (300 to 700 MPa) for a short period, typically ranging from a few seconds to several minutes. Food is pressurized by direct and indirect methods utilizing water as a pressure-transmitting medium (Farr, 1990; Mertens and Deplace, 1993). HHP is a nonthermal process, as it only involves minor increases in temperature during pressurization. For a working pressure of 600 MPa, the temperature increment of pure water is only approximately 15°C (Denys et al., 2000).

Equipment and Engineering Principles

High hydrostatic pressure technology is based on the use of pressure to compress food located inside a pressure vessel. The pressure vessel is the most important component of HHP equipment, consisting of a forged mono-lithic cylindrical piece built of alloy steel with high tensile strength. Multi-layer or wire-wound prestressed vessels are used for pressures higher than 600 MPa. Prestressed vessels are purposely designed with residual compressive stress in order to lower the maximum stress level in the vessel wall during pressurization, hence reducing the cost of producing this important piece of equipment (Mertens and Deplace, 1993).

In HHP equipment utilized in food applications, pressure is transmitted by two methods: direct or indirect. In the direct method, a piston is pushed at its larger diameter end by a low-pressure pump, directly pressurizing the pressure medium at its smaller diameter end (Figure 14.4). This method allows very fast compression but requires a pressure-resistant dynamic seal between the piston and the internal vessel surface to avoid leaks and con-tamination of the food. In the indirect method, high-pressure intensifiers are used to pump the pressure medium from the reservoir into the closed vessel until the desired pressure is achieved (Mertens and Deplace, 1993). The applied pressure is isostatically transmitted by a fluid (Pascal's law) (Earn-shaw, 1996). In this way, uniform pressure from all directions compresses the food, which then returns to its original shape when the pressure is released (Olson, 1995).

High hydrostatic pressure is essentially a batch-wise process for prepack-aged foods, whereas for pumpable liquids semicontinuous processes have been developed. In both cases, the pressure treatment is accomplished in cycles. An initial time is required to reach the desired working pressure (come-up time), after which the pressure is maintained for the required processing time (holding time). Finally, the pressure is released, taking only a few seconds (release time) to complete the process (Horie et al., 1992).

Pressure transmission is instantaneous and independent of product size and geometry. Contrary to heat, pressure transmission is not time or mass dependent; thus, the time required for pressure to reach the internal points of the food being processed is minimized (Mertens and Deplace, 1993). However, the pressure action on microbial inactivation *is* time dependent (Hoover et al., 1989).

Effects of High Hydrostatic Pressure on Microbial Inactivation

Pressures between 300 and 600 MPa can inactivate food spoilage and patho-genic microorganisms (Isaacs et al., 1995; Palou et al., 1999). Pressure induces

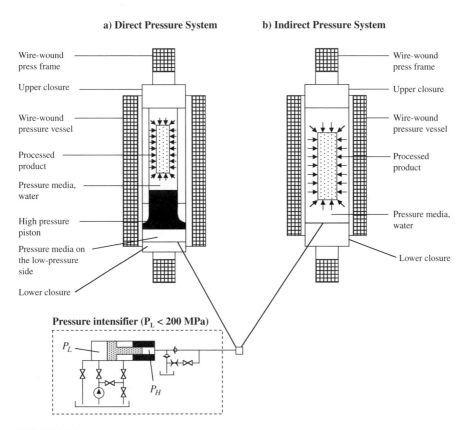

FIGURE 14.4
High-pressure machines with wire-wound vessels and two-pressure transmission system.

a number of changes in the microbial cell membrane, cell morphology, and biochemical reactions that ultimately can cause microbial inactivation (Hoover et al., 1989). Cell membranes are the primary site of pressure damage done to microbial cells (Farr, 1990; Mackey et al., 1994; Burl et al., 2000; Ritz et al., 2000). The microbial membranes play an important role in the transport and respiration functions; thus, a great change in membrane permeability can cause the death of cells (Lechowich, 1993). Changes in cell morphology involve the collapse of intercellular gas vacuoles, anomalous cell elongation, and cessation of movement in the case of motile microorganisms (ZoBell and Cobet, 1964; Mackey et al., 1994). Minor changes in the biochemistry of living cells also play an important role in microbial inactivation. Products that differ in the number of ionizable groups are strongly influenced by pressure; therefore, water and acid molecules increase their ionization under pressure (Earnshaw, 1996). The pH of water and buffered solutions decreases during pressurization by 0.2 to 0.3 units per 100 MPa of applied pressure (Funtenberger et al., 1995), which in turn may enhance the effect of pressure on microbial inactivation (Palou et al., 1999). Pressure may inhibit the availability of energy in cells by affecting energy-producing enzymatic reactions, promoting microbial inactivation. Pressure above 150 MPa induces partial unfolding and dissociation of protein structures, due to modifications of hydrogen and electrostatic bonds, including hydrophobic interactions (Jaenike, 1987; Cheftel, 1991).

The extent of microbial inactivation achieved depends on the type and number of microorganisms, magnitude and duration of HHP treatment, temperature and composition of the suspension media, or food (Hoover et al., 1989; Palou et al., 1999). In general, yeasts and molds are more easily inactivated by pressure than bacteria. Among bacteria, vegetative forms are more susceptible than spores. Spores of bacteria are extremely resistant to pressure action (Nakayama et al., 1996), whereas ascospores of heat-resistant molds are inactivated at pressures of about 300 MPa (Butz et al., 1996). Inactivation of spores increases when combining pressure with other preservation factors, such as heat or antimicrobials (Roberts and Hoover, 1996). Results from combining pressure with low pH are contradictory. Although some authors claim that spore inactivation increases with low pH (Roberts and Hoover, 1996), others have found that lowering pH does not affect spore inactivation, even at extremely low pH values (Sale et al., 1970). Moderate pressures can be utilized to induce spores to germinate and then inactivate them by means of higher pressures or other preservation techniques (Gould, 1973; Wuytack and Michiels, 2001). Because of their relative pressure sensibility, vegetative cells are prime targets for preservation by high-pressure technology, particularly for products with food composition factors such as high acidity (e.g., fruit juices), ensuring that pressure-resistant spores are unable to grow (Gould, 1995). Gram-positive bacteria are more resistant than Gram-negative types (Mackey et al., 1994), and bacteria in the stationary growing phase are more resistant than bacteria in the logarithmic growing phase (ZoBell, 1970; Mackey et al., 1995). When combined with other preservative factors such as water activity, pH, antimicrobials, or temperature, pressure action may have an antagonistic, additive, or synergistic effect. Foods with low water activity (a_w) due to high sugar concentrations decrease the sensibility of *Rodotorula rubra* and *Zygosaccharomices bailii* to pressure (antagonistic effect) (Oxen and Knorr, 1993; Palou et al., 1997b). Combinations of EDTA with nisin and lysozyme increased the lethality of *Escherichia coli* in an additive manner (Hauben et al., 1996). Low pH, antimicrobial peptides, and the use of combined moderate temperatures promote pressure efficacy (synergistic effect) for microbial inactivation (Mackey et al., 1995; Pandya et al., 1995; Masschalack et al., 2001).

Generally, an increase in pressure increases microbial inactivation; however, increasing the treatment time does not necessarily increase the microbial death rate. The types of HHP inactivation kinetics observed with different microorganisms are quite variable (Palou et al., 1999). Most researchers have observed a first-order kinetics (Hashizume et al., 1995; Palou et al., 1997a; Mussa et al., 1999; Erkmen and Karaman, 2001). Other authors report a biphasic inactivation with the first phase as described by a first-order kinetics, followed by a tailing effect. The tailing effect occurs as a result of different pressure sensitivities among microorganisms from the same population (Earnshaw et al., 1995; Reyns et al., 2000). An empirical model to describe the microbial inactivation was originally proposed by Peleg (1995) for pulsed electric fields and was adapted by Palou et al. (1998) to interpret *Zygosaccharomices bailii* inactivation by HHP. Because this model is empirical, it can be used only to compare inactivation patterns and pressure sensitivities of microorganisms under the same processing conditions (Palou et al., 1998).

Current Status and Limitations

High hydrostatic pressure is not a novel technology in the food industry, although interest has been renewed in the last decade. Hite conducted the first studies on HHP in 1899 (Hite, 1899), but it was not until the 1980s that Japan, the U.S., and Europe began working again on this promising technology. Japan launched the first commercial HHP processed products (Earnshaw, 1996). In 1990, Medi-ya Food Co. (Japan) launched the first high-pressure-treated products worldwide, including jams of various flavors. Other HHP products marketed worldwide are fruit sauces and desserts (Medi-ya Food Co.), mandarin juice (Wakayama Co.), grapefruit juice (Pokka Co.), and avocado paste (guacamole) produced in Mexico (Avomex) (Palou et al., 1999). In the U.S., it is expected that the Food and Drug Administration (FDA) will soon approve the use of HHP for processing high-acid foods.

It is unlikely that pressure processing of food will replace canning or freezing technology because it is relatively expensive; however, applications could be found for pressure processing for high-quality products where thermal processing is not suitable and in cases where this technology could confer added value in terms of nutritional or sensorial characteristics (Mertens, 1995; Earnshaw, 1996).

Food Irradiation

Description of the Technology

Food irradiation is a process by which food is exposed to ionizing radiation for the purpose of preservation. In 1896, one year after Roentgen's discovery of x-rays, it was suggested that ionizing radiation was lethal to bacteria (WHO, 1988; Hackwood, 1991). Food irradiation was first patented in England in 1905 and first used in the U.S. in 1921 to inactivate *Trichiniella spiralis* in pork muscle. Since the 1940s, irradiation has been used as a method to supply safe food to U.S. army combat troops (Barbosa-Cánovas et al., 1998). Despite scientific evidence regarding its effectiveness and safety, including regulatory approval for specific uses, consumer acceptance concerns have postponed applications of irradiation in the food industry (Buzby and Morrison, 1999).

Equipment and Engineering Principles

The energy employed in food irradiation technology is referred to as *ionizing irradiation*. The term *ionizing* comes from the fact that certain rays produce electrically charged particles, called *ions*, when these rays strike a material (Urbain, 1986). The propagation of energy through space is known as *radiation* or *radiant energy*. Radiation has a dual nature, of wave and particle (corpuscle). Electromagnetic energy consists of self-propagating electric and magnetic disturbances that travel as waves, involving electric and magnetic vectors characterized by frequency and wavelength. Corpuscular energy can be visualized as a "particle" traveling through space, in which the energy is concentrated into bundles called *photons* (Urbain, 1986). According to quantum theory, the energy content in one photon can be expressed as:

$$E = hw = \frac{hc}{\lambda}$$

where h is Planck's constant, equal to 6.63×10^{-34} J $\cdot$ s; w is the frequency in (Hz); c is a constant representing the velocity of the propagation of the electromagnetic wave (m/sec), closely equal to 3×10^8 m/sec; and λ is the wavelength (m).

Ionizing radiation occurs when one or more electrons are removed from an atom. Electrons orbiting at minimum energy level or ground state can be raised to higher levels, becoming electronically excited (excitation). If enough energy is transferred to an orbital electron, the excited electron may be ejected from the atom (ionization). A minimum amount of absorbed energy, called *ionization potential*, exists for each electron energy level necessary to exit the atom domain. If the energy absorbed by the electron is greater than its ionization potential, the excess energy enters a kinetic state, enabling the electron to leave the atom domain (Urbain, 1986).

Although each electron behaves individually, electrons can be used in large numbers, called *electron beams*, to irradiate food (Barbosa-Cánovas et al., 1997). Electron beams are produced from commercial electron accelerators (WHO, 1988). One advantage of electron beam radiation is that the electron accelerators can be switched off when not in use, leaving no radiation hazard; however, the penetration of electron beams into foods is limited (Kilcast, 1995). Electrostatic forces tend to attract charged particles such as electrons (negatively charged), limiting the electron beam penetration into foods. Incident electrons also strike the orbital electrons of other atoms, limiting the penetration depth of incident electrons. Some of the kinetic energy from the incident electrons can be transferred to an orbital electron, which becomes excited, or leaves the atom, creating a chain of reactions that end when the energy is completely dissipated. The random pathway of the incident electrons also hinders electron penetration (Urbain, 1986). Therefore, electron beam radiation is limited to small food items, such as grains, or to the removal of surface contamination of prepared meals (Kilcast, 1995).

Photons of electromagnetic radiation such as those created by x-rays and gamma rays travel without charge; thus, they do not interact with electrostatic forces while traveling through the food, and can penetrate deeper than electron beams. One mode of producing x-rays is through the use of electron beams. If the highly accelerated electrons penetrate a thin foil of certain metals, such as tungsten, tantalum, or any other material capable of withstanding high heat, x-rays are produced (Figure 14.5). The fact that electron production can be switched off makes this method of x-ray production an ionization source of interest (Radomyski et al., 1994). Radioactive isotopes such as cobalt 60 or cesium 137 are used to produce gamma rays. Although radiation isotopes cannot be switched off, they have the advantage of producing gamma rays, which offer the largest penetration among all ionization sources (Kilcast, 1995).

Dose is the most important parameter in food irradiation. The quantity of energy absorbed by the food is measured in grays (Gy). One gray equals one joule per kilogram of matter. A gray (equal to 100 rad) is a very small quantity; therefore, the dose is expressed in kilograys (kGy). Up to 10 kGy is still a very small amount of energy, equal to the amount of heat required to raise the temperature of water 2.4°C (WHO, 1988). For this reason irradiation is considered a "cold" or nonthermal technology (Farkas, 1988; Barbosa-Cánovas et al., 1998; Foley et al., 2001).

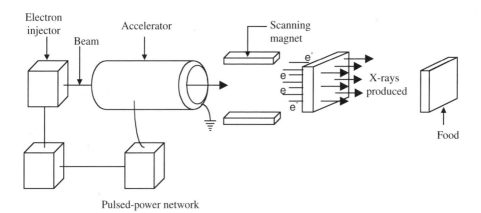

FIGURE 14.5
Schematic representation of an x-ray generator utilizing a linear induction electron accelerator.

Effects of Irradiation on Microbial Inactivation

Irradiation is currently used as a tool to control naturally occurring processes such as ripening or senescence of raw fruits and vegetables and is an effective way to inactivate spoilage and pathogenic microorganisms. Irradiation causes microbial death by inhibiting DNA synthesis (Farkas, 1988). Other mechanisms involved in irradiation microbial inactivation are cell membrane alteration, denaturation of enzymes, alterations in ribonucleic acid (RNA) synthesis, effects on phosphorylation, and DNA compositional changes (Urbain, 1986).

According to the dose used and the goal of the treatment, food irradiation can be classified into three categories (Wilkinson and Gould, 1996):

1. *Radurization* is a process comparable with thermal pasteurization. The goal of radurization is to reduce the number of spoilage micro-organisms, using doses generally below 10 kGy.

2. *Radicidation* is a process in which the irradiation dose is enough to reduce specific non-spore-forming microbial pathogens. Doses generally range from 2.5 to 10 kGy, depending on the food being treated.

3. *Radappertization* is a process designed to inactivate spore-forming pathogenic bacteria, similar to thermal sterilization. Irradiation doses must be between 10 and 50 kGy.

Irradiation microbial inactivation kinetics has been reported to be log linear (Urbain, 1986; Farkas, 1988; Radomysky et al., 1994; Monk et al., 1995; Tarte et al., 1996; Serrano et al., 1997; Buchanan et al., 1998; Thayer et al., 1998; Gerwen et al., 1999). The irradiation D value is a good parameter to predict the effectiveness of irradiation, as it represents the dose required to inactivate target cells by 90% (Wilkinson and Gould, 1996; Gerwen et al., 1999). The D value of a particular microorganism depends on both species and strain, as well as on certain extracellular environmental conditions such as pH, temperature, and chemical composition of the irradiated food (Urbain, 1986; Monk et al., 1995; Barbosa-Cánovas et al., 1998). In comparing the irradiation D value of bacterial spores with the D value of vegetative bacterial cells, viruses are generally found to be the most radiation-resistant microorganisms, followed by spores and yeasts (Wilkinson and Gould, 1996). Gerwen et al. (1999) compared the D values of several bacterial spores with the D values of several bacteria in the vegetative stage. The average D value of the bacterial spores studied (2.48 kGy) was significantly higher than the average D value of vegetative bacterial cells tested (0.762 kGy), which demonstrates that spores are usually more resistant to irradiation than vegetative bacterial cells. Molds and Gram-positive vegetative bacteria are more tolerant than Gram-negative bacteria (Wilkinson and Gould, 1996). The irradiation D value of *Escherichia coli* O157:H7 increased in a range from 0.12 to 0.21 kGy for nonacid-adapted cells to 0.22 to 0.31 kGy for acid-adapted cells when tested in five apple juice brands containing different suspended solids (Buchanan et al., 1998). Resistance of *Salmonella typhimurium* increased at reduced irradiation temperatures (Thayer and Boyd, 1991). Water content also affects microbial sensitivity to irradiation. The lower the water activity of the food product, the more resistant the microorganisms become (Radomyski et al., 1994). The presence of oxygen enhances the indirect action of radiation, increasing its effectiveness and causing a reduction in the D value (Urbain, 1986).

Chemical compounds with nutritional or flavor functions can also be affected by ionizing irradiation; however, the chemical changes produced by irradiation in food components, at recommended doses, are small when compared with other preservation technologies such as heat (Luchsinger, et al., 1996; Kilkast, 1995). Irradiation doses below 1 kGy cause an insignificant loss of nutrients. At higher irradiation doses, such as those required for food sterilization, some vitamins such as A, B_1, C, E, and K can degrade to some extent (Urbain, 1986). The loss of vitamins (thiamin, riboflavin, and α-tocopherol) in various meats due to gamma irradiation is negligible insofar as the diet of U.S. consumers is concerned (Lakritz et al., 1998). Vitamin losses seem to be more than compensated for by the advantages of irradiation processing in controlling bacteriological contamination (Fox et al., 1995). Thiamin is the most radiation sensitive of the water-soluble vitamins and is therefore a good indicator of the effect of irradiation treatment (Graham et al., 1998).

Irradiation may cause some changes in the sensory characteristics of food and the functional properties of food components. Irradiation initiates the autoxidation of fats, which gives rise to rancid off flavors (Kilcast, 1995; Byun et al., 1999). The extent of irradiation-induced lipid oxidation depends on factors associated with oxidation such as temperature, oxygen availability, fat composition, and pro-oxidants (Barbosa-Cánovas et al., 1998). Irradiation-induced lipid oxidation was retarded when antioxidants were added to groundbeef patties prior to irradiation (Lee et al., 1999). Lipid oxidation was also diminished when oxygen availability was lowered (Jo et al., 1999; Foley et al., 2001). Proteins with sulfur-containing amino acids can break down after irradiation treatment, yielding unpleasant off flavors (Kilcast, 1995; Ahn et al., 2000). Dairy products such as milk and other commodities are especially prone to develop off flavors even at low radiation doses (Josephson and Peterson, 1982; Kilcast, 1995). Irradiation can break down high-molecular-weight carbohydrates such as pectins and other cell wall materials into smaller units, causing softening of fruits and vegetables (Kader, 1986).

Current Status and Limitations

Irradiation is the process of choice for spice treatment in many countries, including Canada and the U.S. (Farkas, 1988). Although irradiation is an effective method for food preservation, public apprehension about irradiation has delayed its many potential commercial applications in the food industry. Many consumers naturally tend to relate irradiation with nuclear energy and develop a negative attitude toward irradiated foods (Bord, 1991). Concerns about irradiation include safety, nutritional quality, potential harm to employees, and potential danger to people living near an irradiation facility (Bruhn, 1995).

In the past, lack of regulation was the main reason for limiting potential irradiation applications in the food industry. At the international level, the Food and Agricultural Organization (FAO), International Atomic Energy Agency (IAEA), and World Health Organization (WHO) met in 1980 and concluded that irradiated foods are safe and healthful at levels up to 10 kGy (Radomyski et al., 1994). In the U.S., the FDA has approved the unconditional use of radiation on wheat and wheat flour (1963), potatoes (1964), herbs (1983), spices (1986), pork for parasite control (1985), poultry meat (1992) (Wilkinson and Gould, 1996), and more recently on ground beef (1997). Official regulations are now available for a wide range of foodstuffs, but the availability of official regulations has not been reflected in an increase of commercially irradiated products (Bord, 1991).

Ultrasound Processing

Description of the Technology

Ultrasound technology entails the transmission of mechanical waves through materials at frequencies above 18 MHz (Blitz, 1971). Most ultrasound applications are not associated with food preservation; however, when applied with enough intensity, ultrasound has a lethal effect on microorganisms. Therefore, ultrasound can potentially be used as a food preservation technique (Lillard, 1994; Earnshaw, 1998; Betts et al., 1999).

Equipment and Engineering Principles

Ultrasound is generated in a material because of mechanical disturbances caused by continuous vibrations that are characterized by amplitude and frequency (Blitz, 1971; Povey and McClements, 1988). Transducers transform electric energy into mechanical energy, which in turn is amplified and transmitted through a liquid or solid material to generate traveling sonic energy waves (Povey and McClements, 1988; Sala et al., 1995). The energy of vibrations traveling through a material causes alternating compressive and tensile strains due to minute displacements of particles, described by a sinusoidal equation as follows:

$$y = y_0 \sin w\left(t - \frac{x}{c}\right)$$

where y is the particle displacement, y_0 is the particle displacement amplitude, w is the frequency, t is the time, x is the distance of the particle from
the source, and c is the velocity of the light through the material (Blitz, 1971).

When ultrasound waves travel through liquids, they create periodic cycles
of expansion and compression (Figure 14.6). If the negative pressure created
in the liquid during the expansion cycle is low enough to overcome the
tensile strength, it can cause liquid failure and, as a consequence, the formation of bubbles. These bubbles expand and contract and may finish collaps-
ing during a dynamic process known as *cavitation* (Suslick, 1988; Leighton,
1998). Bubbles enlarge and collapse with different intensities, instanta-
neously increasing the local temperature and pressure up to 5000 K and 100
MPa, respectively. Frequency and amplitude, medium viscosity, temperature, and pressure determine the extent of cavitation (Betts et al., 1999).
Another phenomenon resulting from variation in bubble size and collapse
is the development of strong micro-streaming currents. The phenomenon of
acoustic micro-streaming is associated with high flow-velocity gradients and
shear stresses that alter media characteristics (Suslick, 1988).

Ultrasound applications are typically divided into two categories
according to the power and frequency utilized. Low-power/high-fre-
quency ultrasound operates at frequencies in the megahertz range and
with acoustic power ranging from a few W to several tens of mW (Suslick,
1988). Such acoustic waves are capable of traveling through a medium

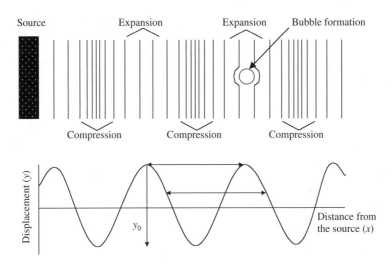

FIGURE 14.6
Schematic representation of layers of a material in which longitudinal sonic energy is propa-
gated.

without altering the material, allowing nondestructive measurements of food processes (Povey and McClements, 1989). High-power/low-frequency ultrasound is operated at frequencies in the kilohertz range, at which the acoustic power may extend from a few milliwatts to kilowatts (Suslick, 1988). Examples of use of this type of ultrasound include surface sanitation, microbial inactivation, and enzyme activity alteration (McClements, 1995; Povey, 1989).

Effects of Ultrasound on Microbial Inactivation

Ultrasound microbial inactivation is mainly attributed to cavitation. Cavitation is associated with shear disruption, localized heating, and free radical formation (Earnshaw, 1998). Sudden changes in temperature and pressure that occur during cavitation are considered the main reasons for cell membrane damage and therefore microbial inactivation. The high-temperature spots created during bubble implosion alone cannot explain microbial inactivation, as their effect is very local (Earnshaw, 1998). It is likely that the lethal effect of microbial inactivation mainly relies on the inability of a microbe to withstand the extreme pressure variations taking place during the cavitation process (Scherba et al., 1991). Microstreaming may also contribute to inactivation, as microcurrents created around the bubbles are of such intensity that they can catalyze chemical reactions and disrupt microorganisms (Suslick, 1988). The third mechanism involved in microbial inactivation with ultrasound is the formation of free radicals and hydrogen peroxide, both of which have bactericidal properties (Riesz and Kondo, 1992).

Efficiency of ultrasound microbial inactivation depends on the microorganism itself as well as on the medium properties and operating conditions. Different microorganisms show different resistances to ultrasound inactivation. Bacterial spores are more resistant to ultrasound than vegetative cells (Raso et al., 1998b). Gram-positive bacteria are more resistant than Gram-negative types, probably due to a thicker cell wall and the firmer adherent layer of peptidoglycans (Earnshaw, 1998). The combination of ultrasound with temperature increases the effectiveness of ultrasound against vegetative cells of *Staphylococcus aureus* (Ordoñez et al., 1984) and *Saccharomyces cerevisiae* (Guerrero et al., 2001) and increases effectiveness against spores of *Bacillus cereus* (Joyce et al., 1960) and *Bacillus subtilis* (Raso et al., 1998b). Combining ultrasound and pressure simultaneously is a very efficient means to increase the lethality of ultrasound waves. Increased antimicrobial efficacy due to pressure is greater at larger ultrasound operating amplitudes (Raso et al., 1998b,c; Pagán et al., 1999).

Microbial inactivation by ultrasound has been reported to follow a first-order kinetics (Raso et al., 1998c; Guerrero et al., 2001); however, shoulders and tails have been reported, suggesting that inactivation kinetics might be nonlinear (Raso et al., 1998c).

Current Status and Limitations

Most likely, ultrasound will become an important tool for measurement in food industry applications such as flow rate, undisolved air, bubble size, solids content, particle size, emulsion stability, and degree of crystallization (Povey and McClements, 1988; McClements, 1995; Mason et al., 1996). Unfortunately, very high intensities are required to sterilize foods using ultrasound alone (Mason et al., 1996). However, in combination with other preservation factors such as heat, pH, pressure, or natural antimicrobials, ultrasound could be used to develop minimally processed foods (Ordoñez et al., 1987; Mason et al., 1996; Guerrero et al., 2001).

Final Remarks

Conventional thermal food processes are among the most widely employed mechanisms in food preservation, primarily because they are very effective in ensuring the required high food safety standards. However, conventional thermal processes frequently lead to undesirable overcooked flavor and loss of desirable fresh flavor, aroma, vitamins, and essential nutrients. These negative effects on food quality can be attributed to the high amount of energy required to heat the food products and to attain the desired microbial inactivation. The high amount of energy required also promotes undesirable changes in nutritional and sensory food characteristics.

Contrary to conventional thermal processes, alternative technologies target the use of energy on the microbial inactivation mechanism itself. Alternative technologies minimize the detrimental effects on food quality and better retain their fresh-like characteristics. In order to be considered successful, however, alternative technologies have to achieve at least the same food safety standards as those of conventional thermal processes.

Understanding the mechanism of microbial inactivation involved in each alternative technology would allow utilizing them in the most suitable conditions. Each alternative technology has specific applications in terms of the type of food that can be processed. Microwave heating, high hydrostatic pressure, and irradiation are useful in processing both liquid and solid foods, whereas pulsed electric fields and ultrasound are more suitable for liquid foods. Also, each of these technologies could be used either alone or in combination with other technologies to optimize product quality, processing time, and microbial inactivation. The future of these alternative technologies is promising; however, technical and legal barriers must be overcome in order to achieve successful industrial implementation.

References

Ahn, D.U., Jo, C., and Olson, D.G. (2000) Analysis of volatile components and the sensory characteristics of irradiated raw pork, *Meat Science*, 54(3), 209–215.

Aronsson, K., Lindgren, M., Johansson, B.R., and Rönner, U. (2001) Inactivation of microorganisms using pulsed electric fields: the influence of process parameters on *Escherichia coli*, *Listeria innocua*, *Leuconostoc mesenteroides* and *Saccharomyces cerevisiae*, *Innovative Food Sci. Emerging Technol.*, 2, 41–54.

Ball, C.O. and Olson, F.C.W. (1957) *Sterilization in Food Technology*, McGraw-Hill, New York.

Barbosa-Cánovas, G.V., Qin, B.L., and Swanson, B.G. (1997) The study of critical values in the treatment of foods by pulsed electric fields, in *Food Engineering (2000)*, P. Fito, E. Ortega-Rodriguez, and G.V. Barbosa-Cánovas, Eds., Chapman & Hall, New York, pp. 141–159.

Barbosa-Cánovas, G.V., Pothakamury, U.R., Palou, E., and Swanson, B.G. (1998) Food irradiation, in *Nonthermal Preservation of Foods*, G.V. Barbosa-Cánovas, U.R. Pothakamury, E. Palou, and B.G. Swanson, Eds., Marcel Dekker, New York, pp. 161–213.

Barbosa-Cánovas, G.V., Góngora-Nieto, M.M., Pothakamury, U.R., and Swanson, B.G. (1999) *Preservation of Foods with Pulsed Electric Fields*, Academic Press, New York.

Barsotti, L., Merle, P., and Cheftel, J.C. (1999) Food processing by pulsed electric fields. I. Physical aspects, *Food Rev. Int.*, 15(2), 163–180.

Betts, G.D., Williams, A., and Oakley, R.M. (1999) Ultrasonic standing waves: inactivation of food-borne microorganisms using power ultrasound, in *Encyclopedia of Food Microbiology*, Vol. 3, R.K. Robinson, C.A. Batt, and P.D. Patel, Eds., Academic Press, New York, pp. 2202–2207.

Blitz, J. (1971) *Ultrasonics: Methods and Applications*, Van Nostrand-Reinhold, New York.

Bord, R.J. (1991) Consumer acceptance of irradiated food in the United States, in *Food Irradiation*, S. Thorne, Ed., Elsevier Science Publishers, New York, pp. 61–86.

Bruhn, C.M. (1995) Consumer attitudes and market response to irradiated food, *J. Food Protection*, 58(2), 175–181.

Buchanan, R.L., Edelson, S.G., Snipes, K., and Boyd, G. (1998) Inactivation of *Escherichia coli* O157:H7 in apple juice by irradiation, *Appl. Environ. Microbiol.*, 64(11), 4533–4535.

Burfoot, D., Griffin, W.J., and James, S.J. (1988) Microwave pasteurization of prepared meals, *J. Food Eng.*, 8, 145–156.

Burl, S., Rommens, A.J.M., and Verrips, C.T. (2000) Mechanistic studies on the inactivation of *Saccharomyces cerevisiae* by high pressure, *Innovative Food Sci. Emerging Technol.*, 1, 99–108.

Bushnell, A.H., Dunn, J.E., and Clark, R.W. (1991) High Pulsed Voltage System for Extending the Shelf Life of Pumpable Food Products, U.S. Patent No. 5,048,404.

Butz, P., Funtenberger, S., Haberditzl, T., and Tauser, B. (1996) High pressure inactivation of *Byssochlamys nivea* ascospores and other heat resistant moulds, *Lebensmittel-Wissenschaft Technologie*, 29, 404–410.

Buzby, J.C. and Morrison, R.M. (1999) Food irradiation: an update, *Food Safety*, 22(2), 21–22.

Byun, M.W., Lee, J.W., Yook, H.S., Lee, K.H., and Kim, K.P. (1999) The improvement of color and shelf life of ham by gamma irradiation, *J. Food Protection*, 62(10), 1162–1166.

Calderón-Miranda, M.L., Swanson, B.G., and Barbosa-Cánovas, G.V. (1999) Inactivation of *Listeria innocua* in liquid whole egg by pulsed electric fields and nisin, *Int. J. Food Microbiol.*, 51, 7–17.

Chang, D.C., Saunders, J.A., Chassy, B.M., and Sowers, A.E. (1992) Overview of electroporation and electrofusion, in *Guide to Electroporation and Electrofusion*, A.J. Chang, B.M. Chassy, and J.A. Saunders, Eds., Academic Press, San Diego, CA, pp. 1–6.

Cheftel, J.C. (1991) Applications des hautes pressions en technologie alimentaire, *Actualites des Industries Alimentaries et Agro-Alimentaries*, 3, 141–153.

Cole, M.B., Davies, K.W., Munro, G., Holyoak, C.D., and Kilsby, D. (1993) A vitalistic model to describe the thermal inactivation of *Listeria monocytogenes*, *J. Indust. Microbiol.*, 12, 232–239.

Coster, H.G.L. and Zimmermann, U. (1975) The mechanism of electrical breakdown in the membranes of *Valonia utricularis*, *J. Membrane Biol.*, 22, 73–90.

Cross, G.A. and Fung, D.Y.C. (1982) The effect of microwaves on nutrient value of foods, *CRC Crit. Rev. Food Sci. Nutr.*, 4:355–381.

Croweley, J.M. (1973) Electrical breakdown of bimolecular lipid membranes as an elecromechanical instability, *Biophys. J.*, 13, 711–724.

Curnutte, B. (1980) Principles of microwave radiation, *J. Food Protection*, 43, 618–624.

Datta, A.K. and Hu, W. (1992) Optimization of quality in microwave heating, *Food Technol.*, 46(12), 53–56.

Decareau, R.V. and Peterson, R.A. (1986) *Microwave Processing and Engineering*, Ellis Horwood, Chichester, England.

Denys, S., Ludikhuyze, L.R., Van Loey, A.M., and Hendrickx, M.E. (2000) Modeling conductive heat transfer and process uniformity during batch high-pressure of foods, *Biotechnol. Progr.*, 16, 92–101.

Dibben, D. (2000) Fundamentals of heat and moisture transport for microwaveable food product and process development, in *Handbook of Microwave Technology for Food Applications*, A.K. Datta and R.C. Anatheswaran, Eds., Marcel Dekker, New York, pp. 1–28.

Dorantes-Alvarez, L., Barbosa-Cánovas, G.V., and Gutiérrez-López, G. (2000) Blanching of fruits and vegetables using microwaves, in *Innovations in Food Processing*, G.V. Barbosa-Cánovas and G.W. Gould, Eds., Technomic, Lancaster, PA, pp. 149–162.

Dreyfuss, M.S. and Chipley, J.R. (1980) Comparison of effects of sublethal microwave radiation and conventional heating on the metabolic activity of *Staphylococcus aureus*, *Appl. Environ. Microbiol.*, 39, 13–16.

Earnshaw, R.G (1996) High pressure food processing, *Nutr. Food Sci.*, 2, 8–11.

Earnshaw, R.G. (1998) Ultrasound: a new opportunity for food preservation, in *Ultrasound in Food Processing*, M.J.W. Povey and T.J. Mason, Eds., Blackie Academic, New York, pp. 183–192.

Earnshaw, R.G., Appleyard, J., and Hurst, R.M. (1995) Understanding physical inactivation processes: combined opportunities using heat, ultrasound and pressure, *International Journal of Food Microbiology*, 28, 197–219.

Erkmen, O. and Karaman, H. (2001) Kinetic studies on the high pressure carbon dioxide inactivation of *Salmonella tiphimurium*, *J. Food Eng.*, 50, 25–28.

Farkas, J. (1988) *Irradiation of Dry Food Ingredients*, CRC Press, Boca Raton, FL.

Farkas, J. (1997) Physical methods of food preservation, in *Food Microbiology: Fundamentals and Frontiers*, M.P. Doyle, L.B. Beuchat, T.J. Montville, Eds., American Society for Microbiology, Washington, D.C., pp. 497–519.

Farr, D. (1990) High pressure technology in the food industry, *Trends Food Sci. Technol.*, 1(7), 14–16.

Fleishman, G.J. (1996) Predicting temperature range in food slabs undergoing long term low power microwave heating, *J. Food Eng.*, 27(4), 337–351.

Foley, D.M., Reher, E., Caporaso, F., Trimboli, S., Musherraf, Z., and Prakash, A. (2001) Elimination of *Listeria monocytogenes* and changes in physical and sensory qualities of a prepared meal following gamma irradiation, *Food Microbiol.*, 18, 193–204.

Fox, J.B., Lakritz, L., Hampson, J., Richardson, R., Ward, K., and Thayer, D.W. (1995) Gamma irradiation effects on thiamin and riboflavin in beef, lamb, pork and turkey, *J. Food Sci.*, 60(3), 596–598, 603.

Fryer, P.J. (1997) Thermal treatment of foods, in *Chemical Engineering for the Food Industry*, P.J. Fryer, D.L. Pyle, and C.D. Rielly, Eds., Blackie Academic, London, pp. 331–382.

Funtenberger, S., Dumay, E., and Cheftel, J.C. (1995) Pressure-induced aggregation of α-lactoglobulin in pH 7.0 buffers, *Lebensmittel-Wissenschaft Technologie*, 28, 410–418.

Gerwen, S.J.C., Rombouts, F.M., Riet, K., and Zwietering, M.H. (1999) A data analysis of the irradiation parameter D10 for bacteria and spores under various conditions, *J. Food Protection*, 62(9), 1024–1032.

Gould, G.W. (1973) Inactivation of spores in food by combined heat and hydrostatic pressure, *Acta Alimentaria*, 2, 377–383.

Gould, G.W. (1995) The microbe as a high pressure target, in D.A. Leward, D.E. Johnson, R.G. Earnshaw, and A.P.M. Hasting, Eds., *High Pressure Processing of Foods*, Nottingham University Press, Nottingham, U.K., pp. 27–35.

Graham, W.D., Stevenson, M.H., and Stewart, E.M. (1998) Effect of irradiation dose and irradiation temperature on the thiamin content of raw and cooked chicken breast meat, *J. Sci. Food Agric.*, 78, 559–564.

Guerrero, S., López-Malo, A., and Alzamora, S.M. (2001) Effect of ultrasound on the survival of *Saccharomyces cerevisiae*: influence of temperature, pH and amplitude, *Innovative Food Sci. Emerging Technol.*, 2, 31–39.

Hackwood, S. (1991) An introduction to the irradiation processing of foods, in *Food Irradiation*, S. Thorne, Ed., Elsevier, New York, pp. 1–18.

Hamilton, W.A. and Sale, A.J.H. (1967) Effects of high electric fields on microorganisms. II. Mechanism of action of the lethal effect, *Biochimica et biophysica acta*, 148, 789–800.

Hashizume, C., Kimura, K., and Hayashi, R. (1995) Kinetic analysis of yeast inactivation by high pressure treatment at low temperatures, *Biosci. Biotechnol. Biochem.*, 59, 1455–1458.

Hauben, K.J.A., Wuytack, E.Y., Soontjens, C.F., and Michelis, C.W. (1996) High-pressure transient sensitization of *Escherichia coli* to lysozyme and nisin by disruption of outer-membrane permeability, *J. Food Protection*, 59(4), 350–355.

Heddleson, R.A. and Doores., S. (1994) Factors affecting microwave heating of foods and microwave induced destruction of foodborne pathogens: a review, *J. Food Protection*, 57(11), 1025–1037.

Hite, B.H. (1899) Effect of pressure in the preservation of milk, *Bull. W. Va. Agric. Exp. Station*, 58, 15–35.

Ho, S.Y. and Mittal, G.S. (1996) Electroporation of cell membranes: a review, *Crit. Rev. Biotechnol.*, 16(4) 349–362.

Ho, S.Y., Mittal, G.S., Cross, J.D., and Griffiths, M.W. (1995) Inactivation of *Pseudomonas fluorescens* by high voltage electric pulses, *J. Food Sci.*, 60(6), 1337–1340, 1343.

Hoover, D.G., Metrick C., Papineau, A.M., Farkas, D.F., and Knorr, D. (1989) Biological effects of high hydrostatic pressure on food microorganisms, *Food Technol.*, 3, 99–107.

Horie, K., Yamamuchi, T., Naoi, T., and Inoue, I. (1992) Present status and future prospects of high pressure food processing equipment, in *Colloque INSERM*, Vol. 224, C. Balany, R. Hayashi, K. Heremans, and P. Mason, Eds., John Libbey Eurotext, Montrouge, France, pp. 521–524.

Hülsheger, H., Potel, J., and Niemann, E.G. (1981) Killing of bacteria with electric pulses of high field strength, *Radiation Environ. Biophys.*, 20, 53–65.

Hülsheger, H., Potel, J., and Neimann, E.G. (1983) Electric field effects on bacteria and yeast cells, *Radiation Environ. Biophys.*, 22, 149–162.

Isaacs, N.S., Chilton, P., and Mackey, B. (1995) Studies on the inactivation by high pressure of micro-organisms, in *High Pressure Processing of Foods*, D.A. Leward, D.E. Johnson, R.G. Earnshaw, and A.P.M. Hasting, Eds., Nottingham University Press, Nottingham, U.K., pp. 65–79.

Jaenike, R. (1987) Cellular components under extremes of pressure and temperature: structure–function relationship of enzymes under pressure, in *Current Perspectives in High Pressure Biology*, H.W. Jannasch, R.E. Marquis, and A.M. Zimmerman, Eds., Academic Press, New York, pp. 257–272.

Jayaham, S., Castle, G.S.P., and Margaritis, A. (1992) Kinetics of sterilization of *Lactobacillus brevis* cells by the application of high voltage pulses, *Biotechnol. Bioeng.*, 40, 1412–1420.

Jeyamkondan, S., Jayas, D.S., and Holley, R.A. (1999) Pulsed electric field processing of foods: a review, *J. Food Protection*, 62(9), 1088–1096.

Jo, C., Lee, J.I., and Ahn, D.U. (1999) Lipid oxidation, color changes and volatiles production in irradiated pork sausage with different fat content and packaging during storage, *Meat Sci.*, 51(4), 355–361.

Josephson, E.S. and Peterson, M.S. (1982) *Preservation of Food by Ionizing Radiation*, CRC Press, Boca Raton, FL.

Joyce, A., Berger, A., and Marr, A.G. (1960) Sonic disruption of spores of *Bacillus cereus*, *J. General Microbiol.*, 22, 147–157.

Kader, A.A. (1986) Potential applications of ionizing radiation in postharvest handling of fresh fruits and vegetables, *Food Technol.*, 40(6), 117–121.

Keefer, R.M. and Ball M.D. (1992) Improving the final quality of microwavable foods, *Microwave World*, 13(2), 14–21.

Khalil, H. and Villota, R. (1986) A comparative study on the thermal inactivation of *B. stearothermophilus* spores in microwave and conventional heating, in *Food Engineering and Process Applications*, Vol 1, *Transport Phenomena*, M. LeMaguer and P. Jelen, Eds., Elsevier, New York, pp. 583–594.

Khalil, H. and Villota, R. (1989) The effect of microwave sublethal heating on the ribonucleic acids of *Staphylococcus aureus*, *J. Food Protection*, 52, 544–548.

Kilcast, D. (1995) Food irradiation: current problems and future potential, *Int. Biodeterioration Biodegradation*, 36(3,4), 279–296.

Lakritz, L., Fox, J.B., and Thayer, D.W. (1998) Thiamin, riboflavin and alpha-tocopherol content of exotic meats and loss due to gamma radiation, *J. Food Protection*, 61(12), 1681–1683.

Lechowich, R.V. (1993) Food safety implications of high hydrostatic pressure as a food processing method, *Food Technol.*, 6, 1970–1972.

Lee, J.W., Yook, H.S., Kim, S.A., Lee, K.H., and Byun, M.W. (1999) Effects of antioxidants and gamma irradiation on the shelf life of beef patties, *J. Food Protection*, 62(6), 619–624.

Leighton, T.G. (1998) The principles of cavitation, in *Ultrasound in Food Processing*, J.W. Povey and T.J. Mason, Eds., Blackie Academic, New York, pp. 151–182.

Lewis, M.J. (1987) Heat transfer mechanisms, in *Physical Properties of Foods and Food Processing Systems*, M.J. Lewis, Ed., Ellis Horwood, Chichester, England, pp. 246–291.

Lillard, H.S. (1994) Decontamination of poultry skin by sonication, *Food Technol.*, 48(12), 72–73.

Lorenz, K. and Decareau, R.V. (1976) Microwave heating of foods: changes in nutrient and chemical composition, *Crit. Rev. Food Sci. Nutr.*, 6, 339–370.

Luchsinger, S.E., Kropf, D.H., Zepeda, C.M., Chambers, E.V., Hollingsworth, M.E., Hunt, M.C., Marsden, J.L., Kastner, C.L., and Kuecker, W.G. (1996) Sensory analysis and consumer acceptance of irradiated boneless pork chops, *J. Food Sci.*, 61(6), 1261–1266.

Lund, D.B. (1975) Heat transfer in foods, in *Principles of Food Science. II. Physical Principles of Food Preservation*, M. Karel, O.R. Fennema, and D.B. Lund, Eds., Marcel Dekker, New York, pp. 1–7.

Mackey, B.M., Forestiere, K., Isaacs, N.S., Stenning, R., and Brooker, B. (1994) The effect of high hydrostatic pressure on *Salmonella thompson* and *Listeria monocytogenes* examined by electron microscopy, *Lett. Appl. Microbiol.*, 19, 429–432.

Mackey, B.M., Forestiere, K., and Isaacs, N. (1995) Factors affecting the resistance of *Listeria monocytogenes* to high hydrostatic pressure, *Food Biotechnol.*, 9, 1–11.

Marquez, V.O., Mittal, G.S., and Griffiths, M.W. (1997) Destruction and inhibition of bacterial spores by high voltage pulsed electric field, *J. Food Sci.*, 62(2), 399–401, 409.

Mason, T.J., Paniwnyk, L., and Lorimer, J.P. (1996) The uses of ultrasound in food technology, *Ultrasonics Sonochem.*, 3:S253-S260.

Masschalack, B., Van Houdlt, R., and Michelis, C.W. (2001) High pressure increases bactericidal activity and spectrum of lactoferrin, lactoferricin and nisin, *Int. J. Food Microbiol.*, 64, 325–332.

McClements, J. (1995) Advances in the application of ultrasound in food analysis and processing, *Trends Food Sci. Technol.*, 6, 293–299.

McDonald, C.J., Lloyd, S.W., Vitale, M.A., Peterson, K., and Innings, E. (2000) Effects of pulsed electric fields on microorganisms in orange juice using electric field strength of 30 and 50 kV/cm, *J. Food Sci.*, 65(6), 984–989.

Meredith, R. (1998) *Engineers' Handbook of Industrial Microwave Heating*, Short Run Press, Exeter, England.

Mertens, B. (1995) Hydrostatic pressure treatment of food: equipment and processing, in *New Methods of Food Preservation*, G.W. Gould, Ed., Blackie Academic, New York, pp. 135–175.

Mertens, B. and Deplace, G. (1993) Engineering aspects of high-pressure technology in the food industry, *Food Technol.*, 6, 1964–1969.

Monk, J.D., Beuchat, L.R., and Doyle, M.P. (1995) Irradiation inactivation of foodborne microorganisms, *J. Food Protection*, 58(2), 197–208.

Mudgett, R.E. (1989) Microwave food processing, *Food Technol.*, 1, 117–126.

Mudgett, R.E. and Schwartzberg, H.G. (1982) Microwave food processing: pasteurization and sterilization, a review, *Food Processing Eng.*, 78, 1–11.

Mussa, D.M., Ramaswamy, H.S., and Smith, J.P. (1999) High-pressure destruction kinetics of *Listeria monocytogenes* on pork, *J. Food Protection*, 62(1), 40–45.

Nakayama, A., Yano, Y., Kobayahi, S., Ishikawa, M., and Sakai, K. (1996) Comparison of pressure resistance of spores of six *Bacillus* strains with their heat resistances, *Appl. Environ. Microbiol.*, 62(10), 3897–3900.

Ohlsson, T. (2000) Minimal processing of foods with thermal methods, in *Innovations in Food Processing*, G.V. Barbosa-Cánovas and G.W. Gould, Eds., Technomic Publishing, Lancaster, PA, pp. 141–148.

Olson, S. (1995) Production equipment for commercial use, in *High Pressure Processing of Foods*, D.A. Leward, D.E. Johnson, R.G. Earnshaw, and A.P.M. Hasting, Eds., Nottingham University Press, Nottingham, U.K., pp. 167–180.

Ordoñez, J.A., Sanz, B., Hernandez, P.E., and López-Lorenzo P. (1984) A note on the effect of combined ultrasonic and heat treatments on the survival of thermoduric streptococci, *J. Appl. Bacteriol.*, 56, 175–177.

Ordoñez, J.A., Anguilera, M.A., Garcia, M.L., and Sanz, B. (1987) Effect of combined ultrasonic and heat treatment (thermoultrasonication) on the survival of a strain of *Staphylococcus aureus*, *J. Dairy Res.*, 54, 61–67.

Oxen, P. and Knorr, D. (1993) Baroprotective effects of high solute concentrations against inactivation of *Rhodotorula rubra*, *Lebensmittel-Wissenschaft Technologie*, 26, 220–223.

Pagán, R., Manas, P., Palop, A., and Sala, F.J. (1999) Resistance of heat-shocked cells of *Listeria monocytogenes* to mano-sonication and mano-thermo-sonication, *Lett. Appl. Microbiol.*, 28, 1–75.

Palou, E., López-Malo, A., Barbosa-Cánovas G.V., and Swanson B.G. (1998) High hydrostatic pressure come-up time and yeast viability, *J. Food Protection*, 61(12), 1657–1660.

Palou, E., López-Malo, A., Barbosa-Cánovas, G.V., Welti-Chanes, J., and Swanson, B.G. (1997a) Effect of water activity on high hydrostatic pressure inhibition of *Zigosaccharomyces bailii*, *Lett. Appl. Microbiol.*, 24, 417–420.

Palou, E., López-Malo, A., Barbosa-Cánovas, G.V., Welti-Chanes, J., and Swanson, B.G. (1997b) Kinetic analysis of *Zigosaccharomyces bailii* by high hydrostatic pressure, *Lebensmittel-Wissenschaft Technologie*, 30:703–708.

Palou, E., López-Malo, A., Barbosa-Cánovas, G.V., and Swanson, B.G. (1999) High pressure treatment in food preservation, in *Handbook of Food Preservation*, M.S. Rahman, Ed., Marcel Dekker, New York, pp. 533–576.

Pandya,Y., Jewett, F.F.J.R., and Hoover, D.G. (1995) Concurrent effects of high hydrostatic pressure, acidity and heat on the destruction and injury of yeasts, *J. Food Protection*, 58, 301–304.

Peleg, M. (1995) A model of microbial survival after exposure to pulsed electric fields, *J. Sci. Food Agric.*, 67, 93–99.

Pol, I.E., Mastwijk, H.C., Bartels, P.V., and Smid, E.J. (2000) Pulsed electric field treatments enhances the bactericidal action of nisin against *Bacillus cereus*, *Appl. Environ. Microbiol.*, 66, 428–430.

Pomeranz, Y. and Meloan, C.E. (1987) *Food Analysis: Theory and Practice*, 2nd ed., Van Nostrand-Reinhold, New York.

Pothakamury, U.R., Vega, H., Zang, Q., Barbosa-Cánovas, G.V., and Swanson, B.G. (1996) Effect of growth stage and processing temperature on the inactivation of *Escherichia coli* by pulsed electric fields, *J. Food Protection*, 59(11), 1167–1171.

Povey, J.W. (1989) Ultrasonics in food engineering. II. Applications, *J. Food Eng.*, 9, 1–20.

Povey, J.W. and McClements, J. (1988) Ultrasonics in food engineering. I. Introduction and experimental methods, *J. Food Eng.*, 8, 217–245.

Qin, B.L., Zhang, Q., Barbosa-Cánovas, G.V., Swanson, B.G., and Pedrow, P.D. (1995) Pulsed electric field treatment chamber design for liquid food pasteurization using a finite element method. *Trans. ASAE*, 38(2), 557–565.

Qin, B.L., Pothakamury, U.R., Barbosa-Cánovas, G.V., and Swanson, B.G. (1996) Nonthermal pasteurization of liquid foods using high intensity pulsed electric fields, *Crit. Rev. Food Sci. Nutr.*, 36(6), 603–627.

Qin, B.L., Barbosa-Cánovas, G.V., Swanson, B.G., Pedrow, P.D., and Olsen, R.G. (1998) Inactivating microorganisms using pulsed electric field continuous treatment systems, *IEEE Trans. Indus. Appl.*, 34(1), 43–50.

Radomyski, T., Murano, E.A., Olson, D.G., and Murano, P.S. (1994) Elimination of pathogens of significance in food by low-dose irradiation: a review, *J. Food Protection*, 57(1), 73–86.

Ramaswamy, H.S. and Pillet-Will, T. (1992) Temperature distribution in microwave-heated food models, *J. Food Quality*, 15, 435–448.

Ramesh, M.N. (1999) Food preservation by heat treatment, in *Handbook of Food Preservation*, M.S. Rahman, Ed., Marcel Dekker, New York, pp. 95–172.

Raso, J., Calderón, M.L., Góngora-Nieto, M.M., Barbosa-Cánovas, G.V., and Swanson, B.G. (1998a) Inactivation of *Zygosacharomyces bailii* in fruit juices by heat, high hydrostatic pressure and pulsed electric fields, *J. Food Sci.*, 63(1), 1042–1044.

Raso, J., Pagán, R., Condón, S., and Sala, F.J. (1998b) Influence of temperature and pressure on the lethality of ultrasound, *Appl. Environ. Microbiol.*, 64(2), 465–471.

Raso, J., Palop, A., Pagán, R., and Condón, S. (1998c) Inactivation of *Bacillus subtilis* spores by combining ultrasonic waves under pressure and mild heat treatment, *J. Appl. Microbiol.*, 85, 849–854.

Raso, J., Alvarez, I., Condón, S., and Trepat, F.J.S. (2000) Predicting inactivation of *Salmonella senftenberg* by pulsed electric fields, *Innovative Food Sci. Emerging Technol.*, 1, 21–29.

Reina, L.D., Jin, Z.T.J., Zang, Q.H., and Yousef, A.E. (1998) Inactivation of *Listeria monocytogenes* in milk by pulsed electric field, *J. Food Protection*, 61(9), 1203–1206.

Reyns, K.M.F.A., Soontjens, C.C.F., Cornelis, K., Weemaes, C.A., Henderickx, E., and Michiels, C.W. (2000) Kinetic analysis and modeling of combined high pressure temperature inactivation of the yeast *Zigosaccharomyces bailii*, *Int. J. Food Microbiol.*, 56, 199–210.

Riesz, P. and Kondo, T. (1992) Free radical formation induced by ultrasound and its biological implications, *Free Radical Biol. Med.*, 13:247–270.

Ritz, M., Freulet, M., Orange, N., and Federighi, M. (2000) Effects of high hydrostatic pressure on membrane proteins of *Salmonella typhimurium*, *Int. J. Food Microbiol.*, 55, 115–119.

Roberts, C.M. and Hoover, D.G. (1996) Sensitivity of *Bacillus coagulans* spores to combinations of high hydrostatic pressure, heat, acidity and nisin. *J. Appl. Bacteriol.*, 81, 363–368.

Ruello, J.H. (1987) Seafood and microwaves: some preliminary observations, *Food Technol. Australia*, 39, 527–530.

Ruhlman, K.T. (1999) Product Examination and Reformulation for Pulsed Electric Field Processing, Master's thesis, Ohio State University, Columbus.

Sala, F.J., Burgos, J., Condón, S., Lopez, P., and Raso, J. (1995) Effects of heat and ultrasound on microorganisms and enzymes, in *New Methods of Food Preservation*, G.W. Gould, Ed., Blackie Academic and Professional, Glasgow, pp. 176–204.

Sale, A.J.H. and Hamilton, W.A. (1967a) Effects of high electric fields on microorganisms. I. Killing bacteria and yeast, *Biochim. Biophys. Acta*, 148, 781–788.

Sale, A.J.H., Gould, G.W., and Hamilton, W.A. (1970) Inactivation of bacterial spores by hydrostatic pressure, *J. General Microbiol.*, 60, 323–334.

Scherba, G., Weigel, R.M., and O'Brien, J.R. (1991) Quantitative assessment of the germicidal efficiency of ultrasonic energy, *Appl. Environ. Microbiol.*, 57, 2079–2084.

Schiffmann, R.F. (1986) Food product development for microwave processing, *Food Technol.*, 40(6), 94–98.

Schiffmann, R.F. (1990) Problems in standardizing microwave oven performance, *Microwave World*, 11(3), 20–24.

Schiffmann, R.F. (2001) Microwave process for the food industry, in *Handbook of Microwave Technology for Food Applications*, A.K. Datta and R.C. Anatheswaran, Eds., Marcel Dekker, New York, pp. 229–337.

Sensoy, I., Zhang, Q.H., and Sastry, S.K. (1997) Inactivation kinetics of *Salmonella dublin* by pulsed electric field, *J. Food Process Eng.*, 20, 367–381.

Serrano, L.E., Murano, E.A., Shenoy, K., and Olson, D.G. (1997) D values of *Salmonella enteritidis* isolates and quality attributes of eggs shell and liquid whole eggs treated with irradiation, *Poultry Sci.*, 76(1), 202–205.

Stanford, M. (1990) Microwave oven characterization and implications for food safety in product development, *Microwave World*, 11(3), 7–9.

Stenstrom, L.A. (1974) Heating of Products in Electromagnetic Field, U.S. Patent Nos. 3,809,845 and 3,814,889.

Suslick, K.S. (1988) *Ultrasound: Its Chemical, Physical and Biological Effects*, VCH Publishers, New York.

Swami, S. and Mudgett, R.E. (1981) Effect of moisture and salt content on the dielectric behavior of liquid and semi-solid foods, *Proc. Microwave Power Symp.*, Ontario, Canada, 16, 48–50.

Tarte, R.R., Murano, E.A., and Olson, D.G. (1996) Survival and injury of *Listeria monocytogenes*, *Listeria innocua*, and *Listeria ivanovii* in ground pork following electron beam irradiation, *J. Food Protection*, 59(6), 596–600.

Thayer, D.W. and Boyd, G. (1991) Effect of ionizing radiation dose, temperature, and atmosphere on the survival of *Salmonella typhimurium* in sterile, mechanically deboned chicken meat, *Poultry Sci.*, 70, 381–388.

Thayer, D.W., Boyd, G., Kim, A., Fox, J.B., and Farrel, H.M. (1998) Fate of gamma irradiated *Listeria monocytogenes* during refrigerated storage on raw or cooked turkey breast meat, *J. Food Protection*, 61(8), 979–987.

Tsong, T.Y. (1990) Review on electroporation of cell membranes and some related phenomena, *Bioelectrochem. Bioenergetics*, 24, 271–295.

Tsong, T.Y. (1991) Electroporation of cell membranes, *Biophys. J.*, (60), 291–306.

Urbain, W.M. (1986) *Food Irradiation*, Academic Press, New York.

Van Zante, H.J. (1973) *The Microwave Oven*, Houghton Mifflin, Boston.

Vega-Mercado, H., Martín-Belloso, O., Chang, F.J., Barbosa-Cánovas, G.V., and Swanson, B.G. (1996a) Inactivation of *Escherichia coli* and *Bacillus subtillis* suspended in pea soup using pulsed electric fields, *J. Food Processing Preservation*, 20(6), 501–510.

Vega-Mercado, H., Pothakamury, U.R., Chang, F.J., Barbosa-Cánovas, G.V., and Swanson, B.G. (1996b) Inactivation of *Escherichia coli* by combining pH, ionic strength and pulsed electric fields hurdles, *Food Res. Int.*, 29(2), 117–121.

Vega-Mercado, H., Góngora-Nieto, M.M., Barbosa-Cánovas, G.V., and Swanson, B.G. (1999) Nonthermal preservation of liquid foods using pulsed electric fields, in *Handbook of Food Preservation*, M.S. Rahman, Ed., Marcel Dekker, New York, pp. 487–520.

Weaver, J.C. and Powel, K.T. (1989) Theory of electorporation, in *Elecroporation and Elecrtrofusion in Cell Biology*, E. Neumann, A.E. Sowers, and C.A. Jordan, Eds., Plenum Press, New York, pp. 111–126.

Wilkinson, V.M. and Gould, G.W. (1996) *Food Irradiation: A Reference Guide*, Reed Educational and Professional Publishing, Oxford, England.

World Health Organization (WHO) (1988) *Food Irradiation: A Technique for Preserving and Improving the Safety of Food*, World Health Organization, Geneva, Switzerland.

Wuytack, E.Y. and Michiels, C.W. (2001) A study on the effects of high pressure and heat on *Bacillus subtilis* spores at low pH, *Int. J. Food Microbiol.*, 64, 333–341.

Yeom, H.W., Streaker, C.B., Zhang, Q.H., and Min, D.B. (2000) Effects of pulsed electric fields on the activities of microorganisms and pectin methyl esterase in orange juice, *J. Food Sci.*, 65(8), 1359–1363.

Yin, Y., Zhang, Q.H., and Sastry, S.K. (1997) High Voltage Pulsed Electric Field Treatment Chambers for the Preservation of Liquid Food Products, U.S. Patent No. 5,690,978.

Yonemoto, Y., Yamashita, T., Muraji, M., Tatbe, W., Ooshima, H., Kato, J., Kimura, A., and Murata, K. (1993) Resistance of yeast and bacterial spores to high voltage electric pulses, *J. Fermentation Bioeng.*, 75, 99–102.

Zhang, H. and Datta, A.K. (2001) Electromagnetics of microwave heating: magnitude and uniformity of energy absorption in an oven, in *Handbook of Microwave Technology for Food Applications*, A.K. Datta and R.C. Anatheswaran, Eds., Marcel Dekker, New York, pp. 33–63.

Zhang, Q., Barbosa-Cánovas, G.V., and Swanson, B.G. (1995) Engineering aspects of pulsed electric field pasteurization, *J. Food Eng.*, 25, 261–281.

Zimmermann, U. (1986) Electrical breakdown, electropermeabilization and electrofusion, *Reviews of physiology, Biochem. Pharmacol.*, 105, 176–250.

ZoBell, C.E. (1970) Pressure effects on morphology and life processes of bacteria, in *High Pressure Effects on Cellular Processes*, A.M. Zimmermann., Ed., Academic Press, New York, pp. 85–130.

ZoBell, C.E. and Cobet, A.B. (1964) Filament formation by *Escherichia coli* at increased hydrostatic pressures, *J. Bacteriol.*, 87(3), 710–719.

15

Emerging Processing and Preservation Technologies for Milk and Dairy Products

Valente B. Alvarez and Taehyun Ji

CONTENTS

Introduction

Pasteurization and, more recently, ultra-high-temperature (UHT) processing are the traditional methods used to eliminate pathogens with minimal detriment to the physical and chemical properties of milk. The main purpose of thermally processing raw milk is to make it safe for human consumption. The demand for better quality dairy products with longer shelf lives has prompted the industry to search for new means to enhance these two important properties. Several recent technologies have shown potential commercial applications for improving quality and extending the shelf life of milk and other dairy products. High-pressure processing (HPP) can inactivate vegetative bacteria in milk. Pulsed electric field (PEF) technology shows some potential in the pasteurization of fluid milk. Irradiation has been an effective method for destroying pathogens but produces undesirable changes in sensory attributes and quality of dairy products. Modified atmosphere packaging (MAP) using CO_2 has an antimicrobial effect and inhibits the growth of some psychrotrophic bacteria in dairy products such as fluid milk, cheese, yogurt, ice cream mixes, and sour cream. Membrane microfiltration has been applied successfully to reduce the microbial load of raw milk; the technique requires less thermal treatment and therefore improves milk quality. The antimicrobial effect of natural components of milk, such as lactoperoxidase, lactoferrin, and xanthine oxidase, can also preserve dairy foods, thus increasing shelf life. The basic concepts of these technologies, their potential uses and limitations, and their possible effects on the quality and shelf life of dairy products are discussed in this chapter.

Processing Technologies

Heat Treatment

Milk and dairy products are heat treated to inactivate pathogenic microorganisms and some undesirable enzymes. This practice improves product safety and prolongs shelf life. The most common heat process is pasteurization, which can be applied at different temperature and time conditions ranging from 63°C for 30 min to 100°C for 0.01 sec (Anon., 2001). Pasteurization produces both irreversible and reversible changes in milk components and causes both desirable and undesirable effects. Heat treatment of milk increases the amount of colloidal phosphate, decreases Ca^{++}, hydrolyzes phosphoric esters, isomerizes lactose, decreases pH, increases titratable acidity, denatures serum proteins, inactivates enzymes, and forms free sulfhydryl groups that drop redox potential and degrades some vitamins (Walstra et al., 1999). Changes of milk components by heat treatment largely depend on the combination of heating intensity and time duration.

High-temperature/short-time (HTST) pasteurization is the most widely used treatment for preserving the quality and extending the shelf life of dairy products. HTST pasteurization destroys all pathogenic microorganisms but not bacterial spores. HTST processing also deteriorates the physical and chemical properties of the product. Plate heat exchanger systems are most commonly used for milk pasteurization in modern industry (Staal, 1986).

During ultra-high-temperature (UHT) treatment, the product is held for a few seconds at a higher temperature (135 to 150°C) than with HTST pasteurization. The two UHT methods are direct heating by steam injection or infusion and indirect heat transfer by heat exchange. The UHT process destroys all viable microorganisms, including bacterial spores, and extends product shelf life. Additionally, UHT causes considerable changes in milk properties such as excessive heated flavor, phosphotase reduction, activation of spores or spore-forming bacteria, gelation, reactivation of lipase, increased viscosity, and decreased nutritive value (Burton, 1998).

High-Pressure Processing

High-pressure processing (HPP) is a novel nonthermal food processing technology that uses extremely high pressure to kill vegetative microorganisms (Farr, 1990; Hjelmqwist, 1998). Inactivation of vegetative microbes through HPP may result from denaturation of deoxyribonucleic acid (DNA) replication and transcription, solidification of lipids, breakage of biomembranes, or leakage of cell contents. Applications of HPP to food extend the shelf life without thermal denaturation; preserve nutritional value with retention of natural flavor, color, texture and taste; and increase food safety (Cheftel, 1995; Hjelmqwist, 1998).

High-pressure processing can operate within a wide range of conditions. High pressure machines can process volumes from 5 mL to 200 L and operating pressures from 30,000 to 130,000 psi (200 to 900 MPa) for typically 30 sec to a few minutes. The typical temperatures are 20 to 40°C, but in some cases the range can go from –20 to 80°C (Farr, 1990; Rovere, 1995). Under HPP, food retains its original shape with minor cellular damage because pressure is applied uniformly in all directions. The effect of HPP on food characteristics depends on the applied pressure, temperature, and duration.

Functional properties of food proteins such as hydration, gelation, and emulsification characteristics are altered by disruption of protein–water interactions and protein–protein interactions. HPP breaks only hydrogen protein bonds, disrupts hydrophobic and electrostatic interactions, makes greater ionization of molecules, and inactivates microorganisms. Covalent bonds are not affected (Messens et al., 1999).

High-pressure processing has been used on dairy products. The process destroys microorganisms in milk and reduces coagulation and ripening time in cheese. HPP technology increases viscosity and apparent elasticity of gel and thus decreases syneresis in yogurt. The effect of HPP on milk depends greatly on the properties and composition of the milk. UHT milk pressurized at 400 MPa and 50°C for 15 min resulted in a reduction of approximately 5 log (CFU/g) for *Escherichia coli*, and 6 log for *Staphylococcus aureus* at 500 MPa (Patterson and Kilpatrick, 1998). Besides destroying microorganisms, HPP may alter the physical characteristics of milk. Pasteurized skim milk was HPP processed up to 700 MPa for 3 min at 20°C and kept constant for 22 min. The treatment increased the dynamic viscosity and surface hydrophobicity of milk. Milk turbidity and lightness decreased (Desobry-Banon et al., 1994). The particle size started to decrease with increased pressure from 230 to 430 MPa and reached minimal size with pressure from 430 MPa and above. Because skim milk is a colloidal emulsion of casein particles, the functional properties of skim milk are largely dependent on the size of casein particles.

High-pressure processing technology was applied to investigate the accelerated ripening of gouda cheese at 14°C, 50 to 400 MPa, for 20 to 100 min (Messens et al., 1999) and cheddar cheese at 25°C, 50 MPa, for 3 days (O'Reilly et al., 2000). The results of those studies showed that HPP treatment using the conditions mentioned above did not accelerate ripening of the cheeses. However, Kolakowski et al. (1998) reported that camembert cheese treated by HPP at 0 to 500 MPa obtained the highest degree of proteolysis at the pressure of 50 MPa for 4 h. The authors also mentioned that the number of microorganisms in gouda and camembert cheeses decreased significantly by HPP at pressures above 400 MPa. Another study concluded that increasing temperature accelerated cheese ripening as well as the risk of microbial spoilage (Fox, 1989). Accelerated cheese ripening was attributed to a combination of HPP and increased ripening temperature (O'Reilly et al., 2000).

Processing of yogurt by HPP technology has also been investigated. The application of HPP at approximately 414 MPa at room temperature can extend shelf life or even produce shelf-stable yogurt (Farkas, 1996). The continuous development of acidity after packaging, which can lead to syneresis, can be reduced by HPP. Tanaka and Hatanaka (1992) subjected yogurt to pressures of 200 to 300 MPa at 10 to 20°C for 10 min. The treatment did not modify the yogurt texture and did not reduce the numbers of viable lactic acid bacteria. Pressure above 300 MPa prevented overacidification; however, the numbers of viable lactic acid bacteria were reduced.

Pulsed Electric Field

Pulsed electric field (PEF) is a nonthermal processing technology that may have the potential to replace traditional thermal pasteurization (Zhang et al., 1995a; Qiu et al., 1998). The use of PEF technology in foods reduces pathogen levels while increasing shelf life; retaining original flavor, color, and nutritional properties; and improving protein functionalities (Dunn, 1996). PEF processing involves the application of a short burst of high-voltage energy to a fluid as it flows between two inert electrodes. The complete PEF system consists of a fluid-handling section, high-voltage pulse generator, and multiple-stage co-field PEF treatment chamber (Yin et al., 1997). The effect of PEF treatment on microorganisms is known as *electroporation*. PEF causes swelling of the cell membrane, resulting in reversible or irreversible ruptures. The high-voltage pulsed electricity discharges that are induced into microbes in the food product develop pores that allow permeation of small molecules in cellular membranes, thus impairing cellular functions (Benz and Zimmermann, 1980; Tsong, 1990; Knorr et al., 1994). Several factors influence PEF process efficiency. The effects of PEF treatment on inactivation of microorganisms largely depend on process conditions such as electric field intensity, pulse width, treatment time and temperature, pulse wave shapes; microbial entities such as type, concentration, and growth stage of microorganisms; and treatment media, such as pH, antimicrobials, ionic compounds conductivity, and medium ionic strength (Hülsheger et al., 1983; Zhang et al., 1995b; Dunn, 1996; Vega-Mercado et al., 1997; Barbosa-Cánovas et al., 2001).

Studies about the effects of PEF on dairy products have been conducted on skim milk, whole milk, and yogurt. Fluid milks containing *Escherichia coli*, *Salmonella dublin*, *Listeria innocua*, and *Listeria monocytogenes* were processed at PEF conditions of 28.6 to 50 kV/cm, 1.5 to 100 μsec, 23 to 100 pulses, and temperatures from 10 to 63°C. Although the PEF conditions were different, the studies reported 2.0 to 4.0 log (CFU/mL) reductions (Dunn and Pearlman, 1987). A study investigated the application of PEF on raw milk by inoculating ultra-high-temperature (UHT) skim milk with *Pseudomonas fluorescens*, *Lactococcus lactis*, and *Bacillus cereus*. The authors reported that the PEF treatment caused 0.3 to 3.0 log reductions of *P. fluorescens*, *L. lactis*, and *B. cereus* in UHT milk and total microorganisms in raw milk. In all cases, PEF had a partial effect on the inactivation and destruction of microorganisms in milk, and the survivability of the cells differed for various organisms (Michalac et al., 2002). Other researchers reported 2.0 to 4.0 log reductions for *Escherichia coli* in skim milk (Qin et al., 1995), *Listeria innocua* in skim milk (Calderon-Miranda et al., 1999), *Listeria monocytogenes (scott A)* in pasteurized whole milk (3.5% milkfat) (Reina et al., 1998), and *Staphylococcus aureus* in simulated milk ultrafiltrate (Pothakamury et al., 1995).

The log reduction of *Saccharomyces cerevisiae*, *Lactobacillus bulgaricus*, and *Streptococcus thermophilus* in yogurt treated by PEF at 23 to 38 kV/cm, 100 μsec, 20 pulses, and 63°C was investigated; the PEF treatment produced a 2.0 log reduction (Dunn and Pearlman, 1987). These results suggest that the use of PEF has some limitations due to the difficulty of inactivating endospores and low conductivity in foods caused by solids. The microbial reduction achieved in the studies is not substantial enough to consider PEF treatments as a substitute for current pasteurization methods of dairy products. However, the efficiency of PEF processing in microbial reduction in milk may be improved by employing longer treatment times and higher electric field strengths. Combination with other nonthermal techniques may be necessary for more efficient PEF treatment for the processing of fluid dairy products.

Irradiation

Food irradiation is the exposure of food to a source of ionizing radiation energy. In general terms, radiation refers to exposure to or illumination by rays or waves of all types. Microwaves, infrared or ultraviolet light, and x-rays are common sources of energy. Gamma (γ) radiation is the most common type of energy used in food irradiation (Satin, 1996). The application of irradiation to foods reduces the pathogenic potential of microorganisms present because nucleic acids and macromolecules of food microorganisms are very sensitive to ionization. Irradiation disrupts the genetic material of living cell, destroys foodborne pathogens, and reduces the number of spoilage microorganisms. Ionization in food caused by irradiation modifies the structure and composition of large and complicated molecules (Thakur and Singh, 1995). Food irradiation also affects food components such as water, carbohydrate, lipid, protein, vitamins, minerals, and other trace elements through reactive ions or free radicals, which combine with other ions to achieve a more stable state (Satin, 1996).

Irradiation of milk and dairy products has been investigated at a wide range of radiation energy intensities (0.07 to 2.5 kGy). Irradiation of milk started to be used during the 1930s in order to increase the vitamin D content in milk. The difference with today's irradiation methods was the use of ultraviolet (UV) light as the source of ionization rather that γ-rays. Fluid milk and evaporated milk were irradiated with UV light (Sadoun et al., 1991; Satin, 1996). In most studies, dairy products have been exposed to high radiation doses suitable for sterilization, causing the development of off flavors, aftertaste, and loss of vitamins A, B_1, and B_2. Irradiation treatment of milk and dairy products has created flavor problems due to sulfur compounds produced from milk protein fraction and oxidative rancidity from lipid fraction. The off-flavor production level in milk and cheese depended on their composition and the conditions of radiation and storage (Wilkinson and Gould, 1996). Irradiation with a dose of 0.25 kGy at room temperature extended the shelf life of pasteurized milk stored at 4°C; however, the milk lost a certain amount of vitamin A, B_1, and B_2 (Sadoun et al., 1991).

Irradiation of other dairy products includes cheese, frozen desserts, and caseinate films. Irradiation at an average dose of 2.5 kGy was an effective method for destroying pathogenic bacteria *Listeria monocytogenes* and *Salmonella* in camembert cheese without affecting enzyme activity and flavor (Boisseau, 1994). Gamma irradiation at a dose of 40 kGy at –78°C was sufficient to sterilize ice cream and frozen yogurt, but not for mozzarella or cheddar cheeses (Hashisaka et al., 1990a). The irradiation caused a decrease in overall acceptability of the product due to off flavor and aftertaste, but it had little effect on product color or texture. Irradiation combined with modified atmosphere packaging or antioxidant was effective at preserving sensory properties of peppermint ice cream packed with helium and strawberry yogurt bars treated with ascorbyl palmitate (Hashisaka et al., 1990b). Water solubility and microbial degradation of caseinate films were reduced by gamma irradiation at 64 kGy dosage. The improved changes were associated with the higher number of cross-links on the film caused by irradiation. (Mezgheni et al., 2000).

Modified Atmosphere Packaging

Modified atmosphere packaging (MAP) is a technology used to extend the shelf life of fresh and processed food products. The MAP process consists of flushing the food packaging with antimicrobial gases just before sealing (Yam and Lee, 1995). Modified atmosphere packaging has been used successfully to extend the shelf life of solid dairy foods such as shredded cheese. In MAP, gas is added directly into liquids or semiliquid foods such as milk, yogurt, cottage cheese, sour cream, and ice cream (Hotchkiss and Chen, 1996). The gas concentration used in MAP can vary from 40 to 1100 ppm, depending on the type of product and packaging characteristics.

Antibacterial agent carbon dioxide and inert nitrogen are the gases used in these technologies. Carbon dioxide is widely used to inhibit the growth of some psychrotrophic bacteria that deteriorate refrigerated food and to restrict the growth of typical aerobic Gram-negative spoilage bacteria. The gas enters microbial cells and lowers the pH, thus retarding the growth of microbes (Hintalian and Hotchkiss, 1987; Buick and Damoglou, 1989). The effectiveness of MAP in foods is improved with exclusion of the oxygen necessary for the growth of aerobic spoilage bacteria. Oxygen also causes oxidative rancidity and color changes (Church, 1994).

Several researchers have investigated MAP in cottage cheese. Flushing CO_2 gas into the headspace of plastic containers filled with cottage cheese inhibited the growth of yeasts, molds, and psychrotrophic bacteria for 112 days; however, flavor and texture were retained for only 45 days (Kosikowski and Brown, 1973). The application of CO_2 gas above 750 mL/L in cottage cheese kept the quality and maintained the color and flavor for 28 days at 4°C (Manier et al., 1994). MAP using 500 mL/L CO_2 gas in cottage cheese inhibited the growth of yeasts and fungi (Fedio et al, 1994). *Pseudomonas* spp. and *Listeria monocytogenes* were inhibited in cottage cheese with 400 mL/L CO_2 (Moir et al., 1993). CO_2 extended the shelf life of cottage cheese cream but caused a slight flavor alteration due to the CO_2 dissolved in the cream (Chen and Hotchkiss, 1993).

Modified atmosphere packaging has also been applied to other cheese products to maintain quality, inhibit microorganism growth, and extend shelf life. MAP with plain CO_2 was applied to whey cheeses stored at 4°C. Chemical composition was retained and lipolysis was inhibited completely (Mannheim and Soffer, 1996). MAP with several ratio combinations was applied to cameros cheese to investigate its effect on shelf life and microbial quality. Gas ratios of 50:50 and 40:60 (CO_2:N_2) were the most effective in reducing proteolysis and lipolysis, which helped in retaining good sensory properties. These gases were also effective in inhibiting the growth of mesophiles, psychrotrophs, enterobacteriaceae, and coliforms, thus extending shelf life (Gonzalez-Fandos et al., 2000).

Effective MAP technology is associated with proper storage temperature, proper amount of gas dissolved, and high-barrier packaging. A disadvantage of MAP is that the gas applied dissipates quickly during modern processing.

Membrane Filtration

Membrane filtration occurs when fluids and solvents are selectively transported and passed through a barrier by applying a driving force across the barrier. The membrane materials that act as the barrier can be organic polymers, metals, ceramics, layers of chemicals, liquids, or gases (Mohr et al., 1989). Membrane filtration has been used for the separation, concentration, demineralization, fractionation, or clarification of liquids. Membrane filtrations are classified by membrane types, pore sizes, and process conditions into microfiltration (MF), nanofiltration (NF), ultrafiltration (UF), and reverse osmosis (RO). These membrane filtrations have been applied in the dairy industry for the removal of bacteria in milk, standardization of milk, concentration of milk protein, fractionation of caseins, cheesemaking, and whey processing for many years (Rosenberg, 1995). Membrane technologies have also been used in the dairy industry for new product development, improving product quality, and enhancing process profitability. The effectiveness of a filtration system depends on the types and characteristics of membrane; therefore, the selection of membrane is important with respect to operating costs, energy requirements, reducing operating time, and increasing product quality.

Microfiltration is applied for the removal of bacteria from milk and cheese milk. It is also used in the preparation of casein-enriched cheese milk and to modify the α- and β-casein ratio of milk (Mistry and Maubois, 1993). MF of skim milk reduced bacteria by about 99.5%, extended shelf life, and retained the milk properties intact (Papachristou and Lafazanis, 1997). Rodriguez et al. (1999) studied the effect of UF and MF technologies on the texture of semihard, low-fat cheese. The use of MF milk improved cheese texture and produced lower retention of whey protein as compared with UF membrane; however, the UF process produced better cheese yields than the MF process.

Nanofiltration membranes have high permeability of salts such as sodium chlorides and potassium chlorides and very low permeability for organic compounds such as lactose, protein, and urea. These characteristics make NF suitable for dairy application. NF concentrates and demineralizes whey at the same time, which traditionally was processed using evaporation or reverse osmosis followed by electrodialysis. With these properties, the use of NF reduces the cost of energy consumption and wastewater disposal considerably (Van der Horst et al., 1995). Demineralization of whey has been accomplished by NF (Van der Horst et al., 1995).

Ultrafiltration and RO are widely used in cheese manufacturing and whey treatment. UF and RO have been used in whey concentration and the development of whey protein concentrates (Rodriguez et al., 1999), and the UF process has been applied to making fresh cheeses (Mahaut and Korolczuk, 1992; Schkoda and Kessler, 1996) and semihard cheeses (De Boer and Nooy, 1980; De Koning et al., 1981; Delbeke, 1987). The use of UF processing in making low-fat cheddar cheese did not significantly improve flavor, body, or texture characteristics (McGregor and White, 1990). Aroma and flavor degradation were reported in cheeses made from UF concentrated milk (Bech, 1993). The advantages of UF milk in cheesemaking are cheese yield increases of about 16 to 20%, reduction of the quantity of coagulant used by 80%, and decrease of biological oxygen demand (BOD) of the whey produced (El-Gazzar and Marth, 1991). RO concentrated 35,000 kg of whey fourfold, reducing the BOD of permeates (Bissett and Schmidtke, 1984).

Natural Components in Milk

Lactoferrin, lactoperoxidase, and xanthine oxidase are naturally present in milk and have some specific properties that have been found to be beneficial for the shelf life and quality of dairy products. These compounds are non-immune antimicrobial proteins that have been investigated by several researchers (Bellamy et al., 1992; Grappin and Beuvier, 1997; Pakkanen and Aalto, 1997; Schanbacher et al., 1998).

Lactoferrin

Lactoferrin is an iron-binding glycoprotein in milk that inhibits the growth of pathogenic bacteria by its high affinity with iron. The antimicrobial mechanism of lactoferrin is more complex than simple binding of iron (Bellamy et al., 1992). Lactoferrin also disrupts bacterial cell membranes by binding bacterial lipopolysaccharide (Nuijens et al., 1996) and modifies membrane permeability by binding porin molecules in the outer membrane (Erdei et al., 1994; Naidu and Arnold, 1994). This interaction with the cell membranes facilitates the bactericidal properties of lactoferrin (Bellamy et al., 1993). When the antimicrobial effects of lactoferrin on *Escherichia coli* (Law and Reiter, 1977), *Salmonella typhimurium*, *Shigella dysenteriae* (Batish et al., 1988), and *Listera monocytogenes* (Payne et al., 1990) were investigated, lactoferrin was shown to inhibit the growth of these microbes.

Lactoperoxidase

Lactoperoxidase is an antibacterial enzyme naturally present in colostrum and milk. This enzyme catalyzes the oxidation of thiocyanate (SCN^-) in the presence of hydrogen peroxide (H_2O_2), producing hypothiocyanite ($OSCN^-$), which is a toxic intermediary oxidation product. This product inhibits bacterial metabolism by oxidation of essential sulfhydryl groups in proteins. This reaction produces a severe change in the cytoplasmic membrane of spoilage bacteria (Reiter, 1978; Pruitt and Reiter, 1985). The use of lactoperoxidase is not approved in the United States because its activation requires a thiocyanide compound, which is unsafe for children (Anon., 1998). The activation of the lactoperoxidase system in refrigerated raw milk retarded the growth of psychrotrophic bacteria for several days (Bjorck, 1978; Haddadin et al., 1996). The lactoperoxidase system has been used to extend the shelf life of raw, pasteurized (Martinez et al., 1988), and UHT-treated milk (Denis and Ramet, 1989) and to preserve cream (Toledo Lopez and Garcia Galindo, 1987), cottage cheese (Earnshaw et al., 1989), mozzarella cheese, and yogurt (Kumar and Mathur, 1989).

Xanthine Oxidase

Xanthine oxidase is a complex metallo-flavo enzyme present in the fat globule membrane (Mondal et al., 2000). The enzyme catalyzes the reaction to produce bactericide superoxide radicals and hydrogen peroxide in the presence of oxygen. Hydrogen peroxide can also be used to activate the lactoperoxidase system (Reiter, 1978). Xanthine oxidase was studied for its activity in dairy products such as raw, evaporated, and powdered milks; ice cream; yogurt; cheese (Cerbulis and Farrell, 1977; Zikakis and Wooters, 1980); and butter and creams (Stannard, 1975; Cerbulis and Farrell, 1977). Nielsen (1999) reported that xanthine oxidase, an oxido-reductase enzyme, catalyzes the oxidation of purine bases and reduces nitrate to nitrite. Nitrate inhibits the germination of spore butyric acid bacteria in cheese.

Conclusions

The development and use of emerging technologies and new processing procedures have been shown to be promising. Most of them provide specific advantages to improve the shelf life and quality of dairy products; however, they also present some disadvantages and limitations. Several of these new technologies are still in experimental stages and may not fall under current Food and Drug Administration (FDA) regulations associated with low-acid canned foods which relate only to processing procedures involving a thermal treatment to render the commercial sterility of the product. Although some of these nonthermal processes may have capabilities to produce shelf-stable products, public health concerns are significant. The FDA specifies that, "Any nonthermal process used to create a commercially sterile food product must result in a food product that has not been prepared, packed, or held under unsanitary conditions whereby it may have become contaminated with filth or whereby it may have been rendered injurious to health." Consequently, the future applications and utilization of nonthermal technologies and new processing procedures will depend greatly on more research results, economics, and legal approval.

References

Anon. (2001) Extending shelf life in dairy foods, in *Innovations in Dairy: Dairy Industry Technology Review, DMI Bull.*, Dairy Management, Inc., Rosemont, IL, pp. 1–6.

Anon. (2001) *Grade A Pasteurized Milk Ordinance*, Publication No. 229, Public Health Service/Food and Drug Administration, Washington, D.C., p. 4.

Barbosa-Cánovas, G.V., Pierson, M.D., Zhang, Q.H., and Schaffner, D.W. (2001) Pulsed electric fields, *J. Food Sci.* (special suppl.), 65–79.

Batish, V.K., Harish, C., Zumdegni, K.C., Bhatia, K.L., and Singh, R.S. (1988) Antibacterial activity of lactoferrin against some common food-borne pathogenic organisms, *Aust. J. Dairy Technol.*, 43, 16–18.

Bech, A.M. (1993) Characterizing ripening in UF-cheese, *Int. Dairy J.*, 3, 329–342.

Bellamy, W.R., Takase, M., Yamauchi, K., Wakabayashi, H., Kawase, K., and Tomita, M. (1992) Identification of the bactericidal domain of lactoferrin, *Biochim. Biophys. Acta*, 121, 130–136.

Bellamy, W.R., Wakabayashi, H., Takase, M., Kawase, K., Shimamura, S., and Tomita, M. (1993) Roll of cell-binding in the antibacterial mechanism of lactoferrin B, *J. Appl. Bacteriol.*, 75, 478–484.

Benz, Z. and Zimmermann, U. (1980) Pulse-length dependence of the electrical breakdown in lipid bilayer membranes, *Biochim. Biophys. Acta*, 597, 637–642.

Bissett, D.W. and Schmidtke, N.W. (1984) Concentrating whey by hyperfiltration at a small Canadian cheese plant, *IDF Bull.*, 184, 96.

Bjorck, L. (1978) Antimicrobial effect of the lactoperoxidase system on psychrotrophic bacteria in milk, *J. Dairy Res.*, 45, 109–118.

Boisseau, P. (1994) Irradiation and the food industry in France, *Food Technol.*, 48(5), 138–140.

Buick, R.K. and Damoglou, A.P. (1989) Effect of modified atmosphere packaging on the microbial development and visible shelf life of mayonnaise-based vegetable salad, *J. Sci. Food Agr.*, 46, 339–347.

Burton, H. (1998) Properties of UHT-processed milk, in *Ultra-High Temperature Processing of Milk and Milk Products*, Burton, H., Ed., Elsevier, New York, pp. 254–291.

Calderon-Miranda, M.L., Barbosa-Canovas, G.V., and Swanson, B.G. (1999) Inactivation of *Listeria innocua* in skim milk by pulsed electric fields and nisin, *Int. J. Food Microbiol.*, 51(1), 19–30.

Cerbulis, J. and Farrell, H.M. (1977) Xanthine oxidase activity in dairy products, *J. Dairy Sci.*, 60(2), 170–176.

Cheftel, J.C. (1995) High-pressure microbial inactivation and food preservation, *Food Sci. Technol. Int.*, 1, 75–80.

Chen, J.H. and Hotchkiss, J.H. (1993) Growth of *Listeria monocytogenes* and *Clostridium sporogenes* in cottage cheese in modified atmosphere packaging, *J. Dairy Sci.*, 76, 972–977.

Church, N. (1994) Developments in modified-atmosphere packaging and related technologies, *Trends Food Sci. Technol.*, 5, 345–352.

De Boer, R. and Nooy, P.F.C. (1980) Low-fat semi-hard cheese from ultrafiltered milk, *North Eur. Dairy J.*, 46, 52–61.

De Koning, P.J., De Boer, R., Both, P., and Nooy, P.F.C. (1981) Comparison of proteolysis in a low-fat semi-hard type of cheese manufactured by standard and by ultrafiltration techniques, *Neth. Milk Dairy J.*, 35, 35–46.

Delbeke, R. (1987) Experiments on making Saint-Paulin cheese by full concentration of milk with ultrafiltration, *Milchwissenschaft*, 42, 222–225.

Denis, F. and Ramet, J.R. (1989) Antimicrobial activity of the lactoperoxidase system of *Listera monocytogenes* in trypticase soy broth, UHT milk and French soft cheese, *J. Food Protection*, 52, 706–711.

Desobry-Banon, S., Richard, F., and Hardy, J. (1994) Study of acid and rennet coagulation of high pressurized milk, *J. Dairy Sci.*, 77(11), 3267–3274.

Dunn, J. (1996) Pulsed light and pulsed electric field for foods and eggs, *Poultry Sci.*, 75, 1133–1136.

Dunn, J.E. and Pearlman, J.S. (1987) Methods and Apparatus for Extending the Shelf-Life of Fluid Food Products, U.S. Patent No. 4,695,472.

Earnshaw, R.G., Banks, J.G., Defrise, D., and Francotte, C. (1989) The preservation of cottage cheese by an activated lactoperoxidase system, *Food Microbiol.*, 6, 285–288.

El-Gazzar, F.E. and Marth, E.H. (1991) Ultrafiltration and reverse osmosis in dairy technology: a review, *J. Food Protection*, 54(10), 801–809.

Erdei, J., Forsgren, A., and Naidu, A.S. (1994) Lactoferrin binds to porins O_{mpf} and O_{mpc} in *Escherichia coli*, *Infection Immunity*, 62, 1236–1240.

Farkas, D.F. (1996) Preservation of foods by ultra-high hydrostatic pressure, *J. Dairy Sci.*, 79(suppl. 1), 102.

Farr, D. (1990) High pressure technology in the food industry, *Trends Food Sci.Technol.*, 1(1), 14–16.

Fedio, W.M., Macleod, A., and Ozimek, L.(1994) The effect of modified atmosphere packaging on growth of microorganisms in cottage cheese, *Milchwissenschaft*, 49, 622–629.

Fox, P.F. (1989) Acceleration of cheese ripening, *Food Biotechnol.*, 2(2), 133–185.

Gonzalez-Fandos, E., Sanz, S., and Olarte, C. (2000) Microbiological, physicochemical and sensory characteristics of cameros cheese packaged under modified atmosphere, *Food Microbiol.*, 17, 407–414.

Grappin, R. and Beuvier, E. (1997) Possible implications of milk pasteurization on the manufacture and sensory quality of ripened cheese: a review, *Int. Dairy J.*, 7, 751–761.

Haddadin, M.S., Ibrahim, S.A., and Robinson, R.K. (1996) Preservation of raw milk by activation of the natural lactoperoxidase systems, *Food Control*, 7(3), 149–152.

Hashisaka, A.E., Matches, J.R., Batters, Y., Hungate, F.P., and Dong F.M. (1990a) Effects of gamma irradiation at $-78°C$ on microbial populations in dairy products, *J. Food Sci.*, 55(5), 1284–1289.

Hashisaka, A.E., Einstein, M.A., Rasco, B.A., Hungate, F.P., and Dong, F.M. (1990b) Sensory analysis of dairy products irradiated with cobalt-60 at $-78°C$, *J. Food Sci.*, 55(2), 404–408, 412.

Hintalian, C.B. and Hotchkiss, J.H. (1987) Comparative growth of spoilage and pathogenic organisms on modified atmosphere-packed cooked beef, *J. Food Protection*, 50, 218–223.

Hjelmqwist, J. (1998) High pressure processing of food — a commercial process, in *Fresh Novel Foods by High Pressure*, K. Autio, Ed., VTT Symp., Vol. 186, VTT Technical Research Centre, Finland, pp. 97–102.

Hotchkiss, J.H. and Chen, J.H. (1996) Microbiological effects of the direct addition of CO_2 to pasteurized milk, *J. Dairy Sci.*, 79(suppl. 1), 87.

Hülsheger, H., Potel, J., and Niemann, E.G. (1983) Electric field effects on bacteria and yeast cells, *Radiat. Environ. Biophys.*, 22, 149–162.

Kolakowski, P., Reps, A., and Babuchowski, A. (1998) Characteristics of pressure ripened cheese, *Polish J. Food Nutri. Sci.*, 7/48(3), 473–483.

Kosikowski, P.A. and Brown, D.P. (1973) Influence of carbon dioxide and nitrogen on microbial population and shelf life of cottage cheese and sour cream, *J. Dairy Sci.*, 56(1), 12–18.

Knorr, D., Geulen M., Grahl, T., and Sitzmann, W. (1994) Food application of high electric field pulses, *Trends Food Sci. Technol.*, 5, 71–75.

Kumar, S. and Mathur, B.N. (1989) Studies on the manufacture of yogurt and mozzarella cheese from milk preserved by LP-system, *Indian J. Dairy Sci.*, 42, 194–197.

Law, B.A. and Reiter, B. (1977) The isolation and bacteriostatic properties of lactoferrin from bovine milk whey, *J. Dairy Res.*, 44, 595–599.

Mahaut, M. and Korolczuk, J. (1992) Effect of whey protein addition and heat treatment of milk on the Viscosity of UF fresh cheeses, *Milchwissenschaft*, 47, 157–159.

Manier, A.B., Marcy, J.E., Bishop, J.R., and Duncan, S.E. (1994) Modified atmosphere packaging to maintain direct set cottage cheese quality, *J. Food Sci.*, 59, 1305–1308.

Mannheim, C.H. and Soffer, T. (1996) Shelf-life extension of cottage cheese by modified atmosphere packaging, *Food Sci. Technol.*, 29(8), 767–771.

Martinez, C.E., Mendoza, P.G., Alacron, F.J., and Gracia, H.S. (1988) Reactivation of the lactoperoxidase system during raw milk storage and its effect on the characteristics of pasteurized milk, *J. Food Protection*, 51, 558–561.

McGregor, J.U. and White, C.H. (1990) Effect of enzyme treatment and ultrafiltration on the quality of low fat cheddar cheese, *J. Dairy Sci.*, 73, 571–578.

Messens, W., Van Camp, J., and Huyghebaert, A. (1997) The use of high pressure to modify the functionality of food protein: a review, *Trends Food Sci. Technol.*, 8, 107–112.

Messens, W., Estepar-Garcia, J., Dewettinck, K., and Huyghebaert, A. (1999) Proteolysis of high-pressure-treated gouda cheese, *Int. Dairy J.*, 9(11), 775–782.

Mezgheni, E., Vachon, C., and Monique, L. (2000) Bacterial use of biofilms crosslinked by gamma irradiation, *Radiat. Phys. Chem.*, 58, 203–205.

Michalac, S., Alvarez, V., Ji, T., and Zhang, Q.H. (2001) Inactivation of selected microorganisms and properties of pulsed electric field processed milk, *J. Food Process. Preser.* (in press).

Mistry, V.V. and Maubois, J.L. (1993) Application of membrane separation technology to cheese production., in *Cheese: Chemistry, Physics and Microbiology*, Vol. 1, *General Aspects*, P.F. Fox, Ed., Elsevier, London, pp. 493–522.

Mohr, C.M., Engelgau, D.E., Leeper, S.A., and Charboneau, B.L. (1989) Membrane technology overview, in *Membrane Applications and Research in Food Processing*, Noyes Data Co., Park Ridge, NJ, pp. 15–39.

Moir, C.J., Eyles, M.J., and Davey, J.A. (1993) Inhibition of *Pseudomonads* in cottage cheese by packaging in atmospheres containing carbon dioxide, *Food Microbiol.*, 10, 345–351.

Mondal, M.S., Sau, A.K., and Samaresh, M. (2000) Mechanism of the inhibition of milk xanthine oxidase activity by metal ions: a transient kinetic study, *Biochim. Biophys. Acta*, 1480, 302–310.

Naidu, A.S. and Arnold, R.R. (1994) Lactoferrin interaction with *Salmonellae* potentiates antibiotic susceptibility *in vitro*, *Diagn. Microbiol. Infect. Dis.*, 20, 69–75.

Nielsen, W. (1999) North European varieties of cheese: Danish cheese varieties, in *Cheese: Chemistry, Physics and Microbiology*, Vol. 2., 2nd ed., P.F. Fox, Ed., Aspen Publishers, Gaithersberg, MD, pp. 245–262.

Nuijens, J.H., Van Berkel, P.H.C., and Schanbacher, F.L. (1996) Structure and biological actions of lactoferrin, *J. Mammary Gland Biol. Neoplasia*, 1, 285–294.

O'Reilly, C.E., O'Connor, P.M., Murphy, P.M., Kelly, A.L., and Beresford, T.P. (2000) The effect of exposure to pressure of 50 MPa on cheddar cheese ripening, *Innovative Food Sci. Emerging Technol.*, 1, 109–117.

Pakkanen, R. and Aalto, J. (1997) Growth factors and antimicrobial factors of bovine colostrum: a review, *Int. Dairy J.*, 7(5), 285–297.

Papachristou E. and Lafazanis, C.T. (1997) Application of membrane technology in the pretreatment of cheese dairy wastes and co-treatment in a municipal conventional biological unit, *Water Sci. Technol.*, 36(2–3), 361–367.

Patterson, M.F. and Kilpatrick, D.J. (1998) The combined effect of high hydrostatic pressure and mild heat on inactivation of pathogens in milk and poultry, *J. Food Protection*, 61(4), 432–436.

Payne, K.D., Davidson, P.M., and Olivier, S.P. (1990) Influence of bovine lactoferrin on the growth of *Listera monocytogenes*, *J. Food Protection*, 53, 468–472.

Pothakamury, U.R., Monsalve-Gonzalez, A., Barbosa-Canovas, G.V., and Swanson, B.G. (1995) Inactivation of *Escherichia coli* and *Staphylococcus aureus* in model foods by pulsed electric field technology, *Food Res. Int.*, 28(2), 167–171.

Pruitt, K.M. and Reiter, B. (1985) Biochemistry of peroxidase system, in *The Lactoperoxidase System: Chemistry and Biological Significance*, K.M. Pruitt and J. Tenovuo, Eds., Marcel Dekker, New York, pp. 143–178.

Qin, B.L., Zhang, Q.H., Barbosa-Canovas, G.V., Swanson, B.G., and Pedrow, P.D. (1995) Pulsed electric field treatment chamber design for liquid food pasteurization using a finite element method, *Trans. ASAE*, 38(2), 557–565.

Qiu, X., Sharma, S., Tuhela, L., Jia, M., and Zhang, Q.H. (1998) An integrated PEF pilot plant for continuous nonthermal pasteurization of fresh orange juice, *Trans. ASAE*, 41(4), 1069–1074.

Reina, L.D., Jin, Z.T., Yousef, A.E., and Zhang, Q.H. (1998) Inactivation of *Listeria monocytogenes* in milk by pulsed electric field, *J. Food Protection*, 61(9), 1203–1206.

Reiter, B. (1978) Review of the progress of dairy science: antimicrobial systems in milk, *J. Dairy Res.*, 45, 131–147.

Rodriguez, J., Requena, T., Fontecha, J., Goudédranche, H., and Juarez, M. (1999) Effect of different membrane separation technologies (ultrafiltration and microfiltration) on the texture and microstructure of semihard low-fat cheeses, *J. Agr. Food Chem.*, 47(2), 558–565.

Rosenberg, M. (1995) Current and future applications for membrane processes in the dairy industry: a review, *Trends Food Sci. Technol.*, 6, 12–19.

Rovere, P. (1995) Next to time and temperature, a new and revolutionary operative parameter: high pressure, the third dimension of food technology, *Technologie Alimentari Bull.*, pp.1–8.

Sadoun, D., Coucercelle, C., Strasser, A., Egler, A., and Hasselmann, C. (1991) Low dose irradiation of liquid milk, *Milchwissenschaft*, 46(5), 295–299.

Satin, M. (1996) Food irradiation and pasteurization, in *Food Irradiation: A Guidebook*, 2nd ed., Technomic, Lancaster, PA, pp. 1–41.

Schanbacher, F.L., Talhouk, R.S., Murray, F.A., Gherman, L.I., and Willett, L.B. (1998) Milk-borne bioactive peptides, *Int. Dairy J.*, 8, 393–403.

Schkoda, P. and Kessler, H.G. (1996) Manufacture of fresh cheese from ultrafiltered milk with reduced amount of acid whey, *IDF Bull.*, 311, 33–35.

Staal, P. (1986) IDF and the development of milk pasteurization, *IDF Bull.*, 200, 4–8.

Stannard, D.J. (1975) The use of marker enzymes to assay the churning of milk, *J. Dairy Res.*, 42, 241–246.

Tanaka, T. and Hatanaka, K. (1992) Application of hydrostatic pressure yogurt to prevent its after acidification, *Nippon Shokuhin Kogyo Gakkaishi*, 39, 173–177.

Thakur, B.R. and Singh, R.K. (1995) Combination process in food irradiation: a review, *Trends Food Sci. Technol.*, 6, 7–11.

Toledo Lopez, V.M. and Garcia Galindo, H.S. (1987) The effect of the lactoperoxidase system on the shelf-life of fresh cream and cultured cream, *Boletin Tecnico LABAL*, 8, 37–48.

Tsong, T.Y. (1990) Electroporation of cell membranes and some related phenomena: a review, *Biochem. Bioenerg.*, 24, 271–295.

Van der Horst, H.C., Timmer, J.M.K., Robbertsen, T., and Leenders, J. (1995) Use of nonofiltration for concentration and demineralization in the dairy industry: model for mass transport, *J. Membrane Sci.*, 104, 205–218.

Vega-Mercado, H., Pothakamury, U.R., Chang, F.J., Barbosa-Canovas, G.V., and Swanson, B.G. (1997) Inactivation of *E. coli* by combining pH, ionic strength and pulsed electric field hurdles, *Food Res. Int.*, 29(2), 117–121.

Walstra, P., Geurts, T.J., Noomen, A., Jellema, A., and van Boekel, M.A.J.S. (1999) Heat treatment, in *Dairy Technology: Principles of Milk Properties and Processes*, Marcel Dekker, New York, pp. 189–240.

Wilkinson, V.M. and Gould, G.W. (1996) *Food Irradiation: A Reference Guide*, Butterworth-Heinemann, Oxford, England, pp. 37–38.

Yam, K.L. and Lee, D.S. (1995) Design of modified atmosphere packaging for fresh produce, in *Active Food Packaging*, M.L. Rooney, Ed., Chapman & Hall, New York, pp. 55–73.

Yin, Y., Zhang, Q.H., and Sastry, S.K. (1997) High Voltage Pulsed Electric Field Treatment Chambers for the Preservation of Liquid Food Products, U.S. Patent No. 5,690,978.

Zhang, Q., Qin, B.L., Barbosa-Cánovas, G.V., and Swanson, B.G. (1995a) Inactivation of *E. coli* for food pasteurization by high-intensity short-duration pulsed electric fields, *J. Food Process. Preser.*, 19, 103–118.

Zhang, Q., Barbosa-Cánovas, G.V., and Swanson, B.G. (1995b) Engineering aspects of pulsed electric field pasteurization, *J. Food Eng.*, 25, 261–281.

Zikakis, J.P. and Wooters, S.C. (1980) Activity of xanthine oxidase in dairy products, *J. Dairy Sci.*, 63(6), 893–904.

Index

A

Abscisic acid, 180
Acetone, butanol, and ethanol (ABE) extraction, 226–228
Adsorbent systems, 223–224
Aerobic bioreactors, 105–118
    bubble size distribution, 111
    hold-up structure in viscous fluids, 105–108
    impeller type effects, 116
    small bubble effect, 109–110
    time dependency model, 107–108
Affinity columns, 97
Agricultural biotechnology, *See* Plant biotechnology
Agrobacterium tumefaciens, 73, 74
Alamine 336, 222
Albumin, 14, 15–18
Alkaline soil-tolerant crops, 69
Allergenic albumin, 17
Alpha-crystalline gel, 255
Alternative food processing technologies, 279–304, *See also* Microbial inactivation technologies
Amaranth, 18, 151
Amarantin, 18–19
Amino acids
    lupine composition, 247
    microbial production, 51–52
    plant proteins, 13–19, 71
Aminocyclopropane-1-carboxylic acid (ACC), 45–46
$<\alpha>$-Amylase, stability of, 161, 168
Amylose, 24–26
Anion exchange chromatography, 132–135
Anthocyanin regulators, 74
Antibiotics resistance, 88
Antibody expression, 30–32
Anticancer monoclonal antibodies, 31–32
Antisense-gene applications
    browning control, 73
    fruit ripening control, 45, 72
    lipid engineering, 70

AOT reverse micelle system, 194–196, 204, 207, 209
$<\beta>$-Apiosidase, 121
Arabidopsis tocopherols, 41
Arabinofuranosidase, 121, 123, 128–13
$<\alpha>$-L-Arachidonic acid, 23
Armoracia rusticana, 201
Aroma and flavor enhancing compounds, 219–229
    downstream processes, 220–221
    grape glycosides, 120–121
    in situ product removal, 221–228
        hollow-fiber membranes, 226
        pervaporation, 227–228
        solid-phase extraction, 223–224
        solvent extraction, 221–223
        supercritical fluids, 225
Ashbya gossypii, 49
Aspergillus niger
    glycosidase extract, 121–136
    pectinolytic extract, 156
Aspergillus oryzae, 166
Astaxanthin, 177–182
    applications, 180–182
    production, 177–180, 185
Avocado, 284

B

Bacillus cereus, 317
Bacteria
    amino acid production, 51–52
    antimicrobial milk protein effects, 321–322
    carotenoid production, 52–53, 180, 185–187
    food processing technologies, *See* Microbial inactivation technologies; *specific technologies*
    high pressure processing, 294, 315
    irradiation, 298–299, 318
    lactic acid production, 50–51
    MAP, 319